T0207182

Lecture Notes in Computer Science 13975

Founding Editors

Gerhard Goos
Juris Hartmanis

Editorial Board Members

Elisa Bertino, *Purdue University, West Lafayette, IN, USA*
Wen Gao, *Peking University, Beijing, China*
Bernhard Steffen ⓘ, *TU Dortmund University, Dortmund, Germany*
Moti Yung ⓘ, *Columbia University, New York, NY, USA*

The series Lecture Notes in Computer Science (LNCS), including its subseries Lecture Notes in Artificial Intelligence (LNAI) and Lecture Notes in Bioinformatics (LNBI), has established itself as a medium for the publication of new developments in computer science and information technology research, teaching, and education.

LNCS enjoys close cooperation with the computer science R & D community, the series counts many renowned academics among its volume editors and paper authors, and collaborates with prestigious societies. Its mission is to serve this international community by providing an invaluable service, mainly focused on the publication of conference and workshop proceedings and postproceedings. LNCS commenced publication in 1973.

Alessio Ferrari · Birgit Penzenstadler

Editors

Requirements Engineering: Foundation for Software Quality

29th International Working Conference, REFSQ 2023
Barcelona, Spain, April 17–20, 2023
Proceedings

Editors
Alessio Ferrari (ID)
CNR ISTI
Pisa, Italy

Birgit Penzenstadler (ID)
Chalmers Tekniska Högskola
Gothenburg, Sweden

ISSN 0302-9743   ISSN 1611-3349 (electronic)
Lecture Notes in Computer Science
ISBN 978-3-031-29785-4   ISBN 978-3-031-29786-1 (eBook)
https://doi.org/10.1007/978-3-031-29786-1

© The Editor(s) (if applicable) and The Author(s), under exclusive license
to Springer Nature Switzerland AG 2023
This work is subject to copyright. All rights are reserved by the Publisher, whether the whole or part of
the material is concerned, specifically the rights of translation, reprinting, reuse of illustrations, recitation,
broadcasting, reproduction on microfilms or in any other physical way, and transmission or information
storage and retrieval, electronic adaptation, computer software, or by similar or dissimilar methodology now
known or hereafter developed.
The use of general descriptive names, registered names, trademarks, service marks, etc. in this publication
does not imply, even in the absence of a specific statement, that such names are exempt from the relevant
protective laws and regulations and therefore free for general use.
The publisher, the authors, and the editors are safe to assume that the advice and information in this book
are believed to be true and accurate at the date of publication. Neither the publisher nor the authors or the
editors give a warranty, expressed or implied, with respect to the material contained herein or for any errors
or omissions that may have been made. The publisher remains neutral with regard to jurisdictional claims in
published maps and institutional affiliations.

This Springer imprint is published by the registered company Springer Nature Switzerland AG
The registered company address is: Gewerbestrasse 11, 6330 Cham, Switzerland

Preface

This volume contains the papers presented at REFSQ 2023, the 29th International Working Conference on Requirements Engineering: Foundation for Software Quality, held on April 17–20, 2023 in Barcelona, Spain. We are very happy to report that submission numbers have gone up significantly and we are excited to host another in-person conference after a successful edition in Birmingham, UK, this past year. The REFSQ series was established in 1994, at first as a workshop series, and since 2010 in the "working conference" format, with ample time for presentations and substantial discussions of each contribution. It is often considered among the major international scientific events in Requirements Engineering, and the only one to be permanently located in Europe, with a special connection to European industry and academia. The need for ever-increasing levels of quality in requirements has not diminished in the 29 years since the first REFSQ; on the contrary, requirements are pervasive in the design, implementation, and operation of software systems and related services that impact the lives of billions. The special theme for REFSQ 2023 was "Human Values in Requirements Engineering". Requirements Engineering (RE) is at the boundary of humans and technology, and values play a crucial role in the interplay between developers, users and systems. When developing technology, we get to be cognizant of how our values inform our designs, because we unconsciously embed them into our systems. In addition, we need to carefully consider possible conflicts between human values and business values. The theme of this year thus aimed to foster discussion around the following questions:

- How do we take care of human values in RE?
- How do we ensure that the systems we design incorporate the values we want them to stand for?
- How do we validate and measure values?
- How do we make sure that systems serve the human as opposed to having the human adapt to them?
- How much do developer habits and characteristics influence their designs?
- What is the interplay between developer and stakeholder values?
- What is the interplay between human values and business values?

We were very happy to observe that the challenge was promptly taken up by the research community, with many submissions focusing on exactly those issues. Several of those contributions were accepted for presentation at the conference, and are now part of this volume. In response to the Call for Papers, we received 84 abstracts, which resulted in 78 full papers, which were single-blind reviewed by three program committee members, extensively discussed among the reviewers, and then brought for additional discussion if needed and a final decision at the plenary program committee meeting that was held (online) on January 17 and 18, 2023. Nine papers for which no consensus had been reached were discussed in special depth, and all of them were accepted on the condition that certain improvements be made (those underwent an additional check by a PC member before final acceptance). Overall, 25 papers were finally accepted for publication, and are now collected in this volume. In particular, based on paper category, the acceptance ratios are as follows:

- Scientific Evaluation (17 pages): 20 submissions, 7 accepted (35%)
- Technical Design (17 pages): 22 submissions, 5 accepted (23%)
- Experience report papers (14 pages): 14 submissions, 5 accepted (36%)
- Vision (10 pages): 3 submissions, 2 accepted (67%)
- Research Preview (10 pages): 14 submissions, 6 accepted (43%)

The acceptance rate of full contributions was thus 29% (12/42). As in previous years, the conference was organized as a three-day symposium (Tuesday to Thursday), with one day devoted to industrial presentations (in a single track), and two days of academic presentations (in two parallel tracks). In addition to paper presentations and related discussions, the program included two keynote talks by Barbara Paech and Klaas-Jan Stol; a Poster & Tools session organised by Sallam Abualhaija and Oliver Karras; a Journal Early-Feedback track, organized by Paola Spoletini and Daniel Amyot; and awards to recognize the best contributions in various categories. On the Monday before the conference, four co-located events were held:

- NLP4RE: 6th Workshop on Natural Language Processing for Requirements Engineering, organized by Sallam Abualhaija, Andreas Vogelsang and Gouri Deshpande;
- RE4AI: 4th International Workshop on Requirements Engineering for Artificial Intelligence, organized by Renata Guizzardi, Jennifer Horkoff, Anna Perini and Angelo Susi;
- ViVA RE!: 1st Workshop on Virtues and Values in Requirements Engineering, organized by Alexander Rachmann and Jens Gulden;
- REFrame: 1st Workshop on Requirements Engineering Frameworks, organized by Andrea Wohlgemuth, Anne Hess and Samuel Fricker;
- REFSQ Doctoral Symposium, organized by Fabiano Dalpiaz and Ana Moreira.

The proceedings of the co-located events and the Poster & Tools track are published in a separate volume via CEUR. We would like to thank all members of the Requirements Engineering community who prepared a contribution for REFSQ 2023: there would be no progress in our discipline without the talent, intelligence and effort that so many brilliant researchers dedicated to the field. We would also like to thank members of the Program Committee and additional reviewers for their invaluable contribution to the selection process. Special thanks are due to all the colleagues that served in various distinguished roles in the organization of REFSQ 2023; your help in assembling a rich program has been invaluable:

- The REFSQ Steering Committee has provided excellent support and guidance throughout the process;
- The previous PC chairs, who happily shared their experiences;
- The Local Organizers Carles Farré and Carme Quer;
- The Steering Committee Chair Anna Perini and Vice-Chair Fabiano Dalpiaz, and
- The head of the Background Organization, Xavier Franch, for making our regular organizational meetings so enjoyable that we almost looked forward to each subsequent one with pleasurable anticipation.

Last but not least, we would like to thank you, the reader. You are the reason for this volume to exist. We hope you will find its contents interesting, useful, stimulating and inspirational.

February 2023

Alessio Ferrari
Birgit Penzenstadler

Organization

Program Committee Chairs

Alessio Ferrari   Consiglio Nazionale delle Ricerche, Italy
Birgit Penzenstadler   Chalmers Tekniska Högskola, Sweden and
     Lappeenranta University of Technology,
     Finland

Local Organization Chairs

Carles Farré   Universitat Politècnica de Catalunya, Spain
Carme Quer   Universitat Politècnica de Catalunya, Spain

Industry Track Chairs

Joan Antoni Pastor   Universitat Politècnica de Catalunya, Spain
Krzysztof Wnuk   Blekinge Institute of Technology, Sweden

Workshop Chairs

Irit Hadar   University of Haifa, Israel
Shola Oyedeji   Lappeenranta University of Technology, Finland

Journal Early Feedback Chairs

Paola Spoletini   Kennesaw State University, USA
Daniel Amyot   University of Ottawa, Canada

Posters and Tools Chairs

Sallam Abualhaija   University of Luxembourg, Luxembourg
Oliver Karras   Leibniz Information Centre for Science and
     Technology, Germany

Doctoral Symposium Chairs

Fabiano Dalpiaz   Utrecht University, The Netherlands
Ana Moreira   NOVA University of Lisbon and NOVA LINCS,
                           Portugal

Most Influential Paper Chair

Martin Glinz   University of Zurich, Switzerland

Social Media and Publicity Chairs

Muhammad Abbas   RISE Research Institutes of Sweden AB, Sweden
Quim Motger   Universitat Politècnica de Catalunya, Spain

Web Chair

Quim Motger   Universitat Politècnica de Catalunya, Spain

Student Volunteer Chair

Claudia Ayala   Universitat Politècnica de Catalunya, Spain

Proceedings Chair

Giorgio O. Spagnolo   Consiglio Nazionale delle Ricerche, Italy

Local Organization Members

Dolors Costal   Universitat Politècnica de Catalunya, Spain
Cristina Gómez   Universitat Politècnica de Catalunya, Spain
Rediana Koçi   Universitat Politècnica de Catalunya, Spain

REFSQ Series Organization

Steering Committee

Anna Perini (Chair)	Fondazione Bruno Kessler, Italy
Fabiano Dalpiaz (Vice-chair)	University of Utrecht, The Netherlands
Xavier Franch (Chair of BO)	Universitat Politècnica de Catalunya, Spain
Alessio Ferrari	Consiglio Nazionale delle Ricerche, Italy
Birgit Penzenstadler	Chalmers Tekniska Högskola, Sweden and Lappeenranta University of Technology, Finland
Paola Spoletini	Kennesaw State University, USA
Nazim Madhavji	Western University, Canada
Michael Goedicke	University of Duisburg-Essen, Germany
Vincenzo Gervasi	University of Pisa, Italy
Andrea Vogelsang	University of Cologne, Germany
Klaus Pohl	University of Duisburg-Essen, Germany
Eric Knauss	Chalmers Tekniska Högskola, Sweden
Daniel Méndez Fernández	Blekinge Institute of Technology, Sweden
Ana Moreira	NOVA University of Lisbon and NOVA LINCS, Portugal

Background Organization

Xavier Franch (Chair)	Universitat Politècnica de Catalunya, Spain
Carme Quer	Universitat Politècnica de Catalunya, Spain
Quim Motger	Universitat Politècnica de Catalunya, Spain

Program Committee

Sallam Abualhaija	University of Luxembourg, Luxembourg
Carina Alves	Universidade Federal de Pernambuco, Brazil
Daniel Amyot	University of Ottawa, Canada
Chetan Arora	Deakin University, Australia
Fatma Başak Aydemir	Boğaziçi University, Turkey
Nelly Bencomo	Durham University, UK
Dan Berry	University of Waterloo, Canada
Stefanie Betz	Furtwangen University, Germany
Travis Breaux	Carnegie Mellon University, USA
Sjaak Brinkkemper	Utrecht University, The Netherlands
Nelly Condori-Fernández	Universidad de Santiago de Compostela, Spain

Fabiano Dalpiaz	Utrecht University, The Netherlands
Maya Daneva	University of Twente, The Netherlands
Joerg Doerr	Fraunhofer, Germany
Xavier Franch	UPC, Spain
Samuel A. Fricker	FHNW, Switzerland
Davide Fucci	University of Hamburg, Germany
Matthias Galster	University of Canterbury, New Zealand
Vincenzo Gervasi	University of Pisa, Italy
Martin Glinz	University of Zurich, Switzerland
Michael Goedicke	Univ. Duisburg-Essen, Germany
Eduard C. Groen	Fraunhofer IESE, Germany
Paul Grünbacher	Johannes Kepler University Linz, Austria
Renata Guizzardi	University of Twente, The Netherlands
Andrea Herrmann	Free Software Engineering Trainer, Germany
Anne Hess	Fraunhofer, Germany
Jennifer Horkoff	Chalmers University of Technology and University of Gothenburg, Sweden
Erik Kamsties	University of Applied Sciences and Arts Dortmund, Germany
Eric Knauss	Chalmers University of Technology and University of Gothenburg, Sweden
Sylwia Kopczyńska	Poznan University of Technology, Poland
Kim Lauenroth	adesso AG, Germany
Emmanuel Letier	University College London, UK
Grischa Liebel	Reykjavik University, Iceland
Nazim Madhavji	University of Western Ontario, Canada
Daniel Méndez Fernández	Blekinge Institute of Technology, Sweden
Luisa Mich	University of Trento, Italy
Lloyd Montgomery	University of Hamburg, Germany
Gunter Mussbacher	McGill University, Canada
John Mylopoulos	University of Ottawa, Canada
Nan Niu	University of Cincinnati, USA
Andreas L. Opdahl	University of Bergen, Norway
Shola Oyedeji	LUT University, Finland
Barbara Paech	Universität Heidelberg, Germany
Elda Paja	IT University of Copenhagen, Denmark
Liliana Pasquale	University College Dublin, Ireland
Oscar Pastor	Universidad Politécnica de Valencia, Spain
Anna Perini	Fondazione Bruno Kessler, Italy
Bjorn Regnell	Lund University, Sweden
Marcela Ruiz	Zurich University of Applied Sciences, Switzerland

Mehrdad Sabetzadeh	University of Ottawa, Canada
Klaus Schmid	University of Hildesheim, Germany
Kurt Schneider	Leibniz Universität Hannover, Germany
Laura Semini	University of Pisa, Italy
Norbert Seyff	FHNW, Switzerland
Paola Spoletini	Kennesaw State University, USA
Jan-Philipp Steghöfer	XITASO GmbH IT & Software Solutions, Germany
Angelo Susi	Fondazione Bruno Kessler, Italy
Colin C. Venters	University of Huddersfield, UK
Michael Vierhauser	Johannes Kepler University Linz, Austria
Andreas Vogelsang	University of Cologne, Germany
Liping Zhao	University of Manchester, UK
Didar Zowghi	CSIRO's Data61, Australia

Additional Reviewers

Anders, Michael	Passaro, Lucia
Fotouhi, Sara	Patkar, Nitish
Heyn, Hans-Martin	Radeck, Leon
Jeswein, Thomas	Ramautar, Vijanti
Koch, Matthias	Rohmann, Astrid
Ly, Delina	Samin, Huma
Molenaar, Sabine	Scherr, Simon André
Oriol Hilari, Marc	van Dijk, Friso

Supporting Institutions, Companies and Groups

International® Requirements Engineering Board

Istituto di Scienza e Tecnologie dell'Informazione "A. Faedo" Consiglio Nazionale delle Ricerche

Contents

RE for Artificial Intelligence

Crowd RE

RE in Practice

Requirements Communication
and Conceptualization

Requirements Engineering Issues Experienced by Software Practitioners: A Study on Stack Exchange

Sávio Freire[1,2] , Felipe Gomes[1] , Larissa Barbosa[1], Thiago Souto Mendes[3] ,
Galdir Reges[4] , Rita S. P. Maciel[1] , Manoel Mendonça[1] ,
and Rodrigo Spínola[4,5(✉)]

[1] Federal University of Bahia, Salvador, Bahia, Brazil
`savio.freire@ifce.edu.br`, {`felipe.gustavo,larissa.leoncio,`
`rita.suzana,manoel.mendonca}@ufba.br`
[2] Federal Institute of Ceará, Morada Nova, Ceará, Brazil
[3] Federal Institute of Bahia, Salvador, Bahia, Brazil
`thiagosouto@ifba.edu.br`
[4] Salvador University, Salvador, Bahia, Brazil
`galdir.reges@unifacs.br`, `spinolaro@vcu.edu`
[5] Virginia Commonwealth University, Richmond, VA, USA

Abstract. **[Context and Motivation]** Requirements engineering (RE) is central to software development. Despite its importance, there are many issues related to its enactment. Question and answer platforms, such as the Stack Exchange, are paramount in contemporary software development. They discuss and bring to light practitioners' viewpoints on software engineering issues. Approaching those platforms focusing on RE deserves investigation because it can reveal current issues experienced by software practitioners and possible solutions for them. **[Question/Problem]** This work investigates RE issues, their causes, effects, and possible solutions as discussed by software practitioners in the Software Engineering Stack Exchange (SWESE). For that, we mine, curate, and analyze a set of 61 discussions related to RE, composed of 414 posts and 770 comments extracted from SWESE. **[Principal Ideas/Results]** We identify 50 issues and their relations with requirements phases. *Customers' unable to describe system requirements* and *the need for detailed specifications* are among the most commonly discussed issues. We also list 20 causes, 23 effects, and 59 solutions for the mined issues. Examples of causes for RE issues are *lack of technical knowledge* and *communication issues*. Examples of the effects of RE issues are *rework* and *unstable requirements*. Solutions encompass practices such as *clearly defining requirements* and using *prototypes*. **[Contribution]** This work organizes the mined RE issues in a Sankey diagram, relating them to RE phases and solutions, which may assist practitioners experiencing them and serve as guidance for future research.

Keywords: Requirements Engineering Issues · Question and Answer Platform · Stack Exchange

© The Author(s), under exclusive license to Springer Nature Switzerland AG 2023
A. Ferrari and B. Penzenstadler (Eds.): REFSQ 2023, LNCS 13975, pp. 3–20, 2023.
https://doi.org/10.1007/978-3-031-29786-1_1

1 Introduction

Requirements engineering (RE) refers to discovering the system's purpose by identifying stakeholders' needs and documenting them to be used in other software development phases [1]. There are many issues inherent to RE processes; for instance, stakeholders' goals may vary and conflict, not be explicit, or maybe challenging to articulate [1, 2]. Knowing such issues and possible solutions to them is necessary to support software teams in improving their chances of success.

Several studies have investigated RE issues experienced by software practitioners [2–9]. After conducting a mapping study, Pekar *et al.* [6] identified that ambiguity is the main issue in requirements. By surveying software practitioners, researchers from the *Naming the Pain in Requirements Engineering* (NaPiRE) project identified issues and their causes and effects [2]. The most commonly mentioned issues and some of their respective causes and effects were *incomplete or hidden requirements*, (cause) *lack of experience of RE team members*, (cause) *weak qualification of RE team members*, (effect) *time overrun*, and (effect) *poor product quality*. Bonfim and Benitti [9] interviewed 19 agile software practitioners to investigate causes, effects, and mitigation practices for requirements debt, i.e., the distance between the optimal requirements specification and the actual system implementation [10]. Examples of identified causes, effects, and mitigation practices are *requirements specification failure*, *requirements not met and realized post-delivery*, and *assessing the impact of technical debt*, respectively.

Despite current work on the area, to the best of our knowledge, no work has analyzed discussions focused on RE in Question and Answer (Q&A) platforms, such as the Stack Exchange. The number of studies performed on it illustrates the importance of such platforms in investigating the state of the practice in software engineering [11]. For instance, there are investigations on the areas of development trends [12], code smells [13], soft skills [14], and technical debt [15–17]. Those platforms are paramount in contemporary software development because countless issues are discussed, bringing to light the practitioners' points of view on possible solutions for those issues [18]. Moreover, Q&A platforms are a rich source of information on professional practices compared to other sources such as surveys, interviews, and literature reviews [19]. They allow researchers to learn directly from the sharing of practitioners' knowledge. Approaching them focusing on RE issues deserves investigation because it can reveal new issues and possible solutions for them.

This work investigates RE issues and their causes, effects, and possible solutions, as discussed by software practitioners in the Software Engineering Stack Exchange (SWESE). We mined, curated, and analyzed a set of 61 discussions related to RE, composed of 414 posts and 770 comments. By analyzing this data set quantitatively and qualitatively, we found 50 RE issues. We also identified 20 causes, 23 effects, and 59 solutions for those issues. Lastly, we organized the relations between issues, RE phases, and solutions in a Sankey diagram.

Software practitioners can use the organized set of issues, causes, effects, and solutions to increase their capability to deal with RE issues. By navigating through the Sankey diagram, software teams can identify which issues commonly affect each RE phase and which solutions they can employ to solve them, learning from other practitioners' experiences. Researchers can use our findings to develop methods and tools to improve the RE process considering practitioners' needs.

This paper is organized into six other sections. Section 2 discusses related work. Section 3 presents the research strategy. Section 4 shows the results reached. Section 5 offers a Sankey diagram encompassing RE issues and their associated solutions and compares our findings to ones reported by related work. Section 6 discusses the threats to validity. Lastly, Sect. 7 presents our work's final remarks and next steps.

2 Related Work

Several studies have investigated issues in RE. Nikula *et al.* [3] surveyed 15 software practitioners from 12 Finnish companies to explore current RE practices, development needs, and preferred ways of technology transfer. Solemon *et al.* [4] surveyed 64 Malaysian software practitioners to identify problems in requirements activities, revealing that the problems are related to organizational structure and process. Liu *et al.* [5] surveyed 377 Chinese software practitioners to recognize their practices in requirements activities. Pekar *et al.* [6] conducted a mapping study to identify requirements issues and solutions. The authors recognized that the main issue is ambiguity and reported a set of nine solutions.

Another related work is the NaPiRE project [2]. By conducting a survey, the project analyzed answers from 228 companies, revealing the ten most cited problems in RE. Among them, *incomplete and/or hidden requirements* was commonly cited. Also, the authors identified leading causes (such as *lack of time* and *lack of experience of RE team members*) and related them to effects.

RE issues have also been investigated from the perspective of requirements and documentation debt. The former refers to the distance between optimal requirements specification and the actual system implementation (e.g., requirements that are only partially implemented). The latter is associated with problems in software project documentation (e.g., missing documentation) [10]. Rios *et al.* [7] investigated the causes, effects, and practices to prevent and repay documentation debt items. The authors analyzed 39 answers from the *InsighTD* project [20] to identify documentation debt causes and effects, and performed interviews with practitioners to recognize preventive and repayment practices for documentation debt items. The authors reported 23 causes (*deadline* was the most cited), 15 effects (*low maintainability*), and three solutions (for example, *keep the documentation updated*) for documentation debt items. Barbosa *et al.* [8] analyzed 78 answers from the *InsighTD* to investigate causes, effects, and practices to prevent and repay requirements and documentation debt items. The authors found 55 causes (*deadline* was the most cited), 33 effects (*delivery delay*), and 18 solutions (*code refactoring*).

Lastly, Bonfim and Benitti [9] interviewed 19 agile software practitioners on causes, effects, and practices to minimize requirements debt. The authors identified eight causes (e.g., *failing to clarify the demand initially received from the customer*), seven effects (e.g., *requirements inconsistency*), and one solution (*assessing the impact of technical debt*).

We recognize that current related work provides valuable information about RE issues. However, a careful analysis of grey literature on the topic is still missing. Grey literature, such as those found in Q&A platforms, is a valuable source of information and commonly provides complementing results compared to those found through other sources of information like surveys, interviews, and literature reviews [19].

Our work investigates RE issues; however, we use a software engineering Q&A forum as a proxy to understand the causes, effects, and solutions for these issues. It uses quantitative and qualitative data analysis to reveal the perception of a broad and diverse set of software development practitioners on the subject. It also analyzes how our results complement the NaPiRe findings [2].

No related work has used a Q&A platform to investigate RE issues considering their causes, effects, and possible solutions. Using this platform as a proxy can allow us to learn from the experience of software practitioners when discussing daily issues.

3   Research Strategy

In this work, we investigate how software engineers experience RE issues. To this end, we seek answers to the following research questions:

- RQ1: What are the main RE issues discussed by software engineers?
- RQ2: What are the causes that lead to RE issues?
- RQ3: What are the effects of RE Issues?
- RQ4: What solutions have been considered?
  The following subsections present the data collection and analysis procedures.

3.1   Data Collection

We use the Stack Exchange data dump (version July 9th, 2022) in our analyses. From this repository, we can access data from any Stack Exchange website, including their posts (questions and answers), comments, and metadata.

In the Q&A structure provided on Stack Exchange, each discussion is composed of the question around which the discussion is centered, the answers to the question, and the comments on both questions and answers. A discussion can have a set of comments presented below the post that goes beyond the discussion, clarifying and enriching the content conveyed through questions and answers [21]. A tag is a keyword or label that categorizes the question. The use of tags makes it easier for others to find and answer the question. To create a new tag, a user needs at least 1500 reputation points in the forum, which is earned by receiving positive feedback from others in questions, answers, and comments.

In this study, we focus on the SWESE website. As SWESE users can refer to a topic in different ways when asking a question or answering it [16], we decided to use the tag 'requirements' to filter the discussions. Thus, we rely on the indication of the author that the question is related to requirements. Although SWESE has some tags related to RE ('requirements', 'requirements-management', 'functional-requirements', and 'minimal-requirements'), we choose 'requirements' as it is a generic tag allowing us to map several issues related to requirements. This resulted in 374 discussions. As we are interested in analyzing requirements-related discussions, we carried out a filtering process, as depicted in Fig. 1.

In **Step 1 - Eliminate incomplete discussions from the data set**; we consider a discussion complete when a question is followed by one or more answers, where there is at least one answer whose author is different from the question's author. After applying this criterion, the initial 374 discussions remained. Concerning **Step 2 - Eliminate untrustworthy discussions**, other studies have found that data from Q&A forums can be affected by noise [22, 23], requiring mitigating this noise using different proxies. As in Gomes et al. [16] and dos Santos et al. [17], we decided to use the discussion score as a filtering proxy. A post score is a Stack Exchange popularity metric in which users, other than the post author, can give an up-vote to the post if they find it useful or a down-vote if they find it not useful. A discussion score is the difference between up-votes and down-votes of all its posts. We decided to filter out discussions with scores lower than ten since we wanted to consider the most relevant discussions according to the community. After this filtering, 155 discussions remained out of the initial 374. These 155 discussions are modified (i.e., received new answers) between 2011 and 2022. Finally, in **Step 3 - Qualitative data analysis**, each of the 155 discussions went through the qualitative data analysis process described in the next section.

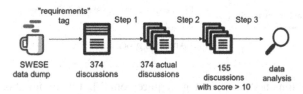

Fig. 1. Data filtering process.

3.2 Data Analysis

The qualitative analysis was composed of three steps, shown in Fig. 2. In **step 1**, the 155 discussions were divided into three subsets (52, 52, and 51 discussions) between three pairs of authors. For each discussion, individually, the researchers filled in the set of questions presented in Fig. 2, reporting the problem (including causes and effects), phase of the requirement process, and solutions reported. This information, when found, was recorded according to the perception of each researcher without any previously established codes. This bottom-up approach was considered most appropriate given the lack of previous analysis of the discussions and the nature of the discussions themselves. The exception to this procedure was the use of the codes for requirement phases, which were based on the ISO 29148 standard [24].

While going through the information-gathering questions, the researchers also looked for false positives, taking into consideration the following rules:

- **Rule 1:** Discussions that, in spite of having the tag 'requirements', did not discuss issues related to requirements were marked as false positives.

- **Rule 2:** This work intends to map real-world issues and doubts faced by practitioners when dealing with requirements in actual software projects. Questions that did not approach a specific issue were marked as false positives.
- **Rule 3:** We considered causes, effects, and solutions present in questions, answers, and comments. When these points were brought by the question's author, or they were confirmed by the author, we marked them as 'Confirmed by author.' The question's author can confirm something by accepting an answer (only one answer can be accepted in Stack Exchange) or by giving a corroborating comment to an answer or comment. When there are not any indications of the author's confirmation, we mark it as 'Not Confirmed by author.'

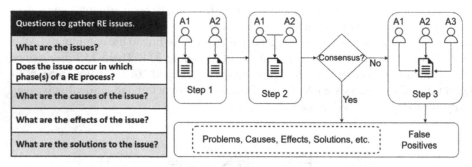

Fig. 2. Analysis procedure.

Thus, discussions that have the tag 'requirements' but were not about a real-world RE issue were considered false positives. For instance, in Fig. 3(a), a student is asking for advice to improve her/his business skills, but the discussion is not related to a real-world situation. On the other hand, the example presented in Fig. 3(b) is considered valid. From the part "our organization", we can see that the author is a practitioner and is reporting a real-world issue. Taking into consideration the whole discussion illustrated in Fig. 3(b), we answered the questions shown in Fig. 2 as follows:

- **What are the issues?** The customer is unable to describe system requirements.
- **Does the issue occur in which phase(s) of a RE process?** Stakeholder needs and requirements definition process.
- **What are the causes of the issue?** The customer is unable to describe system requirements.
- **What are the effects of the issue?** Requirements described by the customer are incomplete or inappropriate.

In Fig. 4, we can see one of the answers related to the question presented in Fig. 3(b). Based on the text, we identified two solutions: *job shadowing or ethnography*, and *prototyping*. By the upvotes on the answer, we can also assume that these are well-accepted approaches by the participants to solve the problem: 19 people considered the answer relevant and helpful.

During the analysis, the researchers also considered the links posted to other web pages, such as articles and tutorials. For instance: *"User stories are requirements. There are a <u>set of characteristics of a good requirement</u> that tend to be well accepted"*. There was also the need to understand the jargon used by practitioners, such as cone of uncertainty, five-whys, and others, that required reading from other sources.

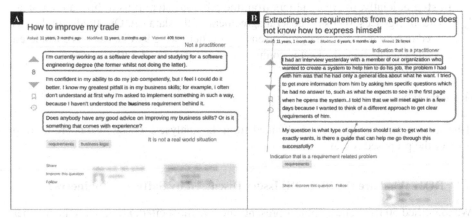

Fig. 3. Example of a (a) false positive and a (b) valid discussion.

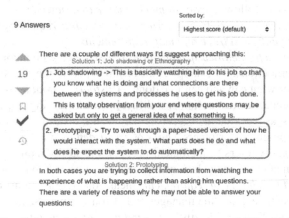

Fig. 4. One of the answers to the question shown in Fig. 3(b).

It is also worth mentioning that during **step 1**, the researchers were encouraged to record the information regarding RE issues freely, without any coding, and to put most data they think are necessary to answer the questions in Fig. 2. We decided to follow this approach because the analysis requires understanding and interpretation by the researchers.

In **step 2**, each pair of researchers performed the coding and then the consensus of their answers. For each discussion, we created terms to represent and group the issues, causes, effects, and solutions. For example, *validating requirements with the customer* was used to code solutions such as: *asking for client feedback* and *performing UAT (User Acceptance Testing) with the client*. In case of agreement of the analysis and coding, the data was added to the results. The divergences went to **step 3**.

In **step 3**, jointly, the two authors from the previous step, plus a third author, analyzed the divergences to ensure general agreement of the data set. Our final sample was composed of 61 out of 155 analyzed discussions. There were 94 discussions considered false positives. The final data set is composed of 414 posts and 770 comments.

4   Results

This section presents the results obtained from the analyses. These results are used to answer the proposed research questions.

4.1   RQ1: What are the Main RE Issues Discussed by Software Engineers?

We identified 50 RE issues. Table 1 presents the five most cited, along with the number of mentions of each issue (**#RI**) and the percentage of the total of mentions (**%RI**). The complete list of issues is available in [25]. *The need for detailed specifications* was the most mentioned issue, followed by *customers' unable to describe system requirements, lack of a well-defined process for gathering and verifying requirements, request for change of requirements*, and *trace and map requirements*.

The issue *the need for detailed specifications* is associated with incomplete or lack of requirements specifications, for example, "the business part of the specs is either incomplete or unaware of what can and can't be done." *Customers' unable to describe system requirements* is related to the customer's difficulty in describing the system purpose and functionalities, as reported in "I receive many weird, invalid or incomplete requests from the actual or potential customers." *Lack of a well-defined process for gathering and verifying requirements* refers to inefficient processes used for requirements activities; for instance, "I am able to see the lack of a systematic approach, but don't know how to proceed!" *Request for change of requirements* means that requirements change over the course of the project, such as "the manager and at times other "senior" keep changing the requirement specification." Lastly, *trace and map requirements* mean that software teams cannot relate requirements to part of the system; for example, "... There are so many stories, it's not immediately clear, for any part of the system which stories relate to it."

Table 1. Five most commonly mentioned RE issues.

Requirement Issue	#RI	%RI
The need for detailed specifications	6	9%
Customers' unable to describe system requirements	5	8%
Lack of a well-defined process for gathering and verifying requirements	3	5%
Request for change of requirements	2	3%
Trace and map requirements	2	3%

Caption:
#RI: Number of mentions of a requirement issue
%RI: Percentage of #RI in relation to the total all mentioned requirements issues (66)

The other best-positioned RE issues are *user stories level of detail, indication of hardware requirements, requirements gathering done by non-technical people, mapping concrete implementation to vague business requirements*, and *user requirements verification*. The issue *user stories level of detail* refers to the level of detail that a user story needs to have to guide the software development, as we can see in "how much detail about a user story can a developer expect?" *Indication of hardware requirements* means the specification of hardware requirements to provide the software deployment, for example, "…they have their own internal IT team, they have asked me on what will be the hardware requirements for the live servers." *Requirements gathering done by non-technical people* is about having a non-technical person collect the system requirements, as reported in "…it is often the case that the members of the development team are not able to get direct access to the client to gather requirements. Is it possible/advisable to give a list of questions to an account manager so that they can gather requirements on your behalf?" *Mapping concrete implementation to vague business requirements* refers to vague business requirements that make it difficult to implement, for example, "How can you link concrete implementation of features to vague business requirements and ensure that the business will be happy with the results, given a lack of technical expertise and buy in from a business?" Lastly, *user requirements verification* is associated with the lack of details provided by the user about their system's needs, for example, "I show the user specifications, prototypes, demos… but still users forget to share some 'insignificant details' about the process or business rules and data."

We also investigated the RE phases affected by the identified issues, as shown in Fig. 5. The phases *stakeholder needs and requirements definition process, requirements management*, and *system [system/software] requirements definition process* are the most commonly impacted. Results also reveal that practitioners have mainly mentioned issues from the requirements conception to their specification.

4.2 RQ2: What are the Causes that Lead to RE Issues?

We found 20 causes that lead to RE issues. Table 2 presents the five most cited causes, along with the number of mentions of each cause (**#C**) and the percentage of the total

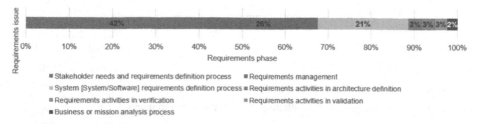

Fig. 5. Relationship between RE issues and RE phases.

of mentions (%C). The complete list of causes is available in [25]. *Lack of technical knowledge* was the most cited cause, followed by *communication issues, requirements from various sources, undefined requirements,* and *difficulty in specifying non-functional requirements.*

Lack of technical knowledge indicates that a software team or managers do not have technical background, such as, "my manager has no background or understanding of computers or software whatsoever." *Communication issues* refer to the lack of access to the customer or users, for example, "…it is often the case that members of the development team are not able to get direct access to the client to gather requirements." *Requirements from various sources* means that users use different ways, such as emails and documents, to send requirements to software teams, as illustrated in "I receive software system's requirements from our potential customers in a very unstructured format from several sources [email, word documents, excel]." *Undefined requirements* are associated with ambiguous, unclear and conflicting requirements, as described in, "I am an inexperienced developer and face the following problem… I am assigned tasks with not clear and conflicting requirements." Lastly, *difficulty in specifying non-functional requirements* is related to doubts about how or where to describe non-functional requirements, for example, "should system configuration settings have a specification assigned to them? On what document?".

Table 2. Five most commonly mentioned causes for RE issues.

Cause	#C	%C
Lack of technical knowledge	26	33%
Communication issues	7	9%
Requirements from various sources	6	8%
Undefined requirements	6	8%
Difficulty in specifying non-functional requirements	4	5%

Caption:
#C: Number of mentions of a cause
%C: Percentage of #C in relation to the total all mentioned causes (79)

The other best-positioned causes are *difficulty in specifying non-functional requirements, incomplete requirements gathering, lack of requirements verification, lack of requirements specification documentation,* and *lack of proper requirements management.* The cause *difficulty in specifying non-functional requirements* is associated with the difficulty of the RE team in specifying non-functional requirements, for example, "Should system configuration settings have a specification assigned to them? On what document? Should this be part of the SRS with the rest of the application's software specifications or should this be recorded as specifications in another domain?" *Incomplete requirements gathering* means that not all information about the requirement has been collected, such as "if you push the current state to production, the app will appear unusable since it is missing a key functionality (that the customer didn't included in the requirements earlier)." *Lack of requirements verification* is related to the lack of officially verify the requirements, as we can see in "I don't have to follow any designs or verify requirements officially." *Lack of requirements specification documentation* is associated with the lack of documentation that has the requirements specification, such as "The previous implementation was done (badly) by a senior developer that left the company and did so without leaving a trace of documentation." Lastly, *lack of proper requirements management* refers to the lack of a proper management of requirements (tracking, specifications etc.), as we can notice in "What is a sane software solution for requirements/software specs?".

4.3   RQ3: What are the Effects of RE Issues?

We found 23 effects of the presence of RE issues. Table 3 presents the five most cited effects, along with the number of mentions of each effect (**#E**) and the percentage of the total of mentions (**%E**). The complete list of effects is available in [25].

Rework, unstable requirements, requirements specification inefficiency, development activity inefficiency, and *requirements management inefficiency* are the most commonly mentioned. *Rework* refers to the need of modifying the source code or requirement specification due to requirements changes, as illustrated in "I code all day and finally get stuck where requirement conflicts and I have to start over again." *Unstable requirements* mean that requirements change out of the software team control; for instance, "I have been through telling the customer what he wants, only to find the requirements change at a later date." *Requirements specification inefficiency* is related to complex or incomplete specifications, as reported in "more often than not, the business part of the specs is either incomplete or unaware of what can and can't be done." *Development activity inefficiency* refers to how lack of complete requirements affects development activities, as described in "most of the times... it's usually only when we start designing and developing that we end up in trouble, as a lot of the spec seems to have holes." Lastly, *requirements gathering inefficiency* indicates that important requirement was not captured during requirements elicitation "...the user realizes a major missing functionality that should be included in the system."

The other best-positioned effects are *requirements gathering inefficiency, project stopped due to lack of requirements, dissatisfaction, inaccuracy in non-functional requirements,* and *difficulty with system evolution/maintenance.* The effect *requirements gathering inefficiency* refers to lack of information about the system requirements, as

Table 3. Five most commonly mentioned effects of RE issues.

Effect	#E	%E
Rework	6	10%
Unstable requirements	5	9%
Requirements specification inefficiency	5	9%
Development activity inefficiency	5	9%
Requirements gathering inefficiency	4	7%

Caption:
#E: Number of mentions of an effect
%E: Percentage of #E in relation to the total all mentioned effects (58)

described in "the user realizes a major missing functionality that should be included in the system…" *Project stopped due to lack of requirements* is associated with users who did not know what functionalities the system may have, delaying its development. *Dissatisfaction* is related to the dissatisfaction of the software team with RE activities, as we can see in "Everything is rosy but the job." *Inaccuracy in non-functional requirements* refers to questions about where to describe non-functional requirements, for example, "In my current role, the team is using Agile Scrum and JIRA to write the user stories to capture the functional requirements etc." Lastly, *difficulty with system evolution/maintenance* is related to the difficulty of software team in maintaining the system due to the lack of system requirements specifications, such as "I don't want to compromise with quality but don't want to re-write everything on some change that I didn't expected."

4.4   RQ4: What Solutions Have Been Considered?

We found 59 possible solutions for RE issues. Table 4 presents the five most cited ones, along with the number of mentions of each solution (**#S**) and the percentage of the total of mentions (**%S**). The complete list of solutions is available in [25].

Clearly defining requirements solution was the most mentioned, followed by *use agile methodology, presenting solutions to the customer, using prototypes*, and *using techniques for requirements estimation*. In the context of this work, *clearly defining requirements* indicates that software teams need to spend time clarifying the requirements, as illustrated in "don't rush into coding the solution. Before typing down a single LOC, spend some time on clarifying the requirements…". *Use agile methodology* refers to following an agile methodology to support the RE process, for example, "what I would suggest in your case is a sort of agile approach." *Presenting solutions to the customer* refers to requesting feedback from the customer for a proposed solution, as described in "write a document proposing 2 or 3 solutions… get the customer to sign off on the ones they want and implement." *Using prototypes* is related to develop incomplete versions of the software to demonstrate its use and support requirements process activities, for example, "build prototypes. Just start drawing screens that don't do anything at first". Lastly, *using techniques for requirements estimation* refers to the use of techniques to estimate the requirement complexity, such as "try to do some collective estimation using planning poker."

Table 4. Five most commonly mentioned solutions for RE issues.

Solution	#S	%S
Clearly defining requirements	18	10%
Use agile methodology	12	6%
Presenting solutions to the customer	12	6%
Using prototypes	10	5%
Using techniques for requirements estimation	7	4%

Caption:
#S: Number of mentions of a solution
%S: Percentage of #S in relation to the total of all mentioned solutions (189)

The other best-positioned solutions are *prioritizing requirements, defining non-functional requirements together with the customer, using requirements management tools, holding meetings with customers,* and *defining done definition.* The solution *prioritizing requirements* refers to use a criterion to prioritize the software requirements, for example, "Give priority to the most required parts of the system and ask you to do the top ones in the given time." *Defining non-functional requirements together with the customer* is related to define the system's restrictions with the customer, as we can see in "you can simply create a user story in JIRA to document a nonfunctional requirement." *Using requirements management tools* is associated with the use of tools to support different RE activities, such as "use of a wiki for tracking requirements." *Holding meetings with customers* indicates that requirements analysts may have meetings with the customer to define the scope, collect the requirements, and have details about them, for example, "Suggest that you meet up and spend around an hour going through the details." Lastly, *defining done definition* refers to ensure that the job is fully completed, such as "You should have a definition of done."

5 Discussion

This section presents the relationship between requirements phases, issues, and solutions. It also compares our findings with those reported in related work.

5.1 Relationship Between Requirements Phases, Issues, and Solutions

From answering RQ1 and RQ4, we identify the main RE issues, RE phases affected by them, and their possible solutions. The data allowed us to go further and identify relationships between them when they are present in the same discussion. For example, we detected that the issue of *access to requirements* is related to the *requirements management* phase in a discussion, and one of the reported solutions was to *use agile methodology*. We recorded the triples formed by RE phase - RE issues - RE solutions as relationships and then grouped and counted all of them.

We organized the identified relationships using a Sankey diagram [26]. This diagram is composed of bars, representing the sources and destination of information, and links, showing the magnitude of the flow between those bars. Figure 6 shows a Sankey diagram considering the most common relationships we found on SWESE (the complete version is available in [25]). The flow starts with RE phases (left side) which are related to RE issues (middle), which are associated with possible solutions (right side). The numeric values next to each element (phase, issue, and solution) show the total number of times that a relationship occurs. The thickness of each link varies according to the value of the relationship.

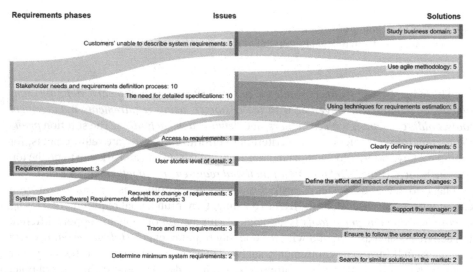

Fig. 6. Sankey diagram for the most affected RE phases and their most mentioned issues and solutions.

By analyzing the diagram, we can see, for instance, that the *requirements management* phase was affected by the following issues: *access to requirements* and *request for change of requirements*. The last issue has the following possible solutions: *support the manager* and *define the effort and impact of requirements change'*.

5.2 Comparison to Related Work

We compared our findings to the ones reported by the NaPiRE project [2]. This comparison was performed by one researcher and reviewed by three others. We realized that some issues, causes, or effects from NaPiRE matched with more than one from our study. For example, the issue of *weak knowledge of customer's application domain* (NaPiRE) corresponds to *requirements team is unaware of business rules* and *absence of business requirements* (our work).

Figure 7 shows the Venn diagrams with comparison results from issues, causes, and effects. We found 30 new issues, such as *requirements gathering done by non-technical people*, *mapping concrete implementation to vague business requirements*, and *user requirements verification*. Also, we recognized five new causes (*lack of requirements verification, poorly implemented requirement, change of development methodology, lack of code review*, and *lack of requirements specification documentation*) and eight new effects (*requirements management inefficiency, project stopped due to lack of requirements, inefficiency in specifying code quality requirements, failed project, need to use a support tool, inefficiency in the agile process, inefficiency in the requirements process, and management wants to start development even without defining requirements*).

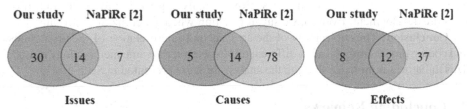

Fig. 7. Venn diagram with a comparison between issues, causes, and effects from our study and the NaPiRe project [2].

The complete comparison between the two works is available in [25], but in summary, our results not only confirm but also extend the set of information already reported in the technical literature.

6   Threats to Validity

This section discusses the threats to validity present in our study and actions taken to mitigate them. The threats follow the classification defined in [27].

Construct: This threat arises from a poor definition of the theoretical basis or the definition of the empirical process. In our case, the questions in Table 1 might not have been enough to gather all the necessary information from the discussions. We mitigated this threat using the NaPiRE protocol [2] to define those questions. Another threat is the process of choosing the discussions about RE issues. We mitigate this threat by performing tests with several terms related to RE, based on [24], before deciding which terms would be used. In addition, we applied both quantitative and qualitative filters to minimize the noise in our analysis. Lastly, a threat emerges from the fact that we only chose one Stack Exchange site (SWESE) and one tag ('requirements'). Stack Exchange has several websites focusing on different aspects of software development process, such as Software Quality Assurance & Testing and Stack Overflow, which can provide different perspectives on software requirements. However, we were interested in investigating issues in software requirements in general, leading us to choose SWESE as proxy and its 'requirements' tag. Even so, we recognize that using only one forum and one tag is a limitation of our work.

Internal: The process used for analyzing and coding the causes, effects, and solutions for RE issues can represent a threat in our study. To avoid this threat, the aforementioned procedures were performed by three pairs of researchers separately. Meetings were held at the beginning of the activities to align the activities among all the researchers. After the analysis, disagreements were resolved in consensus meetings between those researchers. Lastly, the results were reviewed by an experienced researcher.

External: This threat relates to the reasonableness of generalizing our conclusions. We considered a sample from SWESE, a popular Q&A platform where practitioners discuss day-to-day issues. We cannot guarantee that all SWESE participants are knowledgeable software engineers. However, the study only considers real-world practices, and the discussions in the forum are contextualized in the software development area. Thus, the results provide a valid perspective for analyzing RE issues in the context of software development. Even so, we suggest caution when generalizing the results.

Conclusion: The subjectivity of the analysis and interpretation of the discussions and coding process might lead to different results, even if the same methodology is used. This threat was minimized through the analysis process presented in Sect. 3.

7   Concluding Remarks

This work investigates RE issues discussed by software practitioners in the SWESE Q&A forum. Software practitioners can use our results to identify possible solutions for RE issues. For example, if a team is experiencing the issue *the need for detailed specifications*, the team can apply the solution *clearly define the requirements* as suggested in the Sankey diagram. Also, a software team can define strategies to deal with RE issues by consulting the list of causes and effects. By analyzing the causes, practitioners can define preventive practices to curb the presence of RE issues. Having information on effects can drive practitioners to identify the issues present in their projects. Lastly, researchers can use our findings to develop tools, strategies, and methodologies closer to practitioners' needs.

In future work, we intend to: (1) analyze the complete discussion population of SWESE and other Stack Exchange websites, broadening the findings and proving visualizations between issues and their elements (causes, effects, and solutions); (2) conduct interviews with software requirements analysts to assess the findings reported in this work in terms of their level of importance for RE management; and (3) update the proposed diagram considering the results obtained in (2) and empirically assess it with respect to its effectiveness to support the management of RE issues.

References

1. Nuseibeh, B., Easterbrook, S.: Requirements engineering: a roadmap. In: Conference on the Future of Software Engineering, pp. 35–46, ACM, New York, USA (2000)
2. Fernández, D.M., et al.: Naming the pain in requirements engineering. Empir. Softw. Eng. **22**(5), 2298–2338 (2016). https://doi.org/10.1007/s10664-016-9451-7
3. Nikula, U., Sajaniemi, J., Kälviäinen, H.: A state-of-the-practice survey on requirements engineering in small-and medium-sized enterprises. Research Report 951-764-431-0, Telecom Business Research Center Lappeenranta (2000)

4. Solemon, B., Sahibuddin, S., Ghani, A.A.A.: Requirements engineering problems and practices in software companies: an industrial survey. In: Ślęzak, D., Kim, T.-H., Kiumi, A., Jiang, T., Verner, J., Abrahão, S. (eds.) ASEA 2009. CCIS, vol. 59, pp. 70–77. Springer, Heidelberg (2009). https://doi.org/10.1007/978-3-642-10619-4_9

5. Liu, L., Li, T., Peng, F.: Why requirements engineering fails: a survey report from China. In: 18th IEEE International Requirements Engineering Conference, pp. 317–322 (2010)

6. Pekar, V., Felderer, M., Breu, R.: Improvement methods for software requirement specifications: a mapping study. In: 9th International Conference on the Quality of Information and Communications Technology, pp. 242–245 (2014)

7. Rios, N., et al.: Hearing the voice of software practitioners on causes, effects, and practices to deal with documentation debt. In: Madhavji, N., Pasquale, L., Ferrari, A., Gnesi, S. (eds.) REFSQ 2020. LNCS, vol. 12045, pp. 55–70. Springer, Cham (2020). https://doi.org/10.1007/978-3-030-44429-7_4

8. Barbosa, L., et al.: Organizing the TD management landscape for requirements and requirements documentation debt. In: Workshop on Requirements Engineering (2022)

9. Bonfim, V.D., Benitti, F.B.V.: Requirements debt: causes, consequences, and mitigating practices. In: International Conference on Software Engineering & Knowledge Engineering, pp. 13–18, Pittsburgh (2022)

10. Rios, N., Mendonça, M., Spínola, R.: A tertiary study on technical debt: types, management strategies, research trends, and base information for practitioners. Inf. Softw. Technol. **102**, 117–145 (2018)

11. Kamei, F., et al.: Grey literature in software engineering: a critical review. Inf. Softw. Technol. **138**, 106609 (2021)

12. Barua, A., Thomas, S.W., Hassan, A.E.: What are developers talking about? An analysis of topics and trends in stack overflow. Empir. Softw. Eng. **19**, 619–654 (2014)

13. Tahir, A., Dietrich, J., Counsell, S., Licorish, S., Yamashita, A.: A large scale study on how developers discuss code smells and anti-pattern in stack exchange sites. Inf. Softw. Technol. **125**, 106333 (2020)

14. Montandon, J.E., Politowski, C., Silva, L.L., Valente, M.T., Petrillo, F., Guéhéneuc, Y.: What skills do IT companies look for in new developers? A study with stack overflow jobs. Inf. Softw. Technol. **129**, 106429 (2021)

15. Gama, E., Freire, S., Mendonça, M., Spínola, R., Paixao, M., Cortés, M.I.: Using stack overflow to assess technical debt identification on software projects. In: Brazilian Symposium on Software Engineering, pp. 730–739 (2020)

16. Gomes, F., dos Santos, E.P., Freire, S., Mendonça, M., Mendes, T.S., Spínola, R.: Investigating the point of view of project management practitioners on technical debt. In: IEEE/ACM International Conference on Technical Debt, pp. 31–40 (2022)

17. dos Santos, E.P., Gomes, F., Freire, S., Mendonça, M., Mendes, T.S., Spínola, R.: Technical debt on agile projects: managers' point of view at Stack Exchange. In: Brazilian Symposium on Software Quality. ACM, New York (2022)

18. Vasilescu, B., Serebrenik, A., Devanbu, P., Filkov, V.: How social Q&A sites are changing knowledge sharing in open source software communities. In: 17th Computer Supported Cooperative Work and Social Computing, pp. 342–354 (2014)

19. Ahmad, A., Feng, C., Ge, S., Yousif, A.: A survey on mining stack overflow: question and answering (Q&A) community. Data Technol. Appl. **52**, 190–247 (2018)

20. Rios, N., Spínola, R.O., Mendonça, M., Seaman, C.: The practitioners' point of view on the concept of technical debt and its causes and consequences: a design for a global family of industrial surveys and its first results from Brazil. Empir. Softw. Eng. **25**(5), 3216–3287 (2020). https://doi.org/10.1007/s10664-020-09832-9

21. Sengupta, S., Haythornthwaite, C.: Learning with comments: an analysis of comments and community on Stack Overflow. In: Hawaii International Conference on System Sciences, pp. 2898–2907 (2020)
22. Ahasanuzzaman, M., Asaduzzaman, M., Roy, C.K., Schneider, K.A.: Mining duplicate questions of stack overflow. In: IEEE/ACM Working Conference on Mining Software Repositories, pp. 402–412. IEEE (2016)
23. Kavaler, D., Posnett, D., Gibler, C., Chen, H., Devanbu, P., Filkov, V.: Using and asking: APIs used in the android market and asked about in Stack Overflow. In: Jatowt, A., et al. (eds.) SocInfo 2013. LNCS, vol. 8238, pp. 405–418. Springer, Cham (2013). https://doi.org/10.1007/978-3-319-03260-3_35
24. ISO/IEC/IEEE International Standard: Systems and software engineering–life cycle processes–requirements engineering. In: ISO/IEC/IEEE 29148:2018(E), pp. 1–104 (2018)
25. Freire, S., et al.: Requirements engineering issues experienced by software practitioners: a study on stack exchange-complementary material. Zenodo (2023). https://doi.org/10.5281/zenodo.7647916
26. Lupton, R.C., Allwood, J.M.: Hybrid Sankey diagrams: visual analysis of multidimensional data for understanding resource use. Resour. Conserv. Recycl. **124**, 141–151 (2017)
27. Wohlin, C., Runeson, P., Höst, M., Ohlsson, M.C., Regnell, B., Wesslén, A.: Experimentation in Software Engineering. Springer, Cham (2012). https://doi.org/10.1007/978-3-642-29044-2

An Empirical Study of the Intuitive Understanding of a Formal Pattern Language

Elisabeth Henkel[✉][ID], Nico Hauff[ID], Lukas Eber, Vincent Langenfeld[ID], and Andreas Podelski[ID]

Department of Computer Science, University of Freiburg,
Freiburg im Breisgau, Germany
{henkele,hauffn,langenfv,podelski}@informatik.uni-freiburg.de

Abstract. [**Context and motivation**] Formal pattern languages with a restricted English grammar, such as the pattern language of Konrad and Cheng, give us the possibility to combine human intuition and the rigour of a machine. [**Question/problem**] The question arises to what extent the intuitive understanding of such a pattern language is in agreement with its formal semantics. [**Principal ideas/results**] We present an empirical study to address this question. The existence of a formal semantics allows us to use the machine as an objective judge to decide if the intuitive understanding is correct. The study confirms empirically the practical usefulness of HANFORPL in that the intuitive understanding matches the formal semantics in most practically relevant cases. The study reveals that a number of *phrases of interest* represent critical edge cases where even a prior exposure to formal logic is not a guarantee for the correct intuitive understanding. [**Contribution**] We show how the alignment of formal and intuitive semantics can be investigated, and that this alignment can not simply be assumed. Nonetheless, results regarding the understandability of HANFORPL are favourable with high understandability in commonly used patterns. The results of the study will be the basis of improvements in HANFORPL.

Keywords: Pattern Languages · Formal Requirements · Intuitive Understanding · Empirical Study

1 Introduction

The formal representation of requirements is supposed to overcome some of the deficiencies of natural language requirements, especially lack of precision and non-machine readability [2,5,15,16]. However, if requirements are formulated in a formal logic such as temporal logic, they are accessible to only a restricted group of requirement engineers. To overcome the lack of general accessibility, Konrad and Cheng introduced a pattern language to formulate formal requirements as sentences in a restricted English grammar [8]. The intuitive understanding of these sentences is based on the intuitive understanding of natural language, while the formal semantics is derived through corresponding temporal logic formulas.

© The Author(s), under exclusive license to Springer Nature Switzerland AG 2023
A. Ferrari and B. Penzenstadler (Eds.): REFSQ 2023, LNCS 13975, pp. 21–38, 2023.
https://doi.org/10.1007/978-3-031-29786-1_2

For example, we can use its formal semantics to uniquely determine that the requirement below is satisfied by the behaviour depicted in Fig. 1:

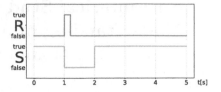

Fig. 1. Example behaviour over the observables R and S.

*Globally, it is always the case that if **R** holds, then **S** holds after **1** time unit.*

It would thus seem that with pattern languages, we are in the ideal situation where we can have both, the precision of formal requirements and the accessibility of natural language. However, while the interface to the computer is fixed by the formal semantics, the interface to the human still relies on the intuitive interpretation of natural language. The question is to what extent we still have the issues of natural language requirements if restricted to the subset of sentences defined in a pattern language. In particular, the question arises to what extent the intuitive understanding of each requirement in the pattern language will be correct.

The existence of a formal semantics for the requirements gives us the unique opportunity to phrase the above question in a mathematically precise sense. We can give a mathematically precise definition of what is the *correct* intuitive understanding of a requirement in the pattern language, namely, through its formal meaning. In contrast, for an informal requirement, it would seem impossible to distinguish one possible intuitive understanding over another one.

For a requirement in the pattern language, the formal meaning is defined as the set of system behaviours that satisfy the corresponding temporal logic formula. Thus, we can base the test of the intuitive understanding of a requirement on a set of example behaviours, some of which satisfy the requirement and some of which do not. The existence of a formal semantics allows us to define an objective judge who decides whether the intuitive understanding is correct: the machine. Both, the requirement and the behaviour have a machine representation, and an algorithm exists to decide whether the behaviour satisfies the requirement. Thus, we only compare the intuitive understanding to the algorithmic decision.

In this paper, we report on an empirical study to investigate the difference between the formal semantics and the intuitive understanding of requirements in a particular example of a pattern language called HANFORPL. The pattern language comes with a framework to specify requirements and behaviours, and to check whether a behaviour satisfies a requirement [1,5]. The study confirms empirically the practical usefulness of HANFORPL in that the intuitive understanding matches the formal semantics in many cases. The study reveals that a number of *phrases of interest* represent critical edge cases where even a prior exposure to formal logic is not a guarantee for the correct intuitive understanding.

Table 1. The table shows all patterns of HANFORPL, with their membership to a group describing overall behaviour (the *Order*, the *Occurrence*, or the *Real-Time*), the names of each pattern, and the pattern text. Due to the available space, we use "..." to omit the shared phrase *it is always the case that*. Names of patterns not already part of the SPS [13] are shown in blue colour.

	Name	Pattern
Order	ConstrainedChain	... if **R** holds, then **S** eventually holds and is succeeded by **T**, where **U** does not hold between **S** and **T**.
	Initialization	... initially **R** holds
	Persistence	... if **R** holds, then it holds persistently
	PrecedenceChain21	... if **R** holds, then **S** previously held and was preceded by **T**
	PrecedenceChain12	... if **R** holds and is succeeded by **S**, then **T** previously held
	Response	... if **R** holds, then **S** eventually holds.
	ResponseChain12	... if **R** holds, then **S** eventually holds and is succeeded by **T**.
	Precedence	... if **R** holds, then **S** previously held.
Occur.	Absence	it is never the case that **R** holds
	ExistenceBoundU	transitions to states in which **R** holds occur at most twice
	Invariance	... if **R** holds, then **S** holds as well
	Universality	... **R** holds
Real-time	DurationBoundL	... once **R** becomes satisfied, it holds for at least **S** time units
	DurationBoundU	... once **R** becomes satisfied, it holds for less than **S** time units
	EdgeResponseBoundU1	... once **R** becomes satisfied and holds for at most **S** time units, then **T** holds afterwards
	EdgeResponseBoundL2	... once **R** becomes satisfied, **S** holds for at least **T** time units
	EdgeResponseDelay	... once **R** becomes satisfied, **S** holds after at most **T** time units
	EdgeResponseDelayBoundL2	... once **R** becomes satisfied, **S** holds after at most **T** time units for at least **U** time units
	InvarianceBoundL2	... if **R** holds, then **S** holds for at least **T** time units
	ReccurrenceBoundL	... **R** holds at least every **S** time units
	ResponseBoundL1	... if **R** holds for at least **S** time units, then **T** holds afterwards
	ResponseBoundL12	... if **R** holds for at least **S** time units, then **T** holds afterwards for at least **U** time units
	ResponseDelay	... if **R** holds, then **S** holds after at most **T** time units
	ResponseDelayBoundL2	... if **R** holds, then **S** holds after at most **T** time units for at least **U** time units
	TriggerResponseBoundL1	... after **R** holds for at least **S** time units and **T** holds, then **U** holds
	TriggerResponseDelayBoundL1	... after **R** holds for at least **S** time units and **T** holds, then **U** holds after at most **V** time units
	UniversalityDelay	... **R** holds after at most **S** time units

2   Hanfor Pattern Language

The HANFOR pattern language (HANFORPL) is based on the patterns of Konrad and Cheng [8] and uses the Duration Calculus semantics of Post [13]. In fact, HANFORPL shares a large portion of patterns with the Specification Pattern System (SPS) from [13].

Each instantiation of a requirement in HANFORPL is a combination of a *scope* defining the general applicability of a pattern, followed by the *pattern* itself. The scopes can be chosen from the following options *Globally*, *After **P***, *After **P** until **Q***, *Before **P***, and *Between **P** and **Q***. The resulting patterns are listed in Table 1. During instantiation placeholders (usually P, Q, R, S, T) have to be replaced by Boolean expressions over observables (using $\neg, \wedge$ for Boolean and $<, =$ for numeric observables).

The semantics of each scope and pattern combination is defined by a logical formula containing the same placeholders. For a more depth introduction to the formal foundations and the pattern semantics in detail, we kindly refer the reader to the cited work.

3   Empirical Study

In this section, we describe the overall goal of our empirical study, our research questions, and the study design.

3.1   Goal and Research Questions

As requirements pattern are used to communicate expected system behaviour, e.g., between customers or different departments, it is necessary that requirements are as understandable as possible to as many stakeholders as possible. That is, the semantics of the pattern defined by formal logics should align with the intuitive understanding of usual stakeholders.

The goal of this study is thus to investigate to what extent the intuitive understanding of formal requirements in HANFORPL is correct in the sense that it matches the formal semantics. This is closely related to the question of the practical usefulness of HANFORPL.

Further, we aim to identify possible reasons for misinterpretation in order to improve HANFORPL in the long term.

Based on previous experience (e.g. [9,11]), we are confident that formally trained people with some training in HANFORPL perform well using the pattern language. With Research Question R1, we want to investigate how well participants without any training in HANFORPL understand the patterns.

However, a basic understanding of formal logics and/or requirements engineering in general may serve as a predictor for the performance dealing with edge cases and uncommon concepts (Research Question R2).

As the requirements pattern are based on natural language sentences, there may be phrases that allow for several sensible interpretations for complex concepts, e.g., formulations referring to timing constraints and quantification. These phrases of interest are investigated in detail in Research Question R3.

R1 How understandable is HANFORPL without former training in the pattern language itself?

R2 Does training in the fields of requirements engineering or formal logics have a positive effect on the understanding of HANFORPL patterns?
    a) Requirements engineering
    b) Formal logics

R3 How is the understanding of HANFORPL impacted by complex concepts, i.e., formulations referring to timing constraints and quantification?

With regard to the last Research Question (R3), we identified several phrases used within HANFORPL to describe concepts like timing constraints and quantification. In the following, we present a list of these *phrases of interest* (highlighted within the according pattern) together with a description of possible interpretations. Additionally, we state which of the possible interpretations matches the *intended meaning*, i.e., the semantic fixed by the corresponding Duration Calculus formula.

(prev) *[...] if **R** holds, then **S*** previously held : For this phrasing, we see two possible points for ambiguity. First, the phrase does not specify whether **S** has to hold persistently or only for a non-zero time interval before any occurrence of **R**. And second, it is not specified whether **S** has to hold at an arbitrary point in time before the occurrence of **R** or directly before **R** holds. The intended meaning is the following: Every occurrence of **R** must at some point be preceded by a non-zero time interval in which **S** held.

(afterw)/(afterw*) *[...] if [...], then **S*** holds afterwards : Analogous to (prev), we identified two possible ambiguities. The phrase does not specify, whether **S** has to hold persistently or only for a non-zero time interval (afterw). Additionally, it is not specified, whether **S** has to hold directly after the trigger event (the [...]-part) or only at an arbitrary point in time after the triggered event (afterw*). The intended meaning is the following: **S** must hold directly after the trigger event for some non-zero time interval.

(aam) *[...]* **R** holds after at most **d** seconds : The phrase does not specify whether **R** has to hold persistently after the **d** seconds have passed (which is the intended meaning), or only has to hold for a non-zero time interval.

(aam-cond) *[...] if [...],* then **S** holds after at most **d** seconds : This wording is the conditioned version of (aam), i.e., it is dependent on the context of a preceding trigger. Analogous, it is not specified whether **S** has to hold persistently or only for a non-zero time interval after **d** seconds have passed. The intended meaning is the following: **S** has to hold for a non-zero time interval. However, due to an oversight while extending the pattern language, this interpretation is clearly inconsistent with the intended meaning provided in (aam).

(obs)/(obs+) *[...]* once **R** becomes satisfied *[...]*: We identified two possible ambiguities in this pattern. The first is regarding the meaning of the phrase *becomes satisfied*. It might be unclear, whether a rising edge of **R** is strictly required in all cases, or whether this phrase also includes system behaviour where **R** initially holds (obs). The second ambiguity concerns the keyword *once*. It might be unclear, whether this means that every occurrence of **R** becoming satisfied should be considered or only the first occurrence (obs+). The intended meaning is the following: all occurrences of rising edges of **R** should be considered.

(rec) *[...]* **R** holds at least every 2 s : The intended meaning of this phrase is that the length of intervals in which **R** does not hold is at most 2 s. However, this wording might be misinterpreted so to mean, that R holds at fixed points in time $t_0 = 0, t_1 = 2, t_2 = 4, \ldots, t_n = 2n$.

Remark. Even though some inconsistencies, e.g., the intended meaning of *holds* in (aam) and (aam-cond), were identified while preparing the study, we decided to make no premature changes for two reasons: First, we are interested to know whether such an inconsistency is noticeable in the results. Second, if it is noticeable, which of the different interpretations is the one that most participants agree with.

3.2 Subject Selection

Participants for the empirical study were selected via convenience sampling of contacts of the authors and second-degree contacts in an original equipment manufacturer (OEM) in the automotive field. Subjects are mostly computer scientists and requirements engineers from the field of software engineering, automotive engineering and formal methods. The empirical study was conducted in the form of an online survey with anonymous participants out of the described group. Participants were asked to complete the survey without any help, but there is no control mechanism against actual cheating. At the beginning of the survey, we asked the participants for demographic information including their age group, their experience in requirements engineering, HANFORPL, and formal logics.

3.3 Object Selection

This first step into the investigation of the understanding of a pattern language is focused on pattern understanding from reading, as it is the basis for further inquiries, e.g., into the generative task of pattern instantiation for formalisation. Therefore, the study (apart from demographic questions) consists of a single repeated task: to decide if pattern instantiations are fulfilled by timing diagrams of system behaviour. Simply checking phrases in isolation (e.g., What is your understanding of the phrase *"holds after at most 2 s"*?) was no option, as their interpretation may differ when embedded into the context of a pattern. This can, for example, be seen when comparing the intended meaning of the two phrases of interest (aam) and (aam-cond) within the patterns *R holds after at most T seconds* and *If R holds, then S holds after at most T seconds*.

Within the survey, we test the participants' understanding of patterns from the HANFORPL. To select a suitable set of patterns, the following criteria are considered: 1) The survey should focus on patterns that are relevant in industrial practice, 2) the survey should include the patterns using phrases of interest, and 3) the survey should be short enough to be filled in without too much interruption to a work day of participants in the industry, i.e., the survey should be completed in about 30 to 40 min.

We considered patterns that were shown to be used frequently for the formalisation of requirements in the automotive context (criterion 1). We then added patterns containing phrases of interest (criterion 2) if not yet included by the

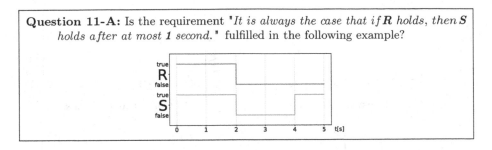

Question 11-A: Is the requirement "*It is always the case that if* **R** *holds, then* **S** *holds after at most* **1** *second.*" fulfilled in the following example?

Fig. 2. The first of the four questions to investigate the understanding of the *ResponseDelay* pattern; correct answer: *yes*.

first selection criterion. For patterns whose meaning is inverse to an already added pattern (e. g. *it is always the case that* **R** *holds* and *it is never the case that* **R** *holds*), we only included the positive formulated pattern in the survey. We do not assume that negative and positive formulations do behave similar, but that using the usually less legible negative formulation is not adding any new insights. Three patterns adding no unique phrases were dropped due to the timing constraint (criterion 3). The selection process resulted in a list of 17 patterns from the HANFORPL (see Table 2).

3.4 Survey Design

The questions should be formulated in a style that avoids errors based on the incomprehensibility of the survey rather than the pattern under investigation. We therefore decided to work with only one type of question, i. e., we asked whether or not a given instantiated requirement in HANFORPL is fulfilled by a given example system behaviour. For each question, the requirement was given as written text, while the example behaviour was depicted as timing diagram. Skipping a question was not permitted. Figure 2 exemplarily shows the first question that was asked to investigate the understanding of the *ResponseDelay* pattern. Consecutively, we asked the same question for three more timing diagrams (Fig. 3). That is, for each of the selected patterns, participants of the study had to match four example system behaviours against an instantiated requirement in HANFORPL, yielding a total of 68 questions.

The order of questions in the survey and therefore the order of the requirements presented to the participants was static. Participants should be eased into the language by a controlled encounter with the different features of the language, from one observable, over several observables, timed quantification and so on. Thereby preventing noise within the answers resulting from being overwhelmed by a first occurrence of too many new concepts at once. Apart from the gradual exposure to the language features, we assume that no relevant training effect is present, as no feedback on the correctness of the answers was given.

To make the survey feasible within a time frame of about 30 to 40 min, the survey includes a high number of example behaviours directly targeting the

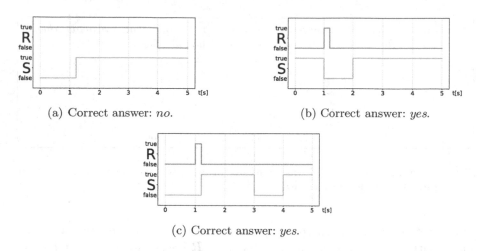

(a) Correct answer: *no*.   (b) Correct answer: *yes*.

(c) Correct answer: *yes*.

Fig. 3. Timing diagrams used to investigate the understanding of the *ResponseDelay* pattern (Questions 11 B - D).

phrases of interest (see Table 3). Correct answers to these questions thus mean, that the general behaviour of the pattern has been understood *and* the phrase of interest was interpreted correctly (with respect to the formal semantics).

The survey does not investigate the understanding of different scopes. This would introduce another level of complexity and hence require more questions to be asked to infer reasons for possible incorrect answers. We therefore implicitly instantiated all requirements with the scope *globally*.

4   Results

The study was completed by 37 participants with an average experience in requirements engineering of 3.3, in HANFORPL of 1.8, and in formal logics of 3.9 on a self assessment scale of 1 (not experienced at all) to 5 (very experienced). The median age group was *41 to 50*. One participant indicated that they clearly misunderstood the given task as part of a feedback email. The described answer set (all *false*) was clearly identifiable, and the participant was removed as an outlier. Table 2 shows the detailed performance of all participants over all patterns and questions.

Participants had to rate their familiarity with HANFORPL in the beginning of the survey (see Fig. 4). To investigate Research Question R1, we separate the participants into two groups: The 26 participants being untrained in HAN-FORPL (answering 1 in the related self assessment question) answered with 75% accuracy (on average 51.1 of 68 questions answered correctly). The 10 participants that received former training in HANFORPL (answering > 1 in the related self assessment question) answered with 79% accuracy (on average 53.7 of 68 questions answered correctly).

Table 2. Survey results per pattern (listed in the order they occur in the survey) and question (columns A,B,C,D).

Pattern Name	Average of correct answers (%)				
	A	B	C	D	Total
Universality	94	100	100	100	99
Invariance	67	97	64	89	79
Initialization	94	100	100	83	94
Persistence	75	100	86	100	90
Precedence	97	53	78	89	79
DurationBoundL	14	100	92	81	72
DurationBoundU	89	14	92	97	73
ReccurrenceBoundL	97	83	86	92	90
UniversalityDelay	53	92	53	86	71
InvarianceBoundL2	64	58	92	61	69
ResponseDelay	47	89	81	89	76
ResponseDelayBoundL1	58	75	92	53	69
ResponseBoundL1	39	94	53	53	60
ResponseBoundL12	50	100	92	83	81
EdgeResponseBoundL2	97	56	17	86	64
EdgeResponseBoundU1	42	72	86	11	53
EdgeResponseDelayBoundL2	100	78	75	56	77

There is a slight, non-significant trend of training in HANFORPL leading to more correct answers (Pearson correlation of $r(34) = 0.292$ with $p = 0.083$). The difference between both groups is statistically not significant (Mann-Whitney-U $U = 103$ with $p = 0.348$). As both groups performed similar, we do not discern between them in the following.

In the beginning of the study, participants had to give a self assessment of their experience in formal logics as well as requirements engineering (relating to R2). We assume that both disciplines give a solid foundation (be it in vocabulary or concepts) for a better understanding of requirements pattern languages.

It turned out, that training in requirements engineering does at best show a weak and statistically not significant trend (Pearson correlation of $r(34) = 0.231$ with $p = 0.175$). Astonishingly, the best and worst participants claimed to have a high understanding for requirements engineering (see Fig. 5).

In contrast, experience in formal logic turned out to have a strong correlation (Pearson correlation of $r(34) = 0.647$ with $p < 0.0001$) with the number of right answers (see Fig. 6).

As the final research question (R3), we investigate the phrases of interest. Detailed results from the relevant questions can be seen in Table 3. For each phrase of interest, its related patterns and questions, the table shows the overall result, as well as the results of participants with prior training in formal logic (answering > 2 in the related self assessment question; $n = 30$) and with little to no training in formal logic (answering ≤ 2; $n = 6$).

Table 3. Correctness results for the phrases of interest. Each row shows the according phrase id, the pattern containing the phrase and which question in the survey prompted that exact behaviour followed by the percentage of correct answers. Column N shows participants with little to no, column L with training in formal logics.

ID	Related pattern	Question	Correct		
			N (6)	L (30)	Overall
prev	Precedence	C	50	83	78
prev	Precedence	D	67	93	89
afterw	ResponseBoundL1	D	0	63	53
afterw	EdgeResponseBoundU1	B	50	77	72
afterw*	ResponseBoundL1	C	33	57	53
afterw*	ResponseBoundL12	A	33	53	50
afterw*	EdgeResponseBoundU1	A	50	40	42
aam	UniversalityDelay	A	33	57	53
aam	UniversalityDelay	C	33	57	53
aam-cond	ResponseDelay	D	83	90	89
aam-cond	ResponseDelayBoundL1	C	83	93	92
obs	DurationBoundL	A	17	13	14
obs	DurationBoundU	B	0	17	14
obs	EdgeResponseBoundL2	C	0	20	17
obs	EdgeResponseBoundU1	D	0	13	11
obs+	DurationBoundL	C	100	90	92
obs+	DurationBoundU	D	100	97	97
obs+	EdgeResponseBoundL2	D	83	87	86
obs+	EdgeResponseDelayBoundL2	B	83	77	78
rec	ReccurrenceBoundL	B	83	83	83

Table 4. Remainder of questions with high error rates not already covered by the phrases of interest. Column N shows participants with little to no, column L with training in formal logics.

ID	Related pattern	Question	Correct		
			N (6)	L (30)	Overall
antec	Invariance	C	17	73	64
antec	InvarianceBoundL2	D	17	70	61
atonce	Precedence	B	50	53	53
atonce	ResponseDelay	A	33	50	47
atonce	ResponseDelayBoundL1	A	33	63	58
atonce	ResponseDelayBoundL1	D	50	53	53
atonce	ResponseBoundL12	A	33	53	50

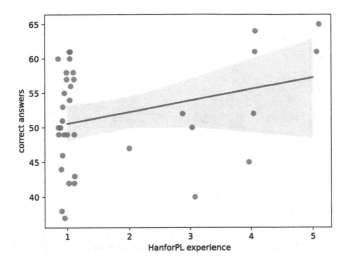

Fig. 4. The influence of former training in HANFORPL (x-Axis) on the number of correct answers given (y-Axis).

Table 4 contains results of the remainder of questions with high error rates. The errors from these questions can be attributed to two kinds of formulations and underlying semantics used in the pattern language. For ease of reading, we define these ad-hoc categories analogous to the phases of interest:

(antec) *[...] if R holds, then S holds as well* : This requirement's semantic is equal to the implication $R \rightarrow S$, i.e., if *R* has to hold, then *S* has to hold as well, but not vice versa.

(atonce) *[...] if R holds, then S holds after at most T time units*: In this example, it is not clear if **S** is expected to be in real succession to **R** (as one would expect for a causal relationship), or if both happening at the same time is also valid behaviour. The latter is the case in HANFORPL.

5 Discussion

The overall results regarding the understanding are positive, showing that most patterns in HANFORPL can be understood even without prior training in the pattern language.

Results of 75% to 79% correct answers of participants untrained and trained in the pattern language entail that generally more than every fifth answer to questions of whether behaviour belongs to the system are erroneous. This interpretation is heavily skewed as the survey is focused on phrases of interest, i.e., on edge cases which are prone to misinterpretation. Thus favouring participants familiar with HANFORPL as well as skewing the distribution of patterns heavily

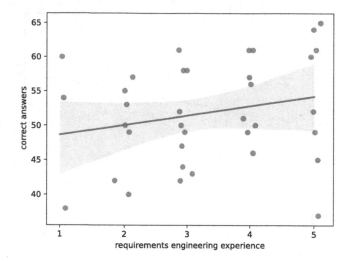

Fig. 5. The influence of experience in requirements engineering (x-Axis) on the number of correct answers given (y-Axis).

towards more complex patterns within the survey, that are used with far lower frequency in practice. Requirements sets usual for industrial practice, as reported by [12], mainly contain patterns that got high success rates. This is especially the case for the *Universality* pattern and common applications of *Invariance-BoundL2* and *ResponseDelay*, i.e., excluding the answers to question A of the latter pattern, (see Table 2). Therefore, we conclude, that HANFORPL turned out to be understandable, even for untrained participants.

Results show, that training in formal logic serves as a good predictor for the comprehension of the requirements pattern. The explanation of this effect could be twofold: For one, formal logics, especially temporal logics (e.g. LTL, MTL or Duration Calculus) have similar interpretation of concepts like referring to, e.g., a future state just requires just a non-zero interval (or one state) except denoted differently. Thus, the everyday understanding of these terms is already aligned with the formal meaning. Second, training in formal logics (in contrast to requirements engineering) may allow for more detachment from the actual physical system, i.e., ignoring the question as to what might happen before or after the timing diagram.

Analysis of individual phrases allows to pinpoint phrases and concepts that are not aligned with their everyday understanding (Table 3).

The results show, that (rec) and (prev) are unproblematic, as questions regarding those phrases of interest were answered correctly by most participants.

For the phrase of interest in (afterw), i.e., the text *S holds afterwards*, participants leaned on the side of *S* only holding for a non-zero interval which matches the intended meaning (with 53% resp. 72% correct answers). For the

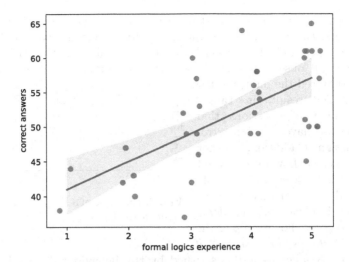

Fig. 6. The influence of experience in formal logics (x-Axis) on the number of correct answers given (y-Axis).

ResponseBoundL1 D question, the divide between logically trained (63% correct) versus untrained (0% correct) shows that there is a different understanding of the phrases depending on training, i.e., all of the latter did assume that S has to hold persistently. Again, disambiguation by including the word *persistently* in pattern where this is the case should solve this case.

Regarding (afterw*), the question whether S has to hold immediately after the trigger event (intended meaning), participants leaned to answer incorrectly (with 53%, 50%, resp. 42% correct answers). This result shows, that the behaviour has to be made explicit. The uncertainty if S has to hold immediately or at some arbitrary point (afterw*) should be addressed by including the word *immediately* as part of the patterns.

All participants performed well on the phrasing of (aam-cond) *if [...], then S holds after at most T seconds*. In contrast, for (aam) only 53% answered correctly, i.e., that the observable has to hold persistently. Thus, the interpretation in (aam-cond) is in alignment with the common understanding, while the *UnversalityDelay* pattern containing the (aam) phrase should be changed to include the phrase *persistently* to be *[...] S holds after at most T seconds persistently.*

The most recent addition to the pattern language is concerned with reaction to changes of observables. Questions related to the phrase *once* **R** *becomes satisfied, [...]* (obs+) were consistently answered correctly, i.e., the requirement has to be evaluated after each time **R** becomes satisfied.

The question if an explicit rising edge is required (obs) and how especially initial behaviour is treated was highly problematic (below 17% correct answers). Answers were systematically given so, that the state of the system before the timing diagram was the missing part to satisfy the change of the observable. As we did not alter the observables, we did not include the negative case. Including the negative case would have been beneficial in analysing if participants just assumed that all observables are *false* in the beginning, or if any state was possible that suited the interpretation.

The detailed results in Table 2 show a number of questions that turned out to have a high error rate. We assigned additional ad-hoc phrases of interest: Low rates of right answers in (antec) (see Table 4) could be attributed to a common error when dealing with implications, the *denying the antecedent*. For example, the pattern *it is always the case that if*

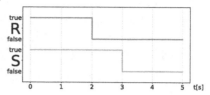

Fig. 7. Example of a *denying the antecedent* error in the survey.

R holds, then S holds as well is satisfied by the behaviour depicted in Fig. 7. Nonetheless, *S* being true without *R* being true in time interval $[2, 3]$ was seen as a violation by 36% of participants, especially those with little to no training in formal logic (only 17% correct in both questions). For nine participants (25%) the error was stable over both questions regarding (antec). This could point to a systematic misunderstanding of implication, or at least a difference in the understanding to the phrasing used for implication in this pattern. The existence of systematic differences of understanding conditionals has been shown by Fischbach et al. [6].

A large number of errors stem from cases in which everything relevant happens at the same point in time (atonce). An example is the requirement *if R holds, then S holds after at most 1 second* together with the behaviour depicted in Fig. 8. One can see that for time interval $[0, 2]$ *R* as well as *S* are true, i.e., the causal relation, although it only needs to be

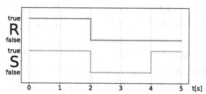

Fig. 8. Problems with immediate satisfaction of a property.

satisfied with a delay of at most one second, is satisfied immediately. This may be again due to a notion of the requirements as more of a physical system, where the trigger results in an action with a real causal delay.

Many of the problems detected in this study should be fixed by small changes regarding the pattern language. As an immediate result of this study, several improvements for the phrases of interest were suggested, as discussed above. These modifications have to be verified carefully so, that the simplicity of the sentences is not lost in an overly complex sequence of adjectives describing each observable.

Similar to the argument in [3,14], a basic understanding of formal logic (better formal methods in general) should be the best mitigation for misalignment in the understanding of formal constructs such as the problems found with (antec).

Additionally, we include clarifications targeted on the misunderstandings found in this survey in our training material.

6   Threats to Validity

6.1   Internal Validity

The threat of *Repeated Testing* is concerned with participants learning over the run of an experiment. As participants were not informed if their answers were correct, they should not have been able to gain information on the correct interpretation of the pattern. An acclimatisation to the pattern language was intended though in order to prevent participants from being overwhelmed with the more complex pattern. As the survey was performed in an industrial context, *Maturation*, i.e., changes over the duration of the survey can influence the results. We tried to keep the survey as short as possible in order to prevent tiring and impatience (or loss of participants) due to more pressing concerns. The threat of *Instrumentation* is concerned with the influence of the experimental material itself on the results. We tried to make the examples of system behaviour as accessible as possible for the use by not formally trained participants [4]. Nonetheless, problems with phrases of interest starting in the beginning of the timing diagram may have suffered from a notion of the system too commonly associated with a real system, i.e., where there is always a previous state even if switched off. To guarantee that the questions themselves do not contain errors, all timing diagrams were automatically verified by the pattern simulator being part of HANFOR.

6.2   Construct Validity

The threat of *Interaction of setting and treatment* is concerned with non-aligning circumstances of experiment and reality. In fact, the experiment is presented in a form focusing on the patterns itself, not on realistic requirements. In a real setting, expressions over observables in pattern instantiations can add another layer of complexity, that is abstracted away, to get data on the pattern themselves. In reality, the correctness numbers can be much lower as requirements get considerably more complex. Nonetheless, the expression language is not likely to have interactions with the phrasing of the surrounding pattern.

6.3   External Validity

The threat of *Interaction of selection and treatment* is concerned with the selection of non-representative participants. Our participants were selected by contacting cooperation partners from different engineering divisions, and the chair mailing list. This way, we tried to spread the risk of convenience sampling over different businesses and person groups likely to be in a position or likely to be in the near future of using a requirements pattern language.

7 Related Work

Winter et al. [15] conduct a survey on the understandability of quantifiers and their negation (such as *all, more than* or *at least*) in natural language requirements. Results show, that there are significant effects on reading speed and error rate between the different quantifiers and their negated forms. Based on the results, advice for writing requirements is given. This recent work shows the relevance of investigations into the understanding of requirements in general. Phrasings are chosen once and reproduced in each instantiation, i.e., any problem introduced to a pattern is multiplied over a requirements specification. Therefore, ensuring understanding by a broad audience is even more relevant.

Giannakopoulou et al. [7] address the problem of pattern understanding by presenting several representations of the instantiated requirement, both graphical and as formal logic, e.g. LTL. This is a necessary support for error recovery by comparison to the intended result, while the pattern language should itself prevent errors in the first place by being aligned with the intuitive understanding of the patterns.

A different approach is taken by Moitra et al. [10], designing the requirements language in the style of a programming language. This surely aligns the intuitive understanding with stakeholders from a computer science background, but may exclude other stakeholders entirely, because of the condensed syntax.

8 Conclusion

In this paper, we demonstrated how an inquiry on the alignment of the formal semantics of the HANFORPL and the intuitive understanding of requirements engineers can help to understand and improve the pattern language. Almost half of the patterns considered in the survey are contained in the SPS by [13]. Parts of the results can therefore be generalized to SPS-like languages.

The analysis results are positive, and the pattern language performed very well in hiding the formal complexity behind intuitively understandable sentences. Nonetheless, the language contains several phrases that lead to near random decisions, and misconceptions of logic can lead to misinterpretations that cannot be mitigated entirely by phrasing. We suggested several improvements through the analysis.

This study was a short, industry friendly foray into the comprehensibility of HANFORPL. In the short term, the pattern of HANFORPL will be improved by the suggested changes. Based on this study, future work will be to design a more thorough investigation of the patterns, especially in conjunction with scopes. The basis of this extended survey could be a mutation based scenario generator to do the tedious work of generating different classes of scenarios. While the examination of each pattern is of immediate use for requirements engineering, the question remains if we can evaluate the meaning of single words (e.g. *after* and *once*), or if their meaning is heavily influenced by the context in which they occur.

Acknowledgements. We thank all participants, and Amalinda Post for forwarding the study.

References

1. Becker, S., et al.: Hanfor: semantic requirements review at scale. In: REFSQ Workshops. CEUR Workshop Proceedings, vol. 2857. CEUR-WS.org (2021)
2. Berry, D.M., Kamsties, E.: The syntactically dangerous all and plural in specifications. IEEE Softw. **22**(1), 55–57 (2005)
3. Bjørner, D., Havelund, K.: 40 years of formal methods. In: Jones, C., Pihlajasaari, P., Sun, J. (eds.) FM 2014. LNCS, vol. 8442, pp. 42–61. Springer, Cham (2014). https://doi.org/10.1007/978-3-319-06410-9_4
4. Dietsch, D., Feo-Arenis, S., Westphal, B., Podelski, A.: Disambiguation of industrial standards through formalization and graphical languages. In: RE, pp. 265–270. IEEE Computer Society (2011)
5. Dietsch, D., Langenfeld, V., Westphal, B.: Formal requirements in an informal world. In: 2020 IEEE Workshop on Formal Requirements (FORMREQ), pp. 14–20. IEEE (2020)
6. Fischbach, J., Frattini, J., Mendez, D., Unterkalmsteiner, M., Femmer, H., Vogelsang, A.: How do practitioners interpret conditionals in requirements? In: Ardito, L., Jedlitschka, A., Morisio, M., Torchiano, M. (eds.) PROFES 2021. LNCS, vol. 13126, pp. 85–102. Springer, Cham (2021). https://doi.org/10.1007/978-3-030-91452-3_6
7. Giannakopoulou, D., Pressburger, T., Mavridou, A., Schumann, J.: Generation of formal requirements from structured natural language. In: Madhavji, N., Pasquale, L., Ferrari, A., Gnesi, S. (eds.) REFSQ 2020. LNCS, vol. 12045, pp. 19–35. Springer, Cham (2020). https://doi.org/10.1007/978-3-030-44429-7_2
8. Konrad, S., Cheng, B.H.C.: Real-time specification patterns. In: ICSE, pp. 372–381. ACM (2005)
9. Langenfeld, V., Dietsch, D., Westphal, B., Hoenicke, J., Post, A.: Scalable analysis of real-time requirements. In: RE, pp. 234–244. IEEE (2019)
10. Moitra, A., Siu, K., Crapo, A.W., et al.: Towards development of complete and conflict-free requirements. In: RE, pp. 286–296. IEEE (2018)
11. Post, A., Hoenicke, J.: Formalization and analysis of real-time requirements: a feasibility study at BOSCH. In: Joshi, R., Müller, P., Podelski, A. (eds.) VSTTE 2012. LNCS, vol. 7152, pp. 225–240. Springer, Heidelberg (2012). https://doi.org/10.1007/978-3-642-27705-4_18
12. Post, A., Menzel, I., Hoenicke, J., Podelski, A.: Automotive behavioral requirements expressed in a specification pattern system: a case study at BOSCH. Requir. Eng. **17**(1), 19–33 (2012)
13. Post, A.C.: Effective correctness criteria for real-time requirements. Ph.D. thesis, University of Freiburg (2012)

14. Westphal, B.: On education and training in formal methods for industrial critical systems. In: Lluch Lafuente, A., Mavridou, A. (eds.) FMICS 2021. LNCS, vol. 12863, pp. 85–103. Springer, Cham (2021). https://doi.org/10.1007/978-3-030-85248-1_6

15. Winter, K., Femmer, H., Vogelsang, A.: How do quantifiers affect the quality of requirements? In: Madhavji, N., Pasquale, L., Ferrari, A., Gnesi, S. (eds.) REFSQ 2020. LNCS, vol. 12045, pp. 3–18. Springer, Cham (2020). https://doi.org/10.1007/978-3-030-44429-7_1

16. Yang, H., Roeck, A.N.D., Gervasi, V., Willis, A., Nuseibeh, B.: Analysing anaphoric ambiguity in natural language requirements. Requir. Eng. **16**(3), 163–189 (2011)

Supporting Shared Understanding in Asynchronous Communication Contexts

Lukas Nagel[1]([✉]) [iD], Oliver Karras[2] [iD], Seyed Mahdi Amiri[1],
and Kurt Schneider[1] [iD]

[1] Software Engineering Group, Leibniz Universität Hannover, Hannover, Germany
{lukas.nagel,kurt.schneider}@inf.uni-hannover.de,
m.amiri@stud.uni-hannover.de
[2] TIB - Leibniz Information Centre for Science and Techology, Hannover, Germany
oliver.karras@tib.eu

Abstract. **[Context and motivation]** The success of software projects depends on developing a system that satisfies the stakeholders' wishes and needs according to their mental models of the intended system. However, stakeholders may have different or misaligned mental models of the same system, resulting in conflicting requirements. For this reason, aligned mental models and thus a shared understanding of the project vision is essential for the success of software projects. **[Question/problem]** While it is already challenging to achieve shared understanding in synchronous contexts, such as meetings, it is even more challenging when only asynchronous contexts, like messaging services, are possible. When multiple stakeholders are involved from different locations and time zones, primarily asynchronous communication occurs. Despite the frequent use of software tools, like *Confluence*, to support asynchronous contexts, their use for the development of a shared understanding has hardly been analyzed. **[Principal ideas/results]** In this paper, we propose five concepts to help stakeholders develop a shared understanding in asynchronous communication contexts. We assess the adaptability of three existing software tools to our concepts, adapt these software tools accordingly, and develop our own prototype that implements all five concepts. In an experiment with 30 participants, we evaluate these four software tools and compare them to a control group that had no support in developing a shared understanding. **[Contribution]** Our results show the suitability of our concepts, as the participants using our concepts were able to achieve a higher level of shared understanding compared to the control group.

Keywords: requirements engineering · shared understanding · asynchronous communication

1 Introduction

A shared understanding of the project vision is paramount to the success of software projects, as its absence can lead to conflicting requirements [31]. Achieving this shared understanding is one of the key challenges in requirements engineering [14]. For this purpose, stakeholders must disclose, discuss, and align

© The Author(s), under exclusive license to Springer Nature Switzerland AG 2023
A. Ferrari and B. Penzenstadler (Eds.): REFSQ 2023, LNCS 13975, pp. 39–55, 2023.
https://doi.org/10.1007/978-3-031-29786-1_3

their mental models of the intended system to achieve a shared understanding [3]. However, stakeholders are often spread across different locations and time zones [22]. In this case, primarily asynchronous communication occurs, as stakeholders can hardly meet for synchronous in-person or even virtual meetings [13]. One way to achieve a shared understanding in asynchronous communication contexts is to distribute a written specification using standards like ISO/IEC/IEEE 29148:2018 [16]. Nevertheless, reading a written specification can be time-consuming due to its low communication richness and effectiveness [1]. For project visions specifically, a richer and more effective way for achieving a shared understanding is the use of so-called vision videos [21].

Vision videos support the development of a shared understanding, as they provide visual reference points to stimulate active discussions among stakeholders to align their mental models [17]. They are primarily used to support the elicitation, documentation and validation of requirements [21]. Nagel et al. [25] have successfully used vision videos to find misaligned mental models in asynchronous settings. However, simply watching a vision video without the opportunity to discuss its contents complicates the resolving of misalignments [25]. For this reason, stakeholders need suitable support for their discussions to achieve a shared understanding in asynchronous communication contexts.

The goal of this paper, which is based on a master's thesis by Amiri [2], is to *develop suitable concepts to support stakeholders in achieving a shared understanding in asynchronous communication contexts.*

In this paper, we propose five corresponding concepts that are designed to solve issues with asynchronous communication extracted from literature. We combine these concepts with vision videos to investigate whether they support stakeholders in achieving a shared understanding. Three existing software tools for asynchronous communication are assessed regarding their adaptability to our concepts and adapted accordingly. We also develop a prototype that implements all five concepts. In an experiment with 30 participants, we evaluate the four software tools and establish a baseline. Our results show evidence for the suitability of our concepts. All software tools support the achievement of a shared understanding. In particular, participants supported by our adaptation of the messaging service Discord and our developed prototype presented a statistically significantly higher level of shared understanding compared to the control group.

This paper is structured as follows: Sect. 2 discusses related work. We present our concepts in Sect. 3 and describe their implementation on existing tools in Sect. 4. In Sect. 5, we provide details on our experiment, whose results are reported in Sect. 6. Section 7 shows threats to validity. Our results are discussed in Sect. 8 before the paper is concluded in Sect. 9.

2   Related Work

Several works address achieving a shared understanding among stakeholders in requirements engineering. Glinz and Fricker [14] discuss the role of shared understanding in software engineering and identify enablers and obstacles. They also introduce implicit and explicit shared understanding. One technique to support the achievement of a shared understanding is the use of vision videos presenting the project vision. The term *vision video* has been defined by Schneider et al. [28] as a video *of a software-based system typically showing a problem, an envisioned solution, and its impact, pretending the solution already exists.* Creighton et al. [9] introduced the use of videos to visualize scenarios by presenting workflows that are not yet implemented. Brill et al. [6] expanded on this idea by investigating potential uses of videos in various phases of requirements engineering. The potential use of vision videos on multimedia platforms like YouTube has already been discussed by Schneider and Bertolli [27]. Karras et al. [19] investigated the use of vision videos on social-media platforms for CrowdRE. The videos motivated crowd members to provide feedback.

Another use case of videos in asynchronous settings is e-learning. Skylar [29] investigated the performance of students in synchronous and asynchronous online courses and found both to be effective. Furthermore, Clark [7] found one of the biggest advantages of asynchronous communication to be the opportunity for reflective thought processes in between messages. A work by Dowling and Lewis [11] discusses further disadvantages of both communication types. They mention the time pressure of synchronous meetings, which might lead to important contributions being missed. However, the temporal linearity of asynchronous communication is missing due to the distribution of comments on the same topic. A response could therefore be separated from the original comment, which hampers discussants following a discussion topic.

Braunschweig and Seaman [5] developed a technique to measure the shared understanding achieved by a group of stakeholders using *Pathfinder Networks* (PFNets). To use this technique, stakeholders fill out a spreadsheet with relatedness ratings of concept pairs. These ratings are then used to create graphs called PFNets as introduced by Dearholt and Schvaneveldt [10]. Shortest paths can be calculated by using the relatedness ratings as edge weights. The PFNets of a pair of stakeholders can be compared by determining the similarity of the neighborhoods of individual concept-nodes. Calculating the average of all concept similarities between two PFNets, a *Network Similarity* (NetSim) value for a stakeholder pair can be obtained.

3   Concepts for Supporting Shared Understanding

For the development of concepts supporting the achievement of a shared understanding among stakeholders communicating asynchronously, we collected common issues of asynchronous communication from existing literature. Based on these issues, we brainstormed concepts to minimize the impact of these issues. The concepts introduced in this paper are based on the master's thesis by Amiri [2]. Table 1 presents an overview of the identified issues and our concepts addressing them. In the following, we explain each concept in more detail.

Questions of Understanding: We adopt the concept of *Questions of Understanding* from the related work by Nagel et al. [24]. These questions ensure that

Table 1. Overview of our Concepts and the Issues they look to solve.

Concepts	Issues
Questions of Understanding	Differing Domain Knowledge [14,25]
	Misunderstandings [11,12,25]
Message Frames	Misunderstandings [11,12,25]
	Missing valuable Ideas [11,25]
	Sequential Ordering of Messages [11]
Req. Engineers as Facilitators	Missing valuable Ideas [11,25]
	Free-Riders [32]
Polls	Missing valuable Ideas [11,25]
	Reaching Final Conclusions [32]
Step-By-Step Design	Coordination of Steps [15,30]

all stakeholders understand the presented content of an artifact correctly and clarify domain-specific terminologies. Differing from prior research, we propose to force stakeholders to answer *Questions of Understanding* before being allowed to take part in a discussion. In this way, we can ensure that all discussion members have a basic understanding of the presented content.

Requirements Engineers as Facilitators: Synchronous meetings are often held under the guidance of a moderator who guides the participants [18,33]. A traditional moderator role cannot be present in asynchronous communication. However, the active and collaborative participation of all stakeholders, that can be motivated by a moderator [33], is still vital for achieving a shared understanding [4]. We therefore propose to have requirements engineers play a facilitating role in asynchronous communication. This can be done by providing some initial questions or reacting to comments made by stakeholders to motivate them to participate even more. However, requirements engineers should remain neutral in discussions so that stakeholders can reach final conclusions on their own.

Message Frames: A logical and sequential ordering of individual sentences is important to enable humans to reach conclusions from conversations [11]. Our concept of *Message Frames* looks to implement this idea on asynchronous communication, where such sequential orderings are hard to follow [11]. *Message Frames* are a filter for incoming information that structures outgoing messages. In asynchronous communication, the order of messages does not necessarily have to follow the order of discussion topics. Stakeholders can start a topic and return to the discussion after other stakeholders have commented with ideas on other topics. When messages regarding the same topic are located in widely different positions in the ordering of messages, it is hard for stakeholders to follow a discussion [11]. This issue is especially prevalent when the number of discussants and messages increase. *Message Frames* summarize comments dealing with the same topic in a logical order. For example, a requirements engineer could summarize all comments regarding the topic of "security" in one *Message Frame*.

This makes it easier for stakeholders to finalize their thoughts on any given topic. *Message Frames* can thereby lead to more explicit shared understanding.

Polls: Polling is one possibility to reach definitive conclusions at the end of a discussion [32]. *Polls* can turn implicit shared understanding into explicit shared understanding [14]. We recommend using the *Paraphrasing Method* [14] to create the polling questions. By paraphrasing the comments made by the participants and asking for their feedback before enabling the polls, requirements engineers can ensure that there are no misunderstandings [14]. Additionally, we propose that stakeholders can suggest additional polling questions themselves. This allows them to directly ask their peers about unresolved uncertainties. A potential side benefit of the use of polls is that they can also be used to gather an initial indication of a group's level of shared understanding. Groups of stakeholders giving the same answer to a polling question are likely to have a higher level of shared understanding than other groups giving more diverse answers.

Step-By-Step Design: Another drawback of asynchronous communication is the difficulty of coordinating the stakeholders [15]. Important steps could be performed in different orders, thereby creating a chasm between individual knowledge bases. Providing an explicit process is one way to counteract this phenomenon [15]. Therefore, we propose an enforcement of such a process. At first, our concepts only allow stakeholders to get familiar with the content of the presented artifact. Their next step is to answer *questions of understanding*, thereby ensuring that they have a common knowledge base. Stakeholders are only allowed to contribute to the discussion once they answer all *questions of understanding* correctly. Furthermore, our concepts also include fixed time frames for the existing steps. One task of moderators in synchronous meetings is to lead participants through the phases of the agenda within a given time [33]. We incorporate this aspect by providing fixed time frames for each step of the process. Stakeholders are thereby kept from delaying their participation. Simultaneously, the fixed time frames also provide requirements engineers with a concrete time at which feedback regarding the presented content will be available.

4 Implementation of Concepts

We developed a prototype that implements all five concepts to evaluate their suitability to our goal. We also assessed the adaptability of existing software tools for asynchronous communication, as preexisting familiarity with these tools could reduce the barrier of entry for stakeholders. An important factor in the choice of software tools was their capability to display a vision video. The video must be directly visible in the software tool so that stakeholders do not need to switch between applications to reduce their cognitive load [20]. We conducted a workshop with three participants to discuss different types of software tools and to choose individual representatives for our experiment. In this workshop, a total of 10 different software tools were discussed. Each participant was asked to identify advantages and disadvantages of the tools. Ultimately, we asked the workshop participants to pick three tools they considered to be best suited to the support of stakeholders in the asynchronous achievement of a shared understanding using vision videos. The multimedia platform *YouTube*, the wiki service *Confluence*, and the messaging service *Discord* were selected as the most suitable existing software tools. More information on the workshop can be found in the master thesis by Amiri [2] which this paper is based on. Table 2 presents an overview of the concepts and the manner in which they were implemented for each tool. The following paragraphs present the implementation of the concepts *Questions of Understanding*, *Polls* and *Step-By-Step Design*. Our concepts *Requirements Engineers as Facilitators* and *Message Frames* were not implemented as technical adaptations of the tools, but as tasks of the requirements engineer's role.

Table 2. Overview of the applicability of our concepts to each tool. Applicabiltiy: ✓ fully, ○ partially, and 🖐 only manually * For YouTube, *Polls* had to be applied using a third party tool.

Concept	YouTube	Confluence	Discord	Prototype
Questions of Understanding	✓	✓	✓	✓
Requirements Engineers as Facilitators	🖐	🖐	🖐	🖐
Message Frames	🖐	🖐	🖐	🖐
Polls	○*	○	○	✓
Step-By-Step Design	○	○	○	✓

YouTube: *YouTube* provides built-in functionality for the presentation of video content. With over 2.1 billion worldwide users[1], most stakeholders should be familiar with the system. *YouTube* offers a comment system which provides functionality to answer previously made comments and to reference other users. *YouTube* also includes a description section in which more context can be given. We used this description section to provide the order of steps and the *Questions of Understanding*. However, there was no way to enforce the *Step-By-Step Design* or to hold *Polls*. While the *Like* and *Dislike* functionality of comments could be used, *YouTube* does not display the exact votes and would be limited to yes or no questions. Third party tools are required for other *Polls* and for the answering of *Questions of Understanding*.

Confluence: *Confluence* includes functionality to organize knowledge on pages and a comment system. Videos can be embedded directly on these pages. We created one page to view the video, one page to answer *Questions of Understanding*, one page for the comment section, and one final page for polling questions. In this way, we partially implement the *Step-By-Step Design*. However, the order of steps could not be enforced. There was also no built-in functionality for *Polls*. Instead, a suite of plugins is available within Atlassian's marketplace.

---

[1] https://www.statista.com/topics/2019/youtube/#dossierKeyfigures.

Discord: *Discord* allows its users to create "Servers" for free. Servers consist of text and voice channels only available to invited users. Voice channels can be joined for conference calls. Text channels offer functionality to write messages, upload files, and embed images. Discord users can reference other messages or other users and pin messages to make them easier findable. Threads can be created that appear as a single message in the original chat history, but can be expanded into a new window with its own set of messages. This allows for a separation of especially important topics. We made use of these threads to implement our concept of *Questions of Understanding* by asking them in a separate thread. For the *Step-By-Step Design*, we pinned a message detailing the order of steps within the text channel. However, we could not enforce the compliance with this order. *Discord* also does not offer built-in functionality for polling. For this reason, we use free plugins that enabled our concept of *Polls*.

Prototype: The existing tools evaluated in this paper offer functionality suited to some of our concepts. However, none of them could be adapted to include all concepts to their full extent. For this reason, we developed a prototype that implements all five concepts. The prototype was implemented as a single page application, a screenshot of which can be found in Fig. 1. The prototype always displays the vision video at the top of the screen (1). Stakeholders can click through the pages of the prototype (2), which represent the *Step-By-Step Design*. Some pages only unlock after performing prior steps. The main area of the prototype displays the selected page's content (3). When providing new comments, stakeholders are required to give a headline to assist requirements engineers in the creation of *Message Frames*.

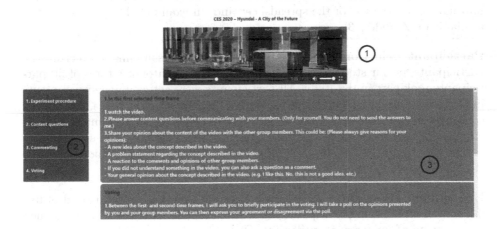

Fig. 1. Screenshot of the prototype presenting the *Experiment Procedure* page.

5   Experiment

A user study was conducted to evaluate our concepts for our research goal (cf. Sect. 1). Based on this goal, we defined the following two research questions.

RQ1: How suited are the six concepts to the support of stakeholders in achieving a shared understanding in asynchronous communication contexts?
RQ2: Which software tools are best suited to provide this support?

5.1   Experiment Design

Material: Our study utilizes a *vision video*[2] on future mobility produced by Hyundai and published on *YouTube* as the basis for all discussion topics. To enable the use of the four software tools mentioned in Sect. 4, members of the treatment groups were provided with new e-mail addresses and user accounts. We thereby preserved their privacy and lowered the barrier of entry.

Furthermore, each participant was provided with a link to a spreadsheet that was used to perform the PFNets method lined out in Sect. 2. Both the link to the spreadsheet and a second link to a questionnaire were distributed at the end of their participation. The questionnaire asked participants about their opinions on the suitability of asynchronous communication for the achievement of a shared understanding and their preference between synchronous and asynchronous communication methods. Another document guiding participants during their participation in the study was also provided. Since none of the participants had previously worked with the PFNets method, this guideline included information on the use of the spreadsheet and an example. The spreadsheet is available on Zenodo [23].

Participant Selection: We performed convenience sampling to recruit the participants for our study. Participation was not mandatory. A total of 30 participants took part in the study. All participants were active university students in Germany. Our only requirement for our participant selection was a functioning computer on which to watch the vision video, answer the questionnaire and fill out the PFNets spreadsheet. We were looking to include potential stakeholders for the topic of future mobility. Therefore, anyone participating in modern traffic is a viable participant.

Experiment Procedure: The study was conducted online over a total of five days, with each group participating on a single day. Participants were assigned to groups based on personal availability. Our only influence on these assignments was limited to the selection of time slots for participants whose availability was suited to multiple groups. The study was performed strictly online due to the Covid-19 pandemic. We performed an experiment session with a control group of 6 participants to establish a baseline. Members of this control group were asked to

---

[2] https://www.youtube.com/watch?v=J_OBgXalGFU.

view the vision video on their own and had no support to discuss with any other group members. They were also explicitly asked to work on the spreadsheets on their own to ensure the validity of their answers. We designed the control group without any means of communication to measure the level of shared understanding that is created by simply watching the same vision video. To the best of our knowledge, no methodology for the achievement of a shared understanding in asynchronous communication contexts exists. Therefore, our study was designed to create a baseline of shared understanding when watching vision videos while also investigating the differences between supporting communication tools.

For members of the treatment groups, the study consisted of two distinct time windows. To ensure a strictly asynchronous setting, no participants were scheduled to take part at the same time. Participants were asked to perform the same set of steps during the two time windows. However, there were some differences in terms of the available functionality as outlined in Sect. 4.

In the first time window, participants were asked to watch the vision video for the first time before answering six *Questions of Understanding*. Participants were explicitly asked to answer these questions first before proceeding. However, this requirement could only be enforced in the prototype. Lastly, participants were allowed to leave comments and add to existing parts of the discussion. Between the two time windows, the experimenter scanned through the comments and created *Message Frames*. Polling questions were also determined.

The second time window started by providing the *Message Frames* before participants answered the polling questions. For the treatment group supported by YouTube, this was done via telephone. Next, each participant was asked to read the submitted comments and respond to them. After all participants had finished the second time window, they were asked to review the results of the *Polls* before answering the questionnaire and filling in the PFNets spreadsheet.

Data Analysis Procedures: To answer our research questions, we created two sets of hypotheses. Each set is designed to answer one research question. The first set of hypotheses aims at finding differences between each of the four treatment groups and the control group:

> **H1$_{i.0}$:** There is no difference in the shared understanding of participants between the control group and the treatment group supported by i.
> $$i \in \{YouTube, Confluence, Discord, Prototype\}$$

The second set deals with the differences between the different supporting tools. For example, we look to find a difference between the treatment group communicating via YouTube and the one being supported by the prototype:

> **H2$_{j.0}$:** There is no difference in the shared understanding of participants between a and b.
> $$j = (a, b) \ with \ a, b \in \{YouTube, Confluence, Discord, Prototype\}, a \neq b$$

To find data on which to base a potential rejection of these null hypotheses, we analyzed the PFNets spreadsheets filled out by our participants according to Braunschweig and Seaman [5]. Their technique resulted in network similarity (NetSim) values for all participant pairs. These were then used to calculate average NetSim values for each group and to calculate the statistical significance of differences in the achieved shared understanding between the groups. The statistical significance was determined by first testing for normal distribution using the Shapiro-Wilk test before applying the Mann-Whitney U test or the t-test, depending on the presence of a normal distribution. We also applied the Bonferroni-Holm correction. In addition, we extracted the results of the *Polls* and gathered answered questionnaires. For the *Polls*, we determined which choice was made by the majority of participants, before averaging the number of participants who were part of this majority for each poll performed in the respective treatment group. This resulted in the average size of the majority vote for each group. We analyzed the answers to the questionnaires descriptively.

6   Results

Our study focuses on measurements for the shared understanding within each group of the experiment. Furthermore, we also obtained information on participants' thoughts on the suitability of asynchronous communication contexts and their general opinion on the software tool they were supported by.

NetSim: We measured the shared understanding within the groups of our experiment using the aforementioned PFNets method. The results are available on Zenodo [23]. As our results were normally distributed for all groups, we used the t-test. The results of the Shapiro-Wilk test can be found in Table 3.

Table 3. Results of Shapiro-Wilk tests. Note that the sample size for a group of 6 participants is 15 as we obtained similarity values for each participant pair.

Tool	W(15)	p	Normal Distribution?
Control	0.889	0.067	Yes
YouTube	0.785	0.965	Yes
Confluence	0.969	0.841	Yes
Discord	0.969	0.842	Yes
Prototype	0.933	0.302	Yes

To test the set Hypotheses H1, we compared the values calculated for the control group with the values measured for each other software tool. We found statistically significant differences between the control group and the treatment groups supported by Discord and our prototype (cf. Table 4).

Hypotheses H2 were tested by determining the statistical significance of differences between the treatment groups. Such differences were found between the group supported by the prototype and all other treatment groups (cf. Table 5).

Table 4. Results for Hypotheses H1. The column *Corrected p* presents the p-values resulting from the Bonferroni-Holm correction.

$H1_{i.0}$	Tool	NetSim			p	Corrected p	Reject $H1_{i.0}$?
		Min	Max	Avg			
N/A	Control	0.118	0.533	0.250	N/A	N/A	N/A
$H1_{1.0}$	YouTube	0.105	0.476	0.297	0.12205	0.12205	No
$H1_{2.0}$	Confluence	0.160	0.467	0.307	0.05776	0.11552	No
$H1_{3.0}$	Discord	0.211	0.556	0.360	0.00401	0.01203	Yes
$H1_{4.0}$	Prototype	0.357	0.538	0.458	<0.00001	<0.001	Yes

Table 5. Results for Hypotheses H2. The column *Corrected p* presents the p-values resulting from the Bonferroni-Holm correction.

$H2_j$	Tool A	Tool B	p	Corrected p	Reject $H2_{j.0}$?
$H2_1$	YouTube	Confluence	0.39055	0.39055	No
$H2_2$	YouTube	Discord	0.05919	0.17757	No
$H2_3$	YouTube	Prototype	0.00002	<0.001	Yes
$H2_4$	Confluence	Discord	0.06814	0.17757	No
$H2_5$	Confluence	Prototype	<0.00001	<0.001	Yes
$H2_6$	Discord	Prototype	0.00211	0.00844	Yes

To gain a better understanding of the magnitude of the differences between the examined groups, we calculated the effect sizes for all comparisons that were positively tested for statistical significance. The results of these calculations can be found in Table 6.

Table 6. Effect sizes for statistically significant differences between groups. We interpret the calculated values according to Cohen [8] and Sawilowsky [26].

Hypothesis	Group A	Group B	Cohen's d	Interpretation
$H1_3$	Control	Discord	1.047	Large
$H1_4$	Control	Prototype	2.354	Huge
$H2_3$	YouTube	Prototype	1.789	Very large
$H2_5$	Confluence	Prototype	1.976	Very large
$H2_6$	Discord	Prototype	1.135	Large

Polls: Polls were created based on the discussion of each group. The groups supported by *YouTube*, *Confluence* and the prototype were asked eight polling questions each, while the group supported by *Discord* answered seven. We found average majority sizes of 72.6% for *YouTube*, 78.8% for *Confluence*, 71.1% for *Discord* and 76.8% for the group supported by the prototype.

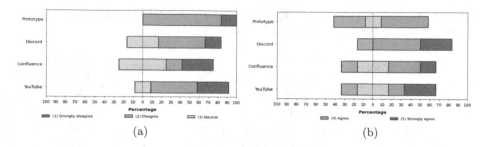

Fig. 2. Answers to the questionnaire regarding the suitability of asynchronous communication (a) and participants' preference between asynchronous and synchronous communication (b).

Questionnaire: The questionnaire consisted of questions regarding the general suitability of asynchronous communication contexts for discussing an artifact. The first question asked participants how suitable they thought asynchronous communication was for the discussion of a vision video's content. No statistically significant differences could be found between the groups. Out of 24 participants, 6 answered neutrally. All other 18 participants indicated that they agreed or strongly agreed that asynchronous communication is suitable. An overview of these results can be found in Fig. 2a. A second question addressed the preference between asynchronous and synchronous communication. Once again, no statistically significant differences could be found. The answers were diverse for all treatment groups. In total, no participant strongly preferred synchronous communication, while 5 participants indicated that they preferred synchronous communication and 5 participants answered neutrally. A total of 9 participants preferred asynchronous communication, with an additional 5 participants strongly preferring asynchronous communication. A visual representation of these results can be found in Fig. 2b. In addition to questions answered on Likert scales, we also asked open questions regarding positive and negative aspects of asynchronous communication. The most often mentioned positives were *having enough time to think*, *developing ideas* and the *temporal flexibility*. Negative aspects included *delayed answers* and *missed comments*, as well as the longer time required for final conclusions. The final question asked for opinions on a statement, indicating *Questions of Understanding* as valuable. Once again, no statistically significant differences could be found between the treatment groups. Only a single participant strongly disagreed, while 2 other participants gave neutral answers. 12 participants agreed with the statement, and a further 9 participants agreed strongly.

7   Threats to Validity

We report the threats to validity of our results according to Wohlin et al. [34].

The *conclusion validity* of our results is threatened by the small sample size. Having only six participants per treatment group increases the risk of statistical noise impacting the results. However, we chose to include three existing software tools in our evaluation rather than increase the sample size for only one or two, as we obtained three clear favorites in the workshop. Another threat to the conclusion validity is the fact that we asked participants who had only discussed the vision asynchronously about their preference between synchronous and asynchronous communication. Nevertheless, it is easy for participants to imagine synchronous discussions and the answers to the open questions of the questionnaire gave concrete reasons for this preference.

One threat to the *internal validity* of our study is the potential of exhausted participants giving incomplete answers. Participants of our study were asked to work in two time windows and asked to fill in multiple documents over the course of a day. We chose this type of study to reliably simulate an asynchronous setting and also gave participants a lengthy break between the time windows. Furthermore, participants could in theory have interacted with one another outside of the asynchronous communication tools. We minimized this threat by creating new accounts without any identifying information for all participants on all software tools used in the study.

A threat to the *construct validity* is the mono-method bias. We chose not to include further metrics to avoid an even higher potential for participant exhaustion. Another threat is that participants might understand the same term differently when filling in the PFNets spreadsheet. We only included terms that were short and clearly visible in the vision video to minimize this threat. Additionally, we only simulated the presence of different time zones by assigning distinct time frames to all participants. An experiment including multiple time zones would have been preferable, but was not feasible.

The *external validity* of our results is that participants knew that they were taking part in an experiment. A study with practitioners in a real-world use case would have been preferable. Another threat is the potential that we might have missed a suitable existing tool. However, we tried to minimize this threat by conducting the workshop and discussing the results with multiple researchers. Furthermore, the experiment was conducted over the course of a single day while a real-world application would likely be performed over the course of multiple days. We accepted this threat as the threat of participant exhaustion might have been increased further, had we conducted a multi-day study.

8   Discussion

The results of our study show clear differences between the achieved level of shared understanding among the participants of the five groups. In particular, we found that all treatment groups supported by one of the four software tools (*YouTube, Confluence, Discord,* and the prototype) achieved a higher average level of shared understanding than the control group. This finding is indicated by the higher average NetSim values, as a higher NetSim value indicates a higher level of shared understanding [5]. When comparing the results of the treatment groups, we found that the group supported by the prototype achieved a statistically significantly higher level of shared understanding than every other treatment group (cf. $H2_{3.1}$, $H2_{5.1}$ and $H2_{6.1}$).

These results substantiate the suitability of our concepts to support stakeholders in achieving a shared understanding in an asynchronous communication context. First, all software tools, even adapted with only a partial implementation of our concepts, result in a higher level of shared understanding than the control group. In accordance with the results of Nagel et al. [25], our results show the importance of enabling discussions between stakeholders in asynchronous settings. Even partial concepts already help to achieve a better understanding, as they improve stakeholders' capabilities to communicate with each other. Second, implementing all concepts to their full extent (as in the prototype) provides a solid basis for achieving a higher level of shared understanding. In all four software tools, we tried to implement each concept as fully as possible. However, for the three existing tools, we had no access to their source code and thus had to make compromises, such as using plugins, to enable the concept as intended. In contrast, the prototype allowed us to implement and combine the concepts to reach their full potential. For this reason, the main difference between the prototype and the adapted software tools is the degree to which the concepts could be implemented. While the results show that even the partial implementations lead to a higher shared understanding than the control group, the prototype achieved the best results overall with effect sizes ranging from large to huge [8,26]. We assume that the main reason for these results is the concept *Step-By-Step Design*. This concept provides a structured framework for all other concepts. For example, the prototype enforces the answering of *Questions of Understanding* before participants can access the comment section due to the *Step-By-Step Design*. In this way, the full implementation of the *Step-By-Step Design* emphasized the importance of these questions and ensured that the participants are familiar with the video content before writing any comment. As a consequence, the concepts were better integrated and combined, resulting in a higher level of shared understanding of stakeholders. Based on these insights, we provide the following answers to our research questions:

Answer to RQ1: The concepts presented in this paper are suited to the support of stakeholders in achieving a shared understanding in an asynchronous communication context. Our participants indicated that *Questions of Understanding* and the *Step-By-Step Design* were especially meaningful.

Answer to RQ2: We found *Discord* to be the most suited existing tool for being adapted to our concepts. However, the group supported by our prototype achieved an even higher level of shared understanding that is statistically significantly different from all other treatment groups. Further development of the prototype to achieve shared understanding in asynchronous communication contexts is a promising endeavor for future research.

Besides the analysis of the shared understanding among the stakeholders in the respective groups, we also investigated the participants' attitude towards the idea of being supported in achieving a shared understanding in asynchronous communication contexts. According to our results, most of them preferred the use of asynchronous communication contexts over synchronous ones. They justified this decision with a higher flexibility to take their time to think about the presented vision and for the development of questions, answers, and ideas for the discussion with the other stakeholders. This finding is in line with the advantages of asynchronous communication contexts found by Dowling and Lewis [11].

However, the generalizability of our results is limited. The groups of participants supported by each software tool are probably smaller than in a real-world setting. In addition, the participants had no real value in understanding the presented vision due to the fictitious experimental context. Nevertheless, our concepts are a promising starting point for future research. On the one hand, future work needs to investigate how each concept individually contributes to a shared understanding, as we only investigated all concepts together. On the other hand, we observed difficulties in the experiment such as language barriers and terminology issues for which we only have the partial solutions of a *Step-By-Step Design* combined with mandatory *Questions of Understanding*.

In summary, our results reveal the value of asynchronous communication contexts. Stakeholders are able to disclose, discuss and align their mental models within an asynchronous context to achieve a shared understanding. An even higher level of shared understanding can be accomplished when using the full extent of our concepts. We conclude that the concepts described in this paper fulfill our goal. In this way, we *developed suitable concepts to support stakeholders in achieving a shared understanding in asynchronous communication contexts*.

9   Conclusion

A shared understanding between stakeholders is vital for successful software projects. The discussion of vision videos present one possible way to achieve such a shared understanding, even in asynchronous settings. However, these discussions depend on asynchronous communication methods. In this paper, we presented concepts to support achieving a shared understanding between stakeholders in asynchronous communication contexts. We adapted existing software tools and developed a prototype according to our concepts and conducted a user study. This study substantiates the suitability of our concepts for supporting shard understanding in asynchronous communication contexts.

In future research, we plan to increase the sample size of our study to obtain more reliable results. We also plan on evaluating our concepts in isolation and to compare our results to the shared understanding created in a synchronous meeting. For the concepts *Requirements Engineers as Facilitators* and *Message Frames* we seek to investigate how requirements engineers can be supported while performing the associated tasks. Furthermore, the PFNets spreadsheet could be extended with terms relating to the topics discussed by the groups. The findings of this paper indicate the potential of our concepts. Further research efforts might lead to a definitive tool supporting the achievement of a shared understanding among stakeholders in asynchronous settings.

Acknowledgement. This work was supported by the Deutsche Forschungsgemeinschaft (DFG) under Grant No.: 289386339, project ViViUse.

References

1. Ambler, S.: Agile Modeling: Effective Practices for Extreme Programming and the Unified Process. Wiley, New York (2002)
2. Amiri, S.M.: Konzeptionierung eines Tools zur Herstellung gemeinsamen Verständnisses durch asynchrone Betrachtung von Vision Videos. Master thesis, Leibniz Universität Hannover (2022)
3. Aranda, J.: A theory of shared understanding for software organizations. University of Toronto (2010)
4. Bittner, E.A.C., Leimeister, J.M.: Why shared understanding matters-engineering a collaboration process for shared understanding to improve collaboration effectiveness in heterogeneous teams. In: 46th Hawaii International Conference on System Sciences. IEEE (2013)
5. Braunschweig, B., Seaman, C.: Measuring shared understanding in software project teams using pathfinder networks. In: 8th ACM/IEEE International Symposium on Empirical Software Engineering and Measurement (2014)
6. Brill, O., Schneider, K., Knauss, E.: Videos vs. use cases: can videos capture more requirements under time pressure? In: International Working Conference on Requirements Engineering: Foundation for Software Quality. Springer, Cham (2010)
7. Clark, C., Strudler, N., Grove, K.: Comparing asynchronous and synchronous video vs. text based discussions in an online teacher education course. Online Learn. **19**(3) (2015)
8. Cohen, J.: Edition 2. Statistical power analysis for the behavioral sciences (1988)
9. Creighton, O., Ott, M., Bruegge, B.: Software cinema - video-based requirements engineering. In: 14th IEEE International Requirements Engineering Conference (2006)
10. Dearholt, D.W., Schvaneveldt, R.W.: Properties of Pathfinder Networks. Ablex Publishing Corp, Norwood (1990)
11. Dowling, K.L., Louis, R.D.S.: Asynchronous implementation of the nominal group technique: is it effective? Decis. Support Syst. **29**(3), 229–248 (2000)
12. Garrison, D.R.: E-Learning in the 21st Century: A Community of Inquiry Framework for Research and Practice. Routledge, New York (2016)
13. Girgensohn, A., Marlow, J., Shipman, F., Wilcox, L.: HyperMeeting: supporting asynchronous meetings with hypervideo. In: 23rd ACM international conference on Multimedia (2015)
14. Glinz, M., Fricker, S.A.: On shared understanding in software engineering: an essay. Comput. Sci. Res. Dev. **30**(3), 363–376 (2015)
15. Hiltz, S.R., Dufner, D., Holmes, M., Poole, S.: Distributed group support systems: social dynamics and design dilemmas. J. Organ. Comput. Electron. Commer. **1**(2), 135–159 (1991)
16. ISO Central Secretary: Systems and software engineering - Life cycle processes - Requirements engineering. Standard ISO/IEC/IEEE 29148:2018 (2018)
17. Karras, O.: Supporting Requirements Communication for Shared Understanding by Applying Vision Videos in Requirements Engineering. Logos (2021)

18. Karras, O., Kiesling, S., Schneider, K.: Supporting requirements elicitation by tool-supported video analysis. In: 24th International Requirements Engineering Conference. IEEE (2016)
19. Karras, O., Kristo, E., Klünder, J.: The potential of using vision videos for CrowdRE: video comments as a source of feedback. In: 29th International Requirements Engineering Conference Workshops. IEEE (2021)
20. Karras, O., Risch, A., Schneider, K.: Interrelating use cases and associated requirements by links: an eye tracking study on the impact of different linking variants on the reading behavior. In: 22nd International Conference on Evaluation and Assessment in Software Engineering (2018)
21. Karras, O., Schneider, K., Fricker, S.A.: Representing software project vision by means of video: a quality model for vision videos. J. Syst. Softw. **162**, 110479 (2020)
22. Lloyd, W.J., Rosson, M.B., Arthur, J.D.: Effectiveness of elicitation techniques in distributed requirements engineering. In: IEEE Joint International Conference on Requirements Engineering. IEEE (2002)
23. Nagel, L., Amiri, S.M.: Supplementary Material - Supporting Shared Understanding in Asynchronous Communication Contexts, February 2023. https://doi.org/10.5281/zenodo.7649336
24. Nagel, L., Karras, O.: Keep your stakeholders engaged: interactive vision videos in requirements engineering. In: IEEE 29th International Requirements Engineering Conference Workshops. IEEE (2021)
25. Nagel, L., Shi, J., Busch, M.: Viewing vision videos online: opportunities for distributed stakeholders. In: IEEE 29th International Requirements Engineering Conference Workshops (2021)
26. Sawilowsky, S.S.: New effect size rules of thumb. J. Mod. Appl. Stat. Methods **8**(2), 26 (2009)
27. Schneider, K., Bertolli, L.M.: Video variants for CrowdRE: how to create linear videos, vision videos, and interactive videos. In: IEEE 27th International Requirements Engineering Conference Workshops. IEEE (2019)
28. Schneider, K., Busch, M., Karras, O., Schrapel, M., Rohs, M.: Refining vision videos. In: Knauss, E., Goedicke, M. (eds.) REFSQ 2019. LNCS, vol. 11412, pp. 135–150. Springer, Cham (2019). https://doi.org/10.1007/978-3-030-15538-4_10
29. Skylar, A.A.: A comparison of asynchronous online text-based lectures and synchronous interactive web conferencing lectures. Issues Teach. Educ. **18**(2), 69–84 (2009)
30. Smith, J.Y., Vanecek, M.T.: Dispersed group decision making using nonsimultaneous computer conferencing: a report of research. J. Manage. Inf. Syst. **7**(2), 71–92 (1990)
31. Van Lamsweerde, A.: Requirements engineering in the year 00: a research perspective. In: 22nd International Conference on Software Engineering (2000)
32. Warkentin, M., Beranek, P.M.: Training to improve virtual team communication. Inf. Syst. J. **9**(4), 271–289 (1999)
33. Wheeler, B.C., Valacich, J.S.: Facilitation, GSS, and training as sources of process restrictiveness and guidance for structured group decision making: an empirical assessment. Inf. Syst. Res. **7**(4), 389–490 (1996)
34. Wohlin, C., Runeson, P., Höst, M., Ohlsson, M.C., Regnell, B., Wesslén, A.: Experimentation in Software Engineering. Springer, Cham (2012). https://doi.org/10.1007/978-3-642-29044-2

Bringing Stakeholders Along for the Ride: Towards Supporting *Intentional* Decisions in Software Evolution

Alicia M. Grubb[1](✉) and Paola Spoletini[2](✉)

[1] Smith College, Northampton, MA, USA
amgrubb@smith.edu
[2] Kennesaw State University, Marietta, GA, USA
pspoleti@kennesaw.edu

Abstract. [Context and Motivation] During elicitation, in addition to collecting requirements, analysts also collect stakeholders' goals and the present and historical interests that motivate their goals. This information can guide the resolution of requirements conflicts, support the evolution of requirements when changes occur (e.g., environmental constraints), and inform decisions in software design. [Problem] Unfortunately, this information is rarely explicitly represented and maintained. When a stakeholder is modeled in the literature, the captured information is only part of that stakeholder's intention (i.e., the goals and the present and historical interests that motivate those goals) and not other requirements documents. In addition, such representations of a stakeholder are not traced and kept aligned with the design and, thus, cannot be used during iterative development and in case of changes. [Principal Idea] To support engineers in making informed decisions during the design, development, and evolution of a system, we propose a framework to collect and maintain intentionality in an efficient and effortless way. [Contributions] To define intentionality, disambiguate it from its use in literature, and position it in relation to similar concepts (i.e., rationale and goals), we conduct a literature review. Based on our derived definition, we present our framework to appropriately include intentionality throughout the stages of a project and the research agenda to realize such a framework.

Keywords: Intentionality · Traceability · Software development

1 Introduction and Motivation

For over two decades, within goal-oriented requirements engineering (GORE), there has been a concerted effort to capture and document the needs of stakeholders and their interdependencies with respect to the system [13]. Thus, stakeholders' needs and motivations have been seen as an important piece of the requirements process and there has been significant work on capturing these needs in the form of models. However, these needs and motivations are often

© The Author(s), under exclusive license to Springer Nature Switzerland AG 2023
A. Ferrari and B. Penzenstadler (Eds.): REFSQ 2023, LNCS 13975, pp. 56–64, 2023.
https://doi.org/10.1007/978-3-031-29786-1_4

lost in the transition from requirements to system design and are not followed downstream. In a systematic literature review of how GORE approaches are used in downstream activities, Horkoff et al. found only one of the top 50 cited papers looked at decision making [13]. This paper, by Wang et al., looks at monitoring the satisfaction of requirements at runtime [18]. In addition to the work by Wang et al., there have been multiple efforts to connect requirements with downstream activities and run-time monitoring [4,7,17], as well as work to automatically link customer wishes with requirements [9]. While the GORE literature offers a foundation for our framework, the importance of stakeholders needs spans the entire field of requirements; thus, we create a generalizable framework, independent of requirements approach.

In this paper, we argue for the inclusion of the underlying motivation of stakeholders, which we call *intentionality* (see Sect. 2 for a definition), throughout the development and maintenance of software, independent of whether or not a GORE methodology is initially used. We begin by understanding in what capacity intentions are being captured and where in the software lifecycle (e.g., elicitation, modeling, design, or development) they are being lost. Additionally, we look at what similar concepts and terminology exist and how they are connected (e.g., design rationale [8]).

Contributions and Organization. We propose a framework, called *Intention Guided Requirements Engineering* (ING-RE), that elicits, models, and maintains stakeholders' intentionality, for the purpose of supporting decision making. Our intellectual contribution explores four research questions:

RQ0: To what extent has intentionality been captured in RE literature?

RQ1: How can analysts retain information about stakeholders' intentionality for use across the software life cycle using existing techniques?

RQ2: To what extent is intentionality currently traced between software artifacts? and how can it be added?

RQ3: How and to what extent can we use stakeholders' intentions to make decisions during software evolution?

In this paper, we make three contributions: (1) In Sect. 2, we complete an initial literature review to define the notion of an intention. (2) In Sect. 3, we present the ING-RE framework and describe how it embeds intentions in the software lifecycle, and how intentions can be used to support the analysis of requirements downstream. (3) In Sect. 4, we present our research agenda and our plan to develop ING-RE. Finally, we conclude in Sect. 5.

2   Defining Intentionality

We begin by addressing RQ0. We conducted an initial light-weight literature review, following the guidelines by Kitchenham [14], to understand how intentions are being captured in the literature. We used Scopus[1] to find papers associated with this concept using the search string:

---

[1] Scopus' coverage is considered optimal when compared to other databases (e.g., IEEE Xplore, ACM Digital library) [15].

TITLE-ABS-KEY ((intent* OR motivation*) AND ("software engineering" OR (requirement* AND engineering)) AND (requirement*) AND (goal*) AND (stakeholder* OR user*)) AND (LIMIT-TO (DOCTYPE , "ar") OR LIMIT-TO (DOCTYPE , "cp")) AND (LIMIT-TO (SUBJAREA , "COMP")) AND (LIMIT-TO (LANGUAGE , "English"))

Scopus returned 135 matches for our search string. The material generated in the literature review is available at: *doi.org/10.35482/csc.001.2023*. With the help of our research assistants (see acknowledgments), we assessed the papers as within scope or out of scope (i.e., Yes, Maybe, No), based on the paper title and abstract. The papers were divided equally among the assistants. All papers labeled as 'Maybe' were discussed as a group with assistants and authors. Then, the assistants checked each others reviews (i.e., checking 40 papers) to ensure they were consistent and reported concerns. Two concerns were raised, but in both the original assessment was deemed appropriate. Of the 135 initial matches, 71 were selected for content analysis. After the content analysis, an additional 20 papers were excluded, 10 because they were not relevant for the context of this work and 10 because they did not contain any fragment of text that, even informally, defined "intention" or other related terms. For each of the remaining 51 papers, we analyzed and extracted fragments of text related to intentionality.

More than 60% of them contain an informal definition of an *intention*. In almost half of these cases, a brief definition such as "intention, i.e., goals, soft-goals, tasks, and resources" is given. In more than 20% of the analyzed papers, intentions are associated with a call to action to reach a targeted state. This is also in line with our analysis of the fragments about goal modeling (where often the stakeholders' intentions are seen as the targeted states and the needs behind the creation of the system goals). Interesting emerging themes from our textual analysis are *emotions* and *emotional goals* [3]—including values, emotional state, and "personality as players" in the motivations of the stakeholders. These concepts are not in contrast with the most common "definition" of intention but highlight the importance of considering the personal dimensions of stakeholders while elucidating their motivations.

Using the definitions and themes that emerged in our literature review and the concept of intentionality defined and discussed philosophically in [1], we define intentionality **as the underlying motivations for the 'wants' and 'needs' of stakeholders that guide the design and development of the system under consideration.** This definition is in line with the assimilation of intention into goal model elements when emotional goals and personal factors are also considered.

Stakeholders' intentions can be mistakenly identified as the rationale for a requirement or as the goals or aims of a project. While these concepts often overlap and even coincide in some circumstances, in general, intentions define different types of *explanations* associated with the users' underlying motivations while the rationale is the final reason that guided the definition of a requirement and captures "the reasoning underlying the creation and use of artifacts" [8].

3   ING-RE: Supporting Intentionality

Intentionality is not always explored during elicitation, especially when analysts concentrate on the system. These motivations add relevant details to the collected needs that can be used during the development process to make more informed decisions. For example, being able to distinguish if a request has its roots in budget considerations or the company's historical behaviors can help stakeholders negotiate between competing needs, as well as help engineers make the most appropriate design choice, create appropriate acceptance tests, and understand the impact of a change in a product. Additionally, when elicited, intentions are rarely documented and, thus, not used to make design decisions.

Concrete Example: Volume Adjustment. A team is upgrading their teleconferencing phone to allow it to work with video conferencing software (e.g., Teams, Zoom). As part of this upgrade, the team is deciding which features to incorporate into a touch-screen and which features to implement with physical components. As part of this analysis, they conduct a focus group with potential users. Table 1 lists the requirements (r_1-r_5) elicited from stakeholders. While each stakeholder has different preferences for the aims of the system, they all want the system to be able to adjust the speaker volume. Taking a closer look, Alex and Jesse's requests have similar underlying motivations. They both want to be able to use the system without having to focus on the interface. Jesse prefers being able to adjust the volume physically until the sound is heard at an appropriate level and wants to adjust it very precisely (r_4); whereas, Alex has an underlying motivation of finding it difficult to learn new systems and multi-task during meetings (r_2). Angelo wants the interface to be usable when others have sticky fingers (r_3). Yet, John, in device manufacturing, wants as few physical components as possible to reduce cost (r_3). When choosing how to design the interface, it is important to consider the underlying motivation of each stakeholder. Even if a strictly digital interface is chosen, these "whys" should be retained and influence the design of the interface.

We propose the ING-RE framework to support the analysts in incorporating stakeholders' intentions from the beginning of the development process (RQ1), adding traceability links throughout the project (RQ2), and using intentions to make more informed decisions when considering proposed changes and competing ideas (RQ3). These aspects are described in the remainder of this section.

Eliciting and Representing Intentions. Ideally, intentionality needs to be elicited when requirements are collected and then maintained throughout the life of the software. Once collected, stakeholders' intentions can be documented alongside the requirements, independent of technique. We propose using a natural language based structured language to limit difficulties in representing intentions, while allowing for some automated analysis of the resulting information. To enable the inclusion of multiple whys, we define I as the set of intention triples $\langle s, m, u \rangle$, where s represents the stakeholder who had the intention, m represents the intention (i.e., motivation), and u represents the status of that intention in the current state of the design.

Table 1. Digital touch panel requirements and their underlying motivation.

Requirements	Intentions triples ⟨Stakeholder, motivation, status⟩
r_1. The interface shall allow for the adjustment of the speaker volume	i_1:⟨John, Input volume may vary., *Queue*⟩
r_2. The speaker volume shall be static and always available	i_2:⟨Alex, Difficulty navigating interfaces under pressure., *Queue*⟩
r_3. The speaker volume shall be adjusted via the touch screen	i_3:⟨Angelo, Ease of daily cleaning., *Queue*⟩, i_4:⟨John, Reduce manufacturing cost., *Queue*⟩
r_4. The speaker volume shall be adjusted via a physical dial	i_5:⟨Jesse, Experienced issues with sound feedback in the past system., *Queue*⟩
r_5. The speaker volume shall be adjusted via physical buttons (i.e., volume up, volume down, mute/unmute)	i_6:⟨Alex, John, Compromise between manufacturing costs and navigation needs., *Queue*⟩

In this paper, we assume that the intention m is expressed as a structured natural language sentence, and the status u assumes values in $\{$*In*, *Past*, *Queue*$\}$: *In* signifies that the intention is currently being considered in the design process, *Past* is used if the intention was considered in the past but is not part of the current set, and *Queue* indicates that the intention has never been considered.

For each of the requirements in our volume adjustment example, we include the ⟨s, m, u⟩ triples elicited during the focus group in the right-hand column of Table 1. An example of a complex intention that can be generated with our triples is $i_3 \wedge i_4$ associated with r_3, indicating that both i_3 and i_4 are considerations for requirement r_3.

Intentions as a Tool for Decision Making. Intentions can be a powerful tool for making decisions in the context of conflicting needs and design choices, as well as when making changes to systems. Equivalent solutions for a requirement might not be equivalent with respect to the motivations for the requirement. We introduced above a representation of intentions and their associated stakeholders as a triple ⟨s, m, u⟩. If this information is matched with tradeoffs for the requirements or design, then we enable engineers to consider intentionality when making these tradeoff decisions both during initial implementation and maintenance.

We develop and evaluate two possible approaches to making trade-off decisions with multiple intentions: (1) pruning requirements and (2) retaining requirements. In our volume adjustment example, r_1 and r_2 are accepted as requirements and added to the specification. Yet, r_3–r_5 are mutually exclusive and the team must select the appropriate requirement to implement. In the first approach r_5 is retained and the intention expression associated with r_5 is $i_6 - \{i_3, i_4, i_5\}$. In the second approach, the requirement intention mappings are preserved as listed in Table 1. In addition to the specification, consisting of r_1, r_2, and r_5, a collection of pruned requirements $\{r_3, r_4\}$ is retained as alternatives to r_5 for further traceability.

Having this information also allows engineers to conduct further elicitation and seek clarification from stakeholders when a change may violate their stated intentions. Automatic analysis of intentions could assist engineers in decision-making without requiring them to gather additional input from stakeholders. For example, if the original concern is budgetary, engineers can consult the project budget directly without reconnecting with the original stakeholder of the intention.

Design models primarily store information about how the system will be built, as well as how stakeholders interact with the system. Design rationale already focuses on storing the reasoning for design decisions, which can be associated with the designer's intentions. Our framework can embed certain design rationale into our representation of intentionality, by treating the designer's intentions synonymously with the intentions of other stakeholders.

Tracing Intentions. Next, we take a wider view and consider how we can build on the extensive traceability literature to implement traceability of intentions across the software lifecycle. Requirements traceability is the "ability to describe and follow the life of a requirement, in both a forward and backward direction" [12], and is critical for maintaining consistency between the different models used in the system lifecycle [11]. Traceability connects entities before and after transformations, and between levels of abstraction. Work has focused on creating traceability links between the requirements for a system and the design artifacts, and ideally the complete implementation [1,16]. This includes the specification of *what* should be built (i.e., features) and any constraints for consideration. While the state-of-the-art in modeling and traceability has looked at tracing non-functional requirements [17], traceability does not include information about the underlying motivation for *why* features exist, even when (partially) elicited during the requirements process [13]. Traces are used to preserve design knowledge, support the integration of changes, and prevent misunderstandings. For these reasons, intentionality, once specified as part of the requirements, will become connected with the design (and implementation) via traceability links.

4   Research Agenda

In this section, we present our agenda for developing the ING-RE framework.

Validate Intention Definition. We defined intentionality based on the results of our initial literature review in Sect. 2. Our initial search string was limited and missed papers about intentionality (e.g., [5,10]). Going forward, we need to validate our definition with a full systematic literature review. Additionally, although the initial motivation for this work stemmed from the GORE literature, we believe that the concept of intentionality has ramifications beyond GORE approaches. We will compare our definition with other intention ontologies in GORE [5] and further juxtapose intentionality with rationale [8].

Theoretical Contribution. In Sect. 3, we presented our initial idea to formalize simple and complex intentions, taking into account the fact that more than one intention can motivate a single aim, and that not all intentions may be considered in the decision process. This initial proposal needs to be further investigated to determine which is the best approach (i.e., pruning or retaining requirements, see Sect. 3) for preserving alternatives in decision making. Our validation efforts will include evaluating both the coverage (i.e., how many common scenarios can be represented) and usability (i.e., the perceptions of analysts and designers when using it) of our framework. We will evaluate coverage through a set of case studies of the most common circumstances, and usability through interviews with experts.

Prototype. We will build a prototype that allows for users to trace and retain intentionality during requirements specification, as well as development. When requirements are selected for a project, their intentions are retained using the set I (see Sect. 3). In the event of a change in the system, engineers can review the affected requirements and underlying intentions. For a given change if the original intention cannot be maintained, then it becomes part of the neglected intentions (or requirements) going forward. This decision process relies on tool support, which we develop, for evaluating the satisfaction of the intention when the requirements they are connected with change. By defining our intention formalism on top of propositional or first order logic, we can build our tooling on top of existing SMT solvers. Given the dominance of the Agile paradigm, we build our prototype on top of an open-source sprint management tool (e.g., MyCollab, Odoo). In the case of approach (2) (see Sect. 3), we extend the paradigm of the product backlog [6] to preserve requirements that were not selected. We extend the prior work on expressing rationales within requirements [2,9].

Classification for Types of Intentions. We observed that while stakeholders' intentions can be articulated in many different ways and come from different contexts, they can be reduced to similar roots (i.e., the type of intention). In our volume adjustment example, John expressed budgetary concerns in i_4. In a different example, a stakeholder may be resistant to adding new features because they have budgetary concerns for the project. Both these examples have similar root-causes, i.e., budgetary concerns. Budget-related intentions are just one of a few types we have identified with our initial brainstorming. Other examples are historic, regulatory, and technology-related intentions. Identifying and classifying most, if not all, types of intentions may suggest what information is needed to evaluate the impact of a change and assist in the decision process. We aim to add this type of classification to the ING-RE framework. We plan to analyze historical problems and brainstorm possible intentions behind the identified needs to collect a data bank of intentions, as well as understand if any associations exist between intention types and NFRs. The procedure for type-specific decision making will be developed in consultation with domain experts, who can reason about how different types of intentions may evolve.

Limitations and Risks. The main limitation of our current proposal is that ING-RE only provides an approach for new systems and does not offer any support for existing systems when intentions were not explicitly collected during elicitation or captured requirements documents were not maintained.

This creates a risk that ING-RE will not reach broad applicability. To mitigate this risk, in our long-term agenda, we plan to study how to extract stakeholders' intentionality from available artifacts. Extracting intentions strictly depends on which artifacts are available in the given system. Additionally, we need to mitigate concerns from analysts about the extra overhead in documenting intentions during the requirements process, which is also part of our long-term agenda.

5   Summary

In this research preview, we proposed the ING-RE framework for capturing and retaining stakeholders' intentions, for use across the software lifecycle. Our literature review demonstrates that intentions are an important concept in requirements engineering, but may not be retained for later use. In our research agenda, we outlined the planned research steps to make ING-RE a reality through the design of lightweight tools and a series of empirical studies. Our aim in this preview is to gain feedback from the RE community and foster collaborations with researchers of existing approaches, in order to complete extensive validation of our framework across multiple paradigms.

Acknowledgments. Bobi Arce Mack, Cyrine Ben Ayed, Annie Karitonze, and Megan H. Varnum assisted in conducting the literature review.

References

1. Agouridas, V., Simons, P.: Antecedence and consequence in design rationale systems. AI EDAM **22**(4), 375–386 (2008)
2. Al-Alshaikh, H.A., Mirza, A.A., Alsalamah, H.A.: Extended rationale-based model for tacit knowledge elicitation in requirements elicitation context. IEEE Access **8**, 60801–60810 (2020). https://doi.org/10.1109/ACCESS.2020.2982837
3. Alatawi, E., Mendoza, A., Miller, T.: Psychologically-driven requirements engineering: A case study in depression care. In: 2018 25th Australasian Software Engineering Conference (ASWEC), pp. 41–50 (2018)
4. Bencomo, N., et al.: Requirements reflection: Requirements as runtime entities. In: Proceedings of ICSE 2010, pp. 199–202 (2010)
5. Bernabé, C.H., Silva Souza, V.E., Almeida Falbo, R., Guizzardi, R.S.S., Silva, C.: GORO 2.0: Evolving an ontology for goal-oriented requirements engineering. In: Guizzardi, G., Gailly, F., Suzana Pitangueira Maciel, R. (eds.) ER 2019. LNCS, vol. 11787, pp. 169–179. Springer, Cham (2019). https://doi.org/10.1007/978-3-030-34146-6_15
6. Bik, N., Lucassen, G., Brinkkemper, S.: A reference method for user story requirements in agile systems development. In: 2017 IEEE 25th International Requirements Engineering Conference Workshops (REW), pp. 292–298 (2017). https://doi.org/10.1109/REW.2017.83
7. Borgida, A., Dalpiaz, F., Horkoff, J., Mylopoulos, J.: Requirements models for design- and runtime: A position paper. In: Proceedings of MiSE 2013, pp. 62–68 (2013)

8. Burge, J.E., Carroll, J.M., McCall, R., Mistrik, I.: Rationale-Based Software Engineering. Springer (2008)

9. Nattoch Dag, J., el al.: Speeding up requirements management in a product software company: Linking customer wishes to product requirements through linguistic engineering. In: Proceedings of RE 2004, pp. 283–294 (2004)

10. Duijf, H., Broersen, J., Meyer, J.J.C.: Conflicting intentions: Rectifying the consistency requirements. Philos. Stud. **176**, 1097–1118 (2019)

11. Galvao, I., Goknil, A.: Survey of traceability approaches in model-driven engineering. In: Proceedings of EDOC 2007, pp. 313–313 (2007)

12. Gotel, O.C., Finkelstein, C.: An analysis of the requirements traceability problem. In: Proceedings of RE 1994 (1994)

13. Horkoff, J., et al.: Using goal models downstream: A systematic roadmap and literature review. Int. J. Inf. Syst. Model. Des. **6**(2), 1–42 (2015)

14. Kitchenham, B.: Procedures for performing systematic reviews. Tech. rep., Keele University and National ICT Australia Ltd. (ISSN: 1353-7776) (2004)

15. Martínez-Fernández, S., et al.: Software engineering for AI-based systems: A survey. ACM TOSEM **31**(2), 1–59 (2022)

16. Nair, S., de la Vara, J.L., Sen, S.: A review of traceability research at the requirements engineering conference[re@21]. In: Proceedings of RE 2013, pp. 222–229 (2013)

17. Vierhauser, M., Cleland-Huang, J., Burge, J., Grünbacher, P.: The interplay of design and runtime traceability for non-functional requirements. In: Proceedings of International Symposium on Software and Systems Traceability (SST), pp. 3–10 (2019)

18. Wang, Y., McIlraith, S.A., Yu, Y., Mylopoulos, J.: An automated approach to monitoring and diagnosing requirements. In: Proceedings of ASE 2007, pp. 293–302 (2007)

Understanding the Role
of Human-Related Factors in Security
Requirements Elicitation

Sanaa Alwidian[1] and Jason Jaskolka[2]([envelope])

[1] Electrical, Computer and Software Engineering, Ontario Tech University, Oshawa, ON, Canada
sanaa.alwidian@ontariotechu.ca
[2] Systems and Computer Engineering, Carleton University, Ottawa, ON, Canada
jason.jaskolka@carleton.ca

Abstract. *Context and motivation:* Many requirements engineering (RE) activities depend not only on the nature of the system itself, but also on human-centric characteristics of the RE teams.

Question/problem: What role do human-related factors of RE teams play in eliciting high-quality security requirements?

Principal ideas/results: This research preview presents our preliminary work in discovering the cognitive factors that represent the intentions and motivations of RE teams to develop secure systems from early stages of the system development, and how these factors impact the quality of the elicited requirements. We outline a framework, with an illustrative example, for describing the variables that affect the decisions of RE teams when they elicit security requirements to address security concerns.

Contribution: The proposed framework helps to characterize the different aspects of human-related factors, and the correlation between the impact of these factors on the quality of the requirements elicitation phase. This is a novel research direction which positions our long-term research agenda, and we urge community contributions in this direction to achieve an enhanced understanding of the role of human-related factors in requirements engineering for security domains.

Keywords: Human factors · Requirements engineering · Security · Cognitive models · Security requirements · Elicitation

1 Introduction

Requirements engineering (RE) is an inherently human-centric process. RE teams are tasked with eliciting requirements that will shape the quality of the developed system. Often, RE teams have varying levels of expertise needed to elicit high quality requirements. By high-quality requirements, we mean those requirements that reflect a *good understanding of stakeholders' needs*. Through the process of requirements elicitation, a good strategy to understand users' needs is to answer the three W's questions: "What are we doing?", "Why are we doing it?", and "Who is going to benefit from what we are doing?". Further, high-quality requirements should be

© The Author(s), under exclusive license to Springer Nature Switzerland AG 2023
A. Ferrari and B. Penzenstadler (Eds.): REFSQ 2023, LNCS 13975, pp. 65–74, 2023.
https://doi.org/10.1007/978-3-031-29786-1_5

unambiguous (i.e., avoid the use of indefinable terms like flexible, user friendly), *attainable* (i.e., feasible to implement within the allocated budget and timeline), *testable* (i.e., testing teams should be able to verify if a requirement has been correctly implemented or not), and *traceable* (i.e., provide the ability to follow the entire life cycle of requirements starting from design up to deployment, and to trace their relationship with other entities.). Finally, high-quality requirements are those that are *simple, specific, concise, and comprehensive* by nature. Security requirements elicitation must also consider the awareness of the team members with respect to threats and vulnerabilities, so that adequate protection goals can be determined for critical system assets [12].

Studying *human-related factors* (HRF) and their impact on the development of a secure system is essential to understanding the role that humans play in these processes. These factors must consider individual human behaviours as well as the social structures that enable collective action by communities of various sizes, and the different types of public and private institutional assemblages that shape societal responses. Considering the various perspectives of stakeholders in the system development life cycle (SDLC) and improving collaboration is noted as a key contributor to more effective security assurance for software systems [12].

In this paper, we focus on characterizing the HRFs (e.g., by studying profiles and cognitive constructs) of *requirements engineers*. To achieve this, we propose a framework comprised of several models to describe the variables that affect the decisions of RE team members when they elicit security requirements to address security concerns. The overall objective of the proposed framework is to discover the cognitive factors that represent the intentions and motivations of RE teams for developing secure systems from early stages of the SDLC, and how these factors impact the quality of the elicited security requirements. Hence, we are not concerned with the usability aspect of the system unless users express usability concerns alongside security concerns as stakeholders.

2   Related Work

Several works have looked to improve understanding of HRFs in the security context by exploring the skills, knowledge, and behaviours of those involved in developing secure systems [1,5]. A conceptual model to structure HRF models in the context of security was proposed by Kruger and Kearney [14]. The model served as the basis to build a tool for assessing information security awareness (ISA). The concepts were organized around the three dimensions: knowledge (what you know), attitude (what you think), and behaviour (what you do). Each dimension was then subdivided in a set of security concerns targeted by the measurement of the ISA in the respective organization. The factors were then used to evaluate the expertise and experience of individuals regarding the respective security concerns.

Wiley et al. [19] explored the relationship between information security awareness, organizational culture, and security culture. However, the literature is still missing studies on how these kinds of relationships contribute to the engineering of secure systems and particularly the requirements engineering stage.

Several works focused on comparing the practices of security and non-security experts [7,10]. Ion et al. [10] studied how security and non-security experts practices impacted the adoption of security solutions, including procedures and technologies. While security experts are most likely to install software updates and use two-factor authentication and a password manager, non-security experts are most likely to use anti-virus, access only known applications, and change passwords frequently. Other studies have explored the role of how human reasoning, judgments, and decisions impacted the adoption of security solutions. [6,8].

Related facets beyond technological solutions such as HRF, including the psychological and behavioural factors that affect the requirements process, should be addressed at both development team member and organizational levels. Other works have explored the interactions among systems developers and engineers to understand how vulnerabilities are introduced in the systems at early stages of development and how to mitigate them under interactions of developers. For example, Astromskis et al. [4] identified patterns of developer behaviour during the development of secure systems. Similarly, Wang et al. [18] examined the communication patterns between developers in the context of security development.

Overall, these works highlight the importance of considering the team's interactions and how they behave during the system development process, and the knowledge they require when performing the corresponding tasks. However, the literature is still missing studies on how these kinds of relationships contribute to the engineering of secure systems and particularly the RE stage.

3 Proposed Framework

We posit that HRFs should be integrated in early stages of system planning. We envision an ecosystem of models to enhance our understanding about the role of HRFs and their impact on security requirements elicitation. Figure 1 highlights the idea of our envisioned framework, its models, and their relationships. The dashed arrow between the security concerns model and the security requirements model indicates a relationship that we wish to exist, but may not when poor security requirements are elicited. It is important to emphasize that although in this paper we focused on the cognitive factors of the requirements engineers only; personality and cognitive factors are also relevant to other actors involved in the development process (e.g., security engineers and stakeholders). We realize that this is an important aspect to be considered in the future.

Profile Characterization Model. This model characterizes the different HRFs of the RE team members, where their combination, along with users' goals and requirements constitute what are referred to as *"user profiles"* [2]. One way in which to build the Profile Characterization Model is to create a set of *RE Team Profiles* by considering one or more of the aforementioned HRFs assessed through questionnaires or observation studies. This will enable the classification of individual RE team members within specific RE team profiles. For illustration, we consider the *"Big Five"* personality model [13] to characterize

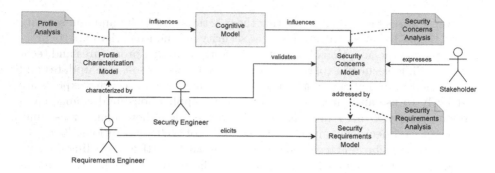

Fig. 1. An overview the proposed framework

the RE team profiles. The Big Five model represents broad categories of personality traits including: Openness, Conscientiousness, Agreeableness, Extraversion, and Neuroticism. In this work, we are not focusing on validating the accuracy or the correctness of profiles. A "profile" is just one example used to characterize the HRFs, and it can be replaced by other more sophisticated modelling mechanisms.

We acknowledge that teams dynamics, and the way team members, with different personalities or profiles, interact with each other would have a major impact on the overall quality of the elicited requirements. We plan to extend the scope of this research in the near future to study the teams dynamics, and their direct relationships to the elicited security requirements. We could consider adopting studies similar to the ones that have been already proposed in the context of software development, (e.g., [9]).

Cognitive Model. This model explains the attitudes, awareness, and intentions for adopting a particular recommendation (protection against security threats in our context). As Fig. 1 illustrates, the cognitive model is related to the Profile Characterization Model in the sense that RE team's cognitive aspects are influenced by the team's HRFs. For example, if we consider an *Openness* RE team profile, then owners of this profile are expected to be open to try new things, and to adapt new attitudes and behaviours towards achieving a particular goal.

One way to construct the cognitive model is by adopting the *Protection Motivation Theory* (PMT) [16] to identify the levels of awareness of the team members regarding security concerns (i.e., threats, vulnerabilities, attacks), and to describe the teams' intentions and motivations for developing secure systems from early stages of the SDLC. The characterization of attitudes, motivations and behaviours of teams is vital so that adequate security requirements and protection goals can be determined for critical system assets. The PMT measures the *coping behaviour* of individuals when they are informed of a threatening event (e.g., unauthorized access to an online banking account) [15]. This behaviour is directly influenced by the individual's willingness to perform a recommended action, referred to as the *coping response* (e.g., use strong passwords). The coping response is the net result of the individual's evaluation of the threat appraisal and coping appraisal.

In PMT, *threat appraisal* refers to an individual's assessment of the level of danger/risk posed by a threat. It consists of the *perceived vulnerability* and *perceived severity*, indicating the individual's assessment of the probability and the severity of the consequences of the threatening event, respectively. *Coping appraisal*, on the other hand, refers to an individual's ability to cope with potential loss/damage resulting from a threat. It consists of *self efficacy*, referring to the individual's confidence in their ability to perform the recommended behaviour (e.g., confidence to choose a strong password), *response efficacy*, representing the efficacy of the recommended behaviour (e.g., benefits of choosing and using a strong password), and *response cost*, indicating the perceived costs (monetary, time, effort, etc.) in adopting the recommended behaviour (e.g., cognitive load that comes with remembering a strong password.).

Security Concerns Model. This model captures the application-specific security concerns (e.g., threats, vulnerabilities, attacks) to the system and its assets. Security concerns are expressed by the stakeholders and validated by a security engineer. We assume that stakeholders express their genuine concerns with respect to the system assets, regardless of their security knowledge and awareness. In our context, the security concerns model can be represented as a list of security concerns, perhaps with priorities indicating the perceived losses and/or criticality of the concerns. The goal of the RE team is to address these concerns by eliciting appropriate security requirements.

Security Requirements Model. This model is elicited by the RE team and captures the set of requirements that prescribe protections against potential threats, vulnerabilities, and attacks to the application-specific system assets. The security requirements model may take the form of scenario-based models, goal-oriented models, Scaled Agile Framework (SAFe) models, UMLsec models, misuse cases, etc. We emphasize that the focus of this paper is not on the processes by which security requirements are elicited, but rather on understanding how HRFs shape the intentions and motivations of the individuals performing these processes. Our next step will be to study the actual impact of these intentions and motivations (which were captured here through profiles) on the elicited requirements through measuring the *quality and adequacy* of these requirements.

4 Illustrative Example

To better understand the relationships between the proposed models (see Fig. 1), we describe one usage scenario. Consider developing an Online Seller of Merchandise (OSM) system similar to the one presented in [17]. The OSM system must deal with several sensitive pieces of *customer information* including the customer's name, address, merchandise item selected, and credit card information. The RE team must elicit security requirements to address the security concerns related to the customer information. Using this scenario, we describe how each of the models in the proposed framework can be built.

Profile Characterization Model. Suppose that RE team profiles are characterized based on the "Big Five" personality model. Suppose also that we have a team composed of three requirements engineers (*RE1*, *RE2*, and *RE3*) to develop the requirements for the OSM system. For illustration, assume that *RE1* highly exhibits the *Openness* trait, *RE2* highly exhibits the *Conscientiousness* trait, and *RE3* highly exhibits the *Agreeableness* trait. According to [11], individuals showing openness tend to be creative, have the desire to try new things, and feel confident to tackle new challenges. On the other hand, individuals showing conscientiousness tend to be careful, show self-discipline, and pay attention to tiny details. Lastly, individuals showing agreeableness are usually considerate, cooperative, and tend to comply with instructions and recommendations. To this end, we model the profiles of the RE team. It is important to note that the big five personality traits are orthogonal dimensions. In our example, we consider *RE1* to belong to the *Openness* profile assuming that this is their most dominant trait. They could still show other personality traits, but just at lower levels. The same reasoning is applied to the other RE team members.

Cognitive Model. The cognitive model is unbounded to any application domain and it is applicable to any type of systems. For our illustrative example, we adopt the PMT constructs listed below.

C1. Perceived Severity: How serious are the consequences if the security concern is realized in the system?

C2. Perceived Vulnerability: How likely is the security concern to be realized in the system?

C3. Self Efficacy: How confident is the RE team in their ability to address the security concerns?

C4. Response Efficacy: To what extent does the RE team know effective security controls to address the security concern?

C5. Response Cost: To what extent does the RE team consider the costs/side-effects associated with solutions to address the security concern?

We consider these cognitive factors in an attempt to predict behaviours of the RE team members, and to differentiate between members who are aware of the importance of eliciting high-quality security requirements from those who do not make such efforts. The outcomes of this analysis can then be potentially used to devise measures that ameliorate quality of the elicited security requirements.

Security Concerns & Security Requirements Models. Assume the project stakeholders expressed the security concerns below related to the customer information assets. The security engineer is responsible for validating these concerns.

Table 1. Mapping security requirements to RE team profiles, cognitive constructs, and security concerns

Security Requirement		RE Profile	PMT Cognitive Construct					Security Concern
Elicited As	Elicited By		C1	C2	C3	C4	C5	
R1	RE1	P1: Openness	L	L	H	L	L	T1
R2	RE2	P2: Conscientiousness	H	H	L	H	H	T2
R3	RE3	P3: Agreeableness	H	H	N	H	H	T3
R4	RE3	P3: Agreeableness	H	H	N	H	H	T4

> T1. An attacker may intercept customer information in transit.
> T2. An attacker may modify customer information in storage.
> T3. An attacker may perform an SQL injection attack to delete customer information from storage.
> T4. A customer may use a weak password for their account.

For the OSM system, the RE team is responsible for eliciting security requirements to address the security concerns (T1–T4) expressed by the stakeholders. For illustration, suppose the following security requirements are elicited by members of the RE team (as noted) belonging to specific profiles described above.

> R1. All data in transit shall be encrypted by proven algorithms. (elicited by RE1)
> R2. The OSM system shall provide an access control mechanism with limited write access to the customer information storage. (elicited by RE2)
> R3. Input queries to the customer information storage shall be validated. (elicited by RE3)
> R4. A password policy for choosing and using strong passwords shall be enforced. (elicited by RE3)

Profile Analysis. To assess the impact of HRFs and their relationships with cognitive constructs, we map the RE team profile to cognitive constructs. For example, RE1, characterized by the *Openness* profile, could underestimate the perceived severity (C1) and perceived vulnerability (C2) of security concerns. At the same time, this team members would correlate highly with self efficacy (C3) because the openness personality traits make them confident about their ability to address security concerns. However, they could be overly optimistic where they may propose solutions that have not been fully thought-out, thus having low correlation to response efficacy (C4) and response cost (C5).

Reasoning in a similar way, we can map the personality traits for the *Conscientiousness* and *Agreeableness* profiles for each cognitive construct. An example result of our hypothetical profile analysis is shown in Table 1. In this example, we used a 3-level rating scale representing low (L), neutral (N), and high (H) correlation. The levels of the actual correlations can only be determined through questionnaire-based studies, which is out of the scope of this paper.

Security Concerns Analysis. To understand the extent to which team members of cognitive levels and profiles are aware of security concerns, we map security concerns with how they were elicited as requirements, and by which RE team member. This gives us an understanding about translating the awareness of security concerns into requirements elicited by a particular RE team member.

To express how security concerns are influenced by cognitive constructs, we suggest a set of generic sample questions that could be asked for each security concern and answered using the same 3-level rating scale discussed previously, as shown in Table 1, where the correlation between cognitive constructs and security concerns is inferred by evaluating the profile of the requirements engineer (RE) who translated the security concern (T) into a security requirement (R).

For instance, security concern $T1$ was handled by $RE1$ and elicited as requirement $R1$. Recall that $RE1$ was profiled with the *Openness* profile ($P1$), which makes them underestimate the consequences and the severity of $T1$. As a result, $RE1$ will score low (L) for $C1$ and $C2$. In addition, $RE1$ with their open-minded, optimistic personality, may not fully consider the potential costs to a solution or care whether it is really effective, thus scoring low (L) for $C4$ and $C5$. Lastly, as mentioned in earlier, $RE1$ will be confident about their ability to address security concerns, hence, they will score high (H) for $C3$, as illustrated in Table 1. It is worthwhile to mention here that it is not necessary for a particular requirements engineer with given profile to always correspond to a specific set of PMT cognitive answers.

Security Requirements Analysis. As illustrated in Table 1, understanding the relationships between an RE team's HRFs and their cognitive levels about protection/security awareness, and understanding how the cognitive levels impact the RE team's perceptions towards security concerns, gives us a framework to reason about the relationship between security concerns (expressed by the stakeholders) and security requirements elicited by the RE team. Specifically, the proposed analyses indicate how an RE team with a particular profile will approach specific security concern and elicit/translate it into a requirement.

Getting back to $RE1$, who elicited the requirement $R1$ to address security concern $T1$, we can see that $R1$ is an over-optimistic requirement, where the algorithms are vaguely prescribed to be the "best available" without much consideration to the costs and effectiveness in the solution. We argue that this over-optimistic requirement was elicited by $RE1$ in this way due to the nature of their HRFs (e.g., profile), which in turn impacted their perception towards the security concern $T1$, and the way they address it as a requirement $R1$. If the same concern $T1$ was elicited by $RE2$ or $RE3$, then we expect that the requirement will be expressed differently, depending on the RE team member's profile.

More detailed analyses investigating the correlations between models in our proposed framework is still needed. We expect that such analyses will result in an improved awareness about the role that HRFs play in the requirements elicitation phase, specifically in the context of developing secure systems.

5 Discussion and Concluding Remarks

The rating scale currently used for describing the relationships between the models may not be expressive enough to draw sufficient conclusions. Hence, we will explore alternatives to better characterize the relationships between models of our proposed framework. Further, we built the models using the "Big Five" model and PMT, but further investigation into other approaches for characterizing HRFs and cognitive constructs is needed. Similarly, other methodologies to characterize RE team profiles (e.g., the HEXACO model [3]) can be used.

To better understand the impact cognitive constructs have on shaping team attitudes and behaviours, we will consider studying the correlations between the different cognitive constructs as done in other studies that adapted the PMT for cybersecurity-related research (e.g., [20]). Last but not least, we currently assume that the stakeholders are capable of expressing their security concerns and that the security engineering is capable of validating these concerns. However, there are other HRFs that influence these activities as well. We need to consider not only the RE team to elicit the security requirements, but also the profiles of other stakeholders who articulate requirements and security engineers who validate these requirements. Exploring ways in which HRFs of the stakeholders impact the quality and validity of their security concerns is another area of interest. In conclusion, this paper previewed our work towards understanding the role of HRFs of RE teams and their impact on eliciting security requirements. Specifically, we sketched a framework based on several models capturing HRFs, cognitive constructs, security concerns and security requirements. Using an illustrative example, we showed how correlations between these models enable the discovery of cognitive factors that represent the intentions and motivations for developing secure systems from early stages of the SDLC. Continued elaboration of the proposed framework is needed to better enhance our understanding of the role that HRFs play in requirements engineering for security domains.

Acknowledgment. This research is supported by Natural Sciences and Engineering Research Council of Canada (NSERC) grant RGPIN-2019-06306.

References

1. Alshaikh, M.: Developing cybersecurity culture to influence employee behavior: A practice perspective. Comput. Secur. **98**, 102003 (2020)
2. Alwidian, S.: Towards integrating human-centric characteristics into the goal-oriented requirements language. In: 12th Model-Driven Requirements Engineering Workshop (2022). (To Appear)
3. Ashton, M.C., Lee, K.: Empirical, theoretical, and practical advantages of the HEXACO model of personality structure. Personal. Soc. Psychol. Rev. **11**(2), 150–166 (2007)
4. Astromskis, S., Bavota, G., Janes, A., Russo, B., Di Penta, M.: Patterns of developers behaviour: A 1000-hour industrial study. J. Syst. Softw. **132**, 85–97 (2017)

5. Bowen, B.M., Devarajan, R., Stolfo, S.: Measuring the human factor of cyber security. In: 2011 IEEE International Conference on Technologies for Homeland Security, pp. 230–235 (2011)
6. Brust-Renck, P.G., Weldon, R.B., Reyna, V.F.: Judgment and decision making. Oxford Research Encyclopedia of Psychology (2021)
7. Compagna, L., Khoury, P.E., Massacci, F., Thomas, R., Zannone, N.: How to capture, model, and verify the knowledge of legal, security, and privacy experts: A pattern-based approach. In: 11th International Conference on Artificial Intelligence and Law, pp. 149–153. ACM (2007)
8. Corbin, J.C., Reyna, V.F., Weldon, R.B., Brainerd, C.J.: How reasoning, judgment, and decision making are colored by gist-based intuition: A fuzzy-trace theory approach. J. Appl. Res. Memory Cognit. 4(4), 344–355 (2015)
9. Gren, L., Torkar, R., Feldt, R.: Group development and group maturity when building agile teams: A qualitative and quantitative investigation at eight large companies. J. Syst. Softw. 124, 104–119 (2017)
10. Ion, I., Reeder, R., Consolvo, S.: "...No one can hack my mind": Comparing expert and non-expert security practices. In: 11th USENIX Conference on Usable Privacy and Security, pp. 327–346. USENIX Association (2015)
11. Jaccard, J.J.: Predicting social behavior from personality traits. J. Res. Personal. 7(4), 358–367 (1974)
12. Jaskolka, J.: Recommendations for effective security assurance of software-dependent systems. In: Arai, K., Kapoor, S., Bhatia, R. (eds.) SAI 2020. AISC, vol. 1230, pp. 511–531. Springer, Cham (2020). https://doi.org/10.1007/978-3-030-52243-8_37
13. John, O.P., Srivastava, S.: The big five trait taxonomy: History, measurement, and theoretical perspectives, pp. 102–138. Guilford Press (1999)
14. Kruger, H., Kearney, W.: A prototype for assessing information security awareness. Comput. Secur. 25(4), 289–296 (2006)
15. Rippetoe, P.A., Rogers, R.W.: Effects of components of protection-motivation theory on adaptive and maladaptive coping with a health threat. J. Personal. Soc. Psychol. 52(3), 596–604 (1987)
16. Rogers, R.W.: A protection motivation theory of fear appeals and attitude change. J. Psychol. 91(1), 93 (1975)
17. Samuel, J., Jaskolka, J., Yee, G.: Analyzing structural security posture to evaluate system design decisions. In: 21st IEEE International Conference on Software Quality, Reliability, and Security, pp. 8–17 (2021)
18. Wang, S., Nagappan, N.: Characterizing and understanding software developer networks in security development. In: 32nd International Symposium on Software Reliability Engineering, pp. 534–545 (2021)
19. Wiley, A., McCormac, A., Calic, D.: More than the individual: Examining the relationship between culture and information security awareness. Comput. Secur. 88, 101640 (2020)
20. Woon, I.M.Y., Tan, G.W., Low, R.: A protection motivation theory approach to home wireless security. In: 2005 International Conference on Information Systems (2005)

Scope Determined (D) and Scope Determining (G) Requirements: A New Categorization of Functional Requirements

Daniel M. Berry[1]([✉]) [iD], Márcia Lucena[2] [iD], Victoria Sakhnini[1] [iD], and Abhishek Dhakla[1] [iD]

[1] Cheriton School of Computer Science, University of Waterloo, Waterloo, Ontario N2L 3G1, Canada
{dberry,vsakhnini,adhakla}@uwaterloo.ca
[2] Department of Computer Science and Applied Mathematics, Universidade Federal do Rio Grande do Norte, Natal, RN, Brazil
marciaj@dimap.ufrn.br

Abstract. Context: Some believe that Requirements Engineering (RE) for a computer-based system (CBS) should be done up front, producing a complete requirements specification before any of the CBS's software (SW) is written. **Problem**: A common complaint is that (1) new requirements *never* stop coming; so upfront RE goes on forever with an ever growing scope. However, data show that (2) the cost to modify written SW to include a new requirement is at least 10 times the cost of writing the SW with the requirement included from the start; so upfront RE saves development costs, particularly if the new requirement is one that was needed to prevent a failure of the implementation of a requirement already included in the scope. The scope of a CBS is the set of requirements that drive its implementation. **Hypothesis**: We believe that both (1) and (2) are correct, but each is about a different category of requirements, (1) scope determininG (G) or (2) scope determineD (D), respectively. **Past Work**: Reexamination of the reported data of some past case studies through the lens of these categories indicates that when a project failed, a large number of its defects were due to missing D requirements, and when a project succeeded, the project focused its RE on finding all of its D requirements. **Conclusions**: The overall aim of the future research is to empirically show that focusing RE for a *chosen* scope, including for a sprint in an agile development, on finding all and only the D requirements for the scope, while deferring any G requirements to later releases or sprints, allows upfront RE (1) that does not go on forever, and (2) that discovers all requirements whose addition after implementation would be wastefully expensive, wasteful because these requirements *are* discoverable during RE if enough time is devoted to looking for them.

Keywords: Agile methods · Defect repair cost · Empirical studies · Exceptions and variations · Requirements specification · Scope · Scope-determined requirement · Scope-determining requirement · Software development lifecycle · Sprint · Upfront requirements engineering · Waterfall methods

© The Author(s), under exclusive license to Springer Nature Switzerland AG 2023
A. Ferrari and B. Penzenstadler (Eds.): REFSQ 2023, LNCS 13975, pp. 75–84, 2023.
https://doi.org/10.1007/978-3-031-29786-1_6

1   Introduction

The current great debate [2, 8, 12, 15, 18, 20, 23, 24, 29, 30, 34, 35] in Requirements Engineering (RE) is whether requirements for a computer-based system (CBS)

1. should be identified upfront before design and coding begin, as in the waterfall lifecycle [31], or
2. should be identified incrementally, interleaved with design and coding of requirements identified so far, as in the spiral or agile lifecycles [1, 10].

Here, "identifying requirements for a CBS up front" means "identifying requirements for the CBS in their entirety".

The argument for identifying requirements upfront is that catching and fixing a requirement defect, i.e., a missing or incorrect requirement, during coding costs 10 times the cost of catching and fixing it during upfront RE [9, 32]. Thus, developing a CBS using waterfall methods, with requirements determined for the entire CBS up front before beginning any coding, leads to the shortest overall development time [4, 6, 9, 11, 16, 28, 33].

The arguments for identifying requirements incrementally are that

- requirements never stop coming [1, 6]; if design and coding do not start until *all* requirements are identified, design and coding will *never* start, and
- many requirements change as more and more of a CBS is developed and as the world changes as a result of the CBS's being used [21, 22]; some requirements that were identified before will be thrown out; and the time spent identifying these thrown-out requirements would be wasted!

Thus, we should develop CBSs using agile methods, with requirements determined for each sprint of coding only at the beginning of the sprint.

Attempts to settle the debate with empirical data have failed. Empirical studies go both ways and are overall inconclusive [2, 12, 20, 24, 29, 34]. Consequently, the choice of CBS development lifecycle, upfront RE or agile, to use in a CBS development project is made on the basis of gut feelings informed by experience and a recognition that if a project does something different from what is established practice, and the project fails, the heads of the project's decision makers will roll.

The reason that data have not decided the debate is that each side is right!

A1. Requirements *do* never stop coming; and many requirements *do* change, resulting in wasted effort.
A2. There *are* a lot of requirements defects that *can* be found and fixed early if one is spending enough time doing RE, and a complete requirements specification (RS) for a CBS *dramatically reduces* the incidence of expensive-to-fix requirement defects that appear in the code for the CBS.

We *believe* that the two competing arguments, A1 and A2, are talking about two different kinds of requirements, respectively:

K1. One kind of requirement emerges only as stakeholders, especially users, identify a previously unknown requirement, as a result of thinking about or using the CBS, sometimes because the CBS has changed the real world and thus, its own requirements [1, 10, 21, 22].

K2. The other kind of requirement can be identified before design and coding if enough time is devoted to RE, and it is wasteful to leave this kind of requirement to be found only later in the lifecycle when it is more expensive to fix [6].

If our belief is correct, then a possible reason that the past empirical studies are inconclusive is that none of them distinguishes these particular different kinds of requirements. They are all lumped together as just requirements.

We have identified a new binary categorization of new requirements being considered for addition to a CBS:

C1. The first category of requirement consists of the *scope determininG (G)*[1] requirements, and
C2. the second category of requirement consists of the *scope determineD (D)* requirements.

Here, the *scope* of a CBS is the set of requirements — a.k.a. use cases or features — it implements. Maybe, the past empirical studies will be more conclusive for each category of requirements studied separately.

This categorization has been identified in the past under different names. For example, among use cases, a variation or exception of another use case is a D requirement, but a new, independent use case is a G requirement. New are the *names* of the categories, which are more suggestive of

– how the categorization of a requirement can be done and
– how knowledge of the categorizations of candidate requirements for a CBS can be used during RE for the CBS and during its subsequent development.

Gause and Weinberg observe that just giving the right name to an old idea can suddenly make the idea operational [13].

This article cites related work from the past, all done for other purposes, which suggest that

– A1 = K1 = C1, and G requirements can be handled incrementally, and
– A2 = K2 = C2, and D requirements are best handled up front.

This article proposes some empirical studies to validate these claims.

In the rest of this article, Sect. 2 describes the categories of D and G requirements in depth. Section 3 defines the completion of a scope as the scope with all its D requirements made explicit. Section 4 makes a few observations about D and G requirements in the wild. Section 5 describes the past empirical work that directly led to the research reported in this article and describes other related past work. Section 6 predicts future work consistent with the long term goals of this research. Section 7 states important implications of successful future work, and Sect. 8 concludes the article.

---

[1] A mnemonic: The first letters that completely distinguish the phrases "scope determining" and "scope determined" are the phrases' last letters!

2   G and D Requirements

Understanding the definitions of G and D requirements is helped by a simple example[2] of their use.

Suppose we have a simple calculator CBS, C, offering only the four operations: addition, subtraction, multiplication, and division. Then the set of requirements[3],

$R =$ {addition, subtraction, multiplication, division},

is a *scope* of C. Then, the requirement,

$r1 =$ exponentiation,

is a G requirement with respect to (w.r.t.) R, because exponentiation is not needed for the correct functioning of any of addition, subtraction, multiplication, and division. Adding $r1$ to R makes a different calculator. That is, the addition of $r1$ is *determininG* a new scope. However, the requirement,

$r2 =$ checking that the division denominator is not zero,

is a D requirement w.r.t. R, because this checking is needed for correct functioning of division. Adding $r2$ to R does *not* make a different calculator; $r2$ is implicitly in C's scope because C's division will break any time the checking fails; $r2$ is *determineD* by the current scope. That is, because $r2$ is already in C's scope, $r2$ is not really added to R.

More precisely, suppose that C is a CBS. A *scope of C* is a set R of C's requirements. Each requirement r can be classified into one of two categories w.r.t. R.

1. r is a *scope determininG (G) requirement* w.r.t. R if r is not needed for correct functioning in C of any element of R
2. r is a *scope determineD (D) requirement* w.r.t. R if r is needed for correct functioning in C of at least one element of R other than r.

When the scope R is understood, "w.r.t. R" can be elided.

In the rest of this article, (1) "r is needed for correct functioning in C of q", (2) "q determines r w.r.t C", and (3) "r is determined w.r.t. C by q" are synonyms. In these sentences, r is a D requirement w.r.t. R.

3   Completion of Scope

That adding to a scope one of its D requirements is not considered changing the scope says that there is some notion of *the completion of a scope*, R, as R with all its D requirements made explicit.

The *completion* w.r.t. C, $\mathcal{C}_C(r)$, of a requirement r, is the set of all requirements R such that each r' in R is determined w.r.t. C by r or by an element of R.

---

[2]   The example does not expose all the nuances that emerge only when the definitions are applied to real-life CBSs.

[3]   Use of mathematical notation, e.g., the variable "C", is *not* signaling any attempt to be formal about inherently informal, real-world concepts [17]. Attempts to stick to only natural language led to confusingly ambiguous sentences, e.g., with two different scopes distinguished by multi-word adjective phrases that confused the authors!

The *completion* w.r.t. C, $\mathcal{C}_C(R)$, of a set of requirements, R, is the union of the completions w.r.t. C of all of R's elements.

In principle, the completion of any set of requirements should be the same, no matter the order in which its elements are considered for completion; testing that this is so is part of future work.

RE for a CBS, C, typically starts when C's customers supply to requirements analysts (RAs) an initial set of features, F. A feature is a requirement, and thus, F is a scope, which is taken, at least initially, as defining C.

The RAs flesh F into its completion, generally requirement by requirement. Because completion adds to a scope only requirements determined by the scope, this fleshing out is not seen as changing the scope of C. Therefore, F, F's completion, and every scope generated during the fleshing out are considered as describing *the scope of* C, $\mathcal{S}(C) = S$. The goal of this fleshing out is to make S explicit, that is to actually contain specifications of all the elements of the completion of F, and to serve as a written specification of C.

There will be an iterative procedure for completion:

Initially $S = F$. Each iteration considers a candidate new requirement, r to add to S, r being identified by any of a variety of elicitation means.

— If r is D w.r.t. S, then $S \cup \{r\}$ becomes S for the next iteration.

— If r is G w.r.t. S, then, *unless it is explicitly decided to expand the scope with* r,

S is unchanged for the next iteration, and r is added to the backlog list. The iteration is complete when $S = \mathcal{C}_C(F)$.

More generally, RE for a scope, R, of a CBS is done when all of the D requirements of R have been found and included in R's RS, which specifies R's completion [6].

If in any iteration, it is decided to expand the scope of C with the new r, then the iteration starts over with $S \cup \{r\}$ as the initial scope. Thus, this procedure takes into account the need to select a scope that is a viable whole, that consists of a set of features that forms a coherent whole, that makes C useful to its users.

To allow the iterative procedure to be used not only for upfront RE but also for each sprint of an agile method, the procedure is allowed to start with *any* scope, *any set of requirements*, not just F, which is intended to be for the whole of C.

4   Observations and Implications

The ability to categorize a requirement as either D or G allows focusing the precious RE effort for any version of a CBS on finding for its scope *all* and *only* those requirements, the scope's D requirements, that are necessary to have a complete RS for the version before its implementation begins. The procedure is to chose a scope, i.e., a set of G requirements, for your CBS. Focus all RE effort on finding all D requirements implied by the requirements in the chosen scope, while ignoring all other G requirements, i.e., those that are orthogonal to the requirements in the chosen scope. While this procedure sounds like the upfront RE in a waterfall method, it can be the initial steps in an agile method sprint for the chosen scope. The test cases that serve as the means to verify the correctness of the code for the sprint can be generated from the requirements that emerge from the procedure, if it is not desired to produce an actual RS.

Once the distinction between D and G requirements is understood, it becomes clear that the addition of a D requirement to the scope currently being implemented is *not* scope creep, because the D requirement was already in the scope even if it were not written in the scope's RS. Only the addition of a G requirement is true scope creep.

Another way to understand a missing D requirement is that it is a case of requirements and requirements documentation debt [3] that may not even reflect a conscious decision to incur the debt.

Still another view arises from use-case-based methods, which distinguish two kinds of use cases, (1) *main use cases* or *basic use cases* and (2) *variation* and *exception use cases*. In retrospect, these kinds of use cases are nothing more than (1) G use cases and (2) D use cases, respectively. When use cases are classified in this way, it becomes clear that all of the D use cases of a G use case have to be considered together and be implemented together with the G use case.

5   Antecedent and Related Work

Some of the relevant literature is cited in Sects. 1–4.

Some papers that author Berry coauthored and that his students wrote in the past show data that are consistent with and even actively support the claims made in this article. Each paper was written before the ideas reported in this article crystalized; it is actually another piece of slowly accumulated evidence leading to these ideas. Nevertheless, since the data were gathered with no notion of D and G requirements, there is no chance that researcher bias towards supporting this article's claims influenced the data gathering or the original conclusions drawn from the data. In these studies, in each challenged or failed project, a large number, if not a majority, of its defects were from missing D requirements [6,7,14,16]. In the one highly successful project, its RE focused on finding *all* D requirements of its scope [4,28].

The NaPiRE effort [26] has developed a survey instrument by which participants can identify what pains them about RE, whether they be artifacts, processes, or whatever. The effort has spawned a number of studies of software development organizations in various different places, including Austria, Brazil, and Germany.

The typical report about a survey [27][4] lists the top 5 or 10 pains. Among the top pains that involve RSs are:

- implicit requirements not made explicit [36]: **D**
- incomplete and/or hidden requirements [36]: **D** #1
- inconsistent requirements [25]: **D**
- missing completeness check [19]: **D**

---

[4]   The data from which the list of pains is obtained are in the papers listed at the cited website [27].

– moving targets (changing goals, business processes and or requirements) [25, 36]: **G**
– underspecified requirements that are too abstract and allow for various interpretations
  [25,36]: **D #2**
– volatile customer's business domain [36]: **G**

The two of these that seem to be consistently listed in the top 2 of the pains that involve requirements specifications are marked "**#1**" and "**#2**", respectively. The other top pains involve RE processes and communication among stakeholders.

For more details on all of these earlier studies, please see Sects. 6 and 7 of a technical report written by the authors of this article [5].

6   Future Work and Long Term Goals

The long-term goal of our future research is to answer the research question (RQ):

RQ: What is the effect on
  1. the development lifecycle of a CBS and
  2. the quality of the developed CBS
  of an RE that focuses on identifying and specifying upfront, all and only the D requirements in the CBS's scope?

A possible answer to the RQ is expressed as falsifiable, testable hypotheses that will be the subject of future research.

As typically done, an agile development discovers *all* requirements the same way: each sprint defines a scope that includes some new requirements, deferring others to later sprint. As typically done, a waterfall development tries to discover *all* requirements up front before its implementation starts.

The cost observations lead to the testable hypotheses:

H1: Regardless of development model,
  1. the quality of a CBS, by any measure, is negatively correlated with and
  2. the cost of developing the CBS is positively correlated with
  the number of D requirements missing from the CBS's scope.
H2: Let S be a scope that is missing some D requirements D'. Regardless of development model, a development from $S \cup D'$ produces a CBS
  1. with better quality and
  2. with lower cost
  than does a development from S.

Some past empirical studies need to be redone taking into account G and D requirements to see if they produce more conclusive results.

7 Implications of Validation

Support for these hypotheses recommends modifying agile methods so that each sprint, with a scope, S, begins with upfront RE that continues as long as necessary to identify all D requirements for S. This modified agile method is agile globally, but within each sprint, it is a waterfall for the scope of the sprint. This modified agile method should produce better CBSs more quickly and with lower cost than do unmodified agile methods. This claim, too, must be validated empirically.

8 Conclusions

This article has identified a new categorization of functional requirements, D and G requirements and has described past case studies showing that focusing a project's RE on finding all of its scope's D requirements has led to higher than expected project success. If future work shows this observation to be true in general, then each sprint of an agile method should include full upfront RE for its scope.

Acknowledgements. The authors thank Luiz Márcio Cysneiros, Sarah Gregory, Irit Hadar, Andrea Herrmann, John Mylopoulos, Mike Panis, Davor Svetinovic, and Anna Zamansky for their comments on previous drafts or in oral presentations of this work.

References

1. Agile Alliance. Principles: The Agile Alliance (2001). http://www.agilealliance.org/
2. Balaji, S., Sundararajan Murugaiyan, M.: WATEERFALLVs [sic] V-MODEL Vs AGILE: A COMPARATIVE STUDY ON SDLC. JITBM **2**(1) (2012)
3. Barbosa, L., Freire, S., et al.: Organizing the TD management landscape for requirements and requirements documentation debt. In: Proceedings of WER (2022). http://wer.inf.puc-rio.br/WERpapers/artigos/artigos_WER22/WER_2022_Camera_ready_paper_28.pdf
4. Berry, D., Daudjee, K., et al.: User's manual as a requirements specification: Case studies. REJ **9**(1), 67–82 (2004)
5. Berry, D., Lucena, M., et al.: Scope determined (D) versus scope determining (G) requirements: A new significant categorization of functional requirements. Tech. rep., Univ. Waterloo (2023). https://cs.uwaterloo.ca/~dberry/FTP_SITE/tech.reports/GvsDprelimTechReport.pdf
6. Berry, D.M., Czarnecki, K., et al.: Requirements determination is unstoppable: An experience report. In: Proceedings of RE, pp. 311–316 (2010)
7. Berry, D.M., Czarnecki, K., et al.: The problem of the lack of benefit of a document to its producer (PotLoBoaDtiP). In: Proceedings of SWSTE, pp. 37–42 (2016)
8. Berry, D.M., Damian, D., et al.: To do or not to do: If the requirements engineering payoff is so good, why aren't more companies doing it? In: Proceedings of RE, p. 447 (2005)

9. Boehm, B.W.: Software Engineering Economics. Prentice-Hall, Englewood Cliffs (1981)
10. Boehm, B.W.: A spiral model of software development and enhancement. SIG-SOFT Softw. Eng. Notes **11**(4), 14–24 (1986)
11. Ellis, K., Berry, D.M.: Quantifying the impact of requirements definition and management process maturity on project outcome in business application development. REJ **18**(3), 223–249 (2013)
12. Gaborov, M., Karuović, D., et al.: Comparative analysis of agile and traditional methodologies in IT project management. jATES **11**(4), 1–24 (2021)
13. Gause, D., Weinberg, G.: Exploring Requirements: Quality Before Design. Dorset House, New York (1989)
14. Gellert, C.: Requirements engineering and management effects on downstream developer performance in a small business findings from a case study in a CMMI/CMM context. Master's thesis, Univ. Waterloo, Canada (2021). http://hdl.handle.net/10012/17777
15. Greenspan, S.J.: Extreme RE: What if there is no time for requirements engineering? In: Proceedings of RE, pp. 282–284 (2001)
16. Isaacs, D., Berry, D.M.: Developers want requirements, but their project manager doesn't; and a possibly transcendent Hawthorne effect. In: Proceedings of EmpiRE (2011)
17. Jackson, M.A.: Problems and requirements. In: Proceedings of ISRE, pp. 2–8 (1995)
18. Jiang, L., Eberlein, A.: An analysis of the history of classical software development and agile development. In: Proceedings of IEEE SMC, pp. 3733–3738 (2009)
19. Kalinowski, M., Curty, P., et al.: Supporting defect causal analysis in practice with cross-company data on causes of requirements engineering problems. In: Proceedings of ICSE SEIP, pp. 223–232 (2016)
20. Kasauli, R., Knauss, E., et al.: Requirements engineering challenges and practices in large-scale agile system development. JSS **172**, 110851 (2021)
21. Lehman, M.M.: Programs, life cycles, and laws of software evolution. Proc. IEEE **68**(9), 1060–1076 (1980)
22. Lehman, M.M.: Laws of software evolution revisited. In: Montangero, C. (ed.) EWSPT 1996. LNCS, vol. 1149, pp. 108–124. Springer, Heidelberg (1996). https://doi.org/10.1007/BFb0017737
23. Lucia, A., Qusef, A.: Requirements engineering in Agile software development. J. Emerg. Technol. Web Intell. **2**(3), 212–220 (2003)
24. Malm, T.: Requirements engineering in agile projects - Comparing a sample to requirements-engineering literature. Master's thesis, Faculty of Social Sciences, Business and Economics, Åbo Akademi Univ., Turku, Finland (2020). https://www.doria.fi/bitstream/handle/10024/177487/malm_tobias.pdf
25. Mendez, D., Tießler, M., et al.: On evidence-based risk management in requirements engineering. In: SWQD: Methods and Tools for Better Software and Systems, pp. 39–59 (2018)
26. Mendez, D., Wagner, S., et al.: Naming the pain in requirements engineering: Contemporary problems, causes, and effects in practice. EMSE **22**, 2298–2338 (2017)
27. NaPiRE: Napire data and publications (Viewed 1 November 2022). http://napire.org/#/data
28. Ou, L.: WD-pic, a new paradigm for picture drawing programs and its development as a case study of the use of its user's manual as its specification. Master's thesis, Univ. Waterloo (2002). https://cs.uwaterloo.ca/~dberry/FTP_SITE/tech.reports/LihuaOuThesis.pdf

29. Rasheed, A., Zafar, B., et al.: Requirement engineering challenges in agile software development. Math. Prob. Eng. **2021** (2021)
30. Rogers, G.: How agile can requirements engineers really be? RE Magazine (2014). https://re-magazine.ireb.org/articles/requirements-engineers
31. Royce, W.W.: Managing the development of large software systems: Concepts and techniques. In: WesCon (1970)
32. Schach, S.R.: Classical and object-oriented software engineering with UML and Java, 4th edn. McGraw-Hill, New York (1998)
33. So, J., Berry, D.M.: Experiences of requirements engineering for two consecutive versions of a product at VLSC. In: Proceedings of RE, pp. 216–221 (2006)
34. Thesing, T., Feldmann, C., Burchardt, M.: Agile versus waterfall project management: Decision model for selecting the appropriate approach to a project. Procedia Comput. Sci. **181**(01), 746–756 (2021)
35. Van Cauwenberghe, P.: Chapter 18: Refactoring or up-front design? (2002). http://wwww.agilecoach.net/html/refactoring_or_upfront.pdf
36. Wagner, S., Mendez, D., et al.: Requirements engineering practice and problems in Agile projects: Results from an international survey. In: Proceedings of CibSE, pp. 85–98 (2017)

NLP and Machine Learning for AI

Using Language Models for Enhancing the Completeness of Natural-Language Requirements

Dipeeka Luitel[✉], Shabnam Hassani[✉], and Mehrdad Sabetzadeh[✉]

University of Ottawa, Ottawa, ON K1N 6N5, Canada
{Dipeeka.Luitel,S.Hassani,M.Sabetzadeh}@uottawa.ca

Abstract. [**Context and motivation**] Incompleteness in natural-language requirements is a challenging problem. [**Question/problem**] A common technique for detecting incompleteness in requirements is checking the requirements against external sources. With the emergence of language models such as BERT, an interesting question is whether language models are useful external sources for finding potential incompleteness in requirements. [**Principal ideas/results**] We mask words in requirements and have BERT's masked language model (MLM) generate contextualized predictions for filling the masked slots. We simulate incompleteness by withholding content from requirements and measure BERT's ability to predict terminology that is present in the withheld content but absent in the content disclosed to BERT. [**Contribution**] BERT can be configured to generate multiple predictions per mask. Our first contribution is to determine how many predictions per mask is an optimal trade-off between effectively discovering omissions in requirements and the level of noise in the predictions. Our second contribution is devising a machine learning-based filter that post-processes predictions made by BERT to further reduce noise. We empirically evaluate our solution over 40 requirements specifications drawn from the PURE dataset [1]. Our results indicate that: (1) predictions made by BERT are highly effective at pinpointing terminology that is missing from requirements, and (2) our filter can substantially reduce noise from the predictions, thus making BERT a more compelling aid for improving completeness in requirements.

Keywords: BERT · Natural Language Processing · Machine Learning

1 Introduction

Improving the completeness of requirements is an important yet challenging problem in requirements engineering (RE) [2]. The RE literature distinguishes two notions of completeness [3]: (1) *Internal* completeness is concerned with requirements being closed with respect to the functions and qualities that one can infer exclusively from the requirements. (2) *External* completeness is concerned with ensuring that requirements are encompassing of all the information that external sources of knowledge suggest the requirements should cover. These external

© The Author(s), under exclusive license to Springer Nature Switzerland AG 2023
A. Ferrari and B. Penzenstadler (Eds.): REFSQ 2023, LNCS 13975, pp. 87–104, 2023.
https://doi.org/10.1007/978-3-031-29786-1_7

sources can be either people (stakeholders) or artifacts, e.g., higher-level requirements and existing system descriptions [4]. External completeness is a relative measure, since the external sources may be incomplete themselves or not all the relevant external sources may be known [3]. Although external completeness cannot be defined in absolute terms, relevant external sources, when available, can be useful for detecting missing requirements-related information.

When requirements and external sources of knowledge are textual, one can leverage natural language processing (NLP) for computer-assisted checking of external completeness. For example, Ferrari et al. [5] use NLP to check completeness against stakeholder-interview transcripts. And, Dalpiaz et al. [6] use NLP alongside visualization to identify differences among stakeholders' viewpoints; these differences are then investigated as potential incompleteness issues.

With (pre-trained) language models, a new opportunity arises for NLP-based improvement of external completeness in requirements: Using self-supervised learning, language models have been trained on very large corpora of textual data, e.g., millions of Wikipedia articles. This raises the prospect that *a language model can serve as an external source of knowledge for completeness checking.* In this paper, we explore a specific instantiation of this idea using BERT [7].

BERT has been trained to predict masked tokens by finding words or phrases that most closely match the surrounding context. To illustrate how BERT can help with completeness checking of requirements, consider the example in Fig. 1. In this example, we have masked one word, denoted [MASK], in each of requirements R1, R2 and R3. We have then had BERT make five predictions for filling each masked slot. For instance, in R1, the masked word is *availability*. The predictions made by BERT are: *performance, efficiency, stability, accuracy,* and *reliability.* As seen from the figure, one of these predictions, namely *stability,* is a word that appears in R6. Similarly, the predictions that BERT makes for the masked words in R2 and R3 (*audit* and *connectivity*) reveal new terminology that is present in R4 and R5 (*network, traffic, comply* and *security*).

In the above example, *if requirements* **R4–R6** *were to be missing,* ***BERT's predictions over* R1–R3** *would provide useful cues about some of the missing terminology.*

Contributions. We need a strategy to study how well BERT predicts relevant terminology that is absent from requirements. To this end, we *simulate* missing information by randomly withholding a portion of a given requirements specification. We disclose the remainder of the specification to BERT for obtaining masked-word predictions. In our example of Fig. 1, the disclosed part would be requirements R1–R3 and the withheld part would be requirements R4–R6. BERT can be configured to generate multiple predictions per mask. Our ***first contribution*** is to determine how many predictions per mask is an optimal trade-off between effectively discovering simulated omissions and the amount of unuseful predictions (noise) that BERT generates.

We observe that a large amount of noise would result if predictions by BERT are to achieve good coverage of requirements omissions. Some of the noise is trivial to filter. For instance, in the example of Fig. 1, one can dismiss the predictions of *service* and *system* (made over R3); these words already appear in

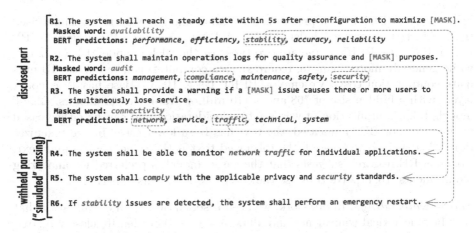

Fig. 1. Illustrative requirements specification split into a *disclosed* and a *withheld* part. The withheld part *simulates* requirements omissions. Masking words in the disclosed part and having BERT make predictions for the masks reveals some terms that appear only in the withheld part.

the disclosed portion, thereby providing no cues about missing terminology. Furthermore, one can dismiss words that carry little meaning, e.g., "any", "other" and "each", should such words appear among the predictions. After applying these obvious filters, one would still be left with considerable noise. Our *second contribution* is a machine learning-based filter that post-processes predictions by BERT to strike a better balance between noise and useful predictions.

Our solution development and evaluation is based on 40 requirements specifications from the PURE dataset [1]. These specifications contain over 23,000 sentences combined. To facilitate replication and further research, we make our implementation and evaluation artifacts publicly available [8].

2   Background

Below, we review the background for our work, covering the NLP pipeline, language models, word embeddings, machine learning and corpus extraction.

NLP Pipeline. Natural language processing (NLP) is usually performed using a pipeline of modules [9]. In this paper, we apply an NLP pipeline composed of tokenizer, sentence splitter, part-of-speech (POS) tagger and lemmatizer modules. The tokenizer demarcates the tokens of the text. The sentence splitter divides the text into sentences. The POS tagger assigns a POS tag to each token in each sentence. The lemmatizer maps each word in the text to its lemma form. For example, the lemma for both "running" and "ran" is "run". We use the annotations produced by the NLP pipeline for several purposes, including the identification and lemmatization of terms in requirements documents as well as processing predictions made by BERT in their surrounding context.

Language Models. Recent NLP approaches heavily rely on deep learning, and in particular, transfer learning [7]. Bidirectional Encoder Representations from

Transformers (BERT) is a pre-trained language model using two unsupervised tasks: Masked Language Model (MLM) and Next Sentence Prediction (NSP). BERT Base and BERT Large are two types of the BERT model. BERT Large, while generally more accurate, requires more computational resources. To mitigate computation costs, we employ BERT Base. BERT Base has 12 encoder layers with a hidden size of 768 and ≈110 million trainable parameters. BERT models take capitalization of words into consideration and can be either cased or uncased. For BERT uncased, the text has been lower-cased before tokenization, whereas in BERT cased, the tokenized text is the same as the input text. Previous RE research suggests that the cased model is preferred for analyzing requirements [10,11]. We thus use the cased model in this paper.

Word Embeddings. In our work, we need a semantic notion of similarity that goes beyond lexical equivalence and allows us to further identify closely related terms; examples would be (i) "key" and "unlock", and (ii) "encryption" and "security". For this, we use cosine similarity over *word embeddings*. Word embeddings are mathematical representations of words as dense numerical vectors capturing syntactic and semantic regularities [12]. We employ GloVe's pre-trained model [13]. This choice is motivated by striking a trade-off between accuracy and efficiency. BERT also generates word embeddings; however, these embeddings are expensive to compute and do not scale well when a large number of pairwise term comparisons is required.

Machine Learning (ML). We use supervised learning, more specifically classification, to identify the relevant predictions made by BERT. Our features for learning and our process for creating labelled data are discussed in Sects. 3 and 4, respectively. Classification models have a tendency to predict the more prevalent class(es) [14]. In our context, non-relevant terms outnumber relevant ones. We under-sample the majority class (i.e., non-relevant) to counter imbalance in our training set and thereby reduce the risk of filtering useful information [15]. To further reduce this risk, we additionally employ cost-sensitive learning (CSL) [14]. CSL enables us to assign a higher penalty to relevant terms being filtered than non-relevant terms being classified as relevant.

Domain-Corpus Extraction. Domain-specific corpora are useful resources for improving the accuracy of automation in RE [16]. When such corpora do not exist a priori, they can be constructed using domain documents from sources such as Wikipedia, books and magazines [16–19]. In our work, we require statistical information from a domain-specific corpus to better determine relevance of terms predicted by BERT. For this purpose, we employ the WikiDoMiner corpus extractor [19]. WikiDoMiner is a fully automated tool that gathers domain knowledge for an input requirements specification by crawling Wikipedia. The tool extracts keywords from the input specification and assembles a set of Wikipedia articles relevant to the terminology and thus the domain of the specification.

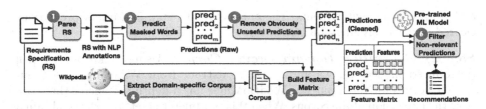

Fig. 2. Approach Overview.

3   Approach

Figure 2 provides an overview of our approach. The input to the approach is a (textual) requirement specification (RS). The approach has six steps. The first step is to parse the RS. The second step is to generate predictions for masked words using BERT. The third step is to remove the predictions that provide little or no additional information. The fourth step is to construct a domain-specific corpus for the given RS. Using this corpus and the results from Step 1, the fifth step is to build a feature matrix for ML-based filtering of non-relevant terms from predictions by BERT. The sixth and last step is to feed the computed feature matrix to a (pre-trained) classifier in an attempt to remove noise (non-relevant words) from the predictions. The output of the approach is a list of recommended terms that are likely relevant to the RS but are currently absent from it.

Step 1) Parsing RS using NLP. The RS is fed to an NLP pipeline. The pipeline first annotates the sentences in the RS using a sentence splitter. A sentence annotation does not necessarily demarcate a grammatical sentence but rather what the sentence splitter finds to be a sentence. Next, each word in each sentence is annotated with a POS tag and the lemma form of the word.

Step 2) Obtaining Predictions from BERT. Our approach loops through each sentence of the annotated RS obtained from Step 1. It masks, one at a time, each word that has a POS tag of "noun" or "verb". The resulting sentence with a single masked word is fed to BERT in order to obtain a configurable number of predictions for the masked word. We focus on nouns and verbs because noun phrases and verb phrases are the main meaning-bearing elements of sentences [4]. In our illustration of Fig. 1, we had BERT generate five predictions per masked word. As we argue empirically in our evaluation of Sect. 4, for our purposes, our recommendation is 15 predictions per masked word. For each prediction, BERT provides a probability score indicating its confidence in the prediction. We retain the probability scores for use in Step 5 of the approach.

Step 3) Removing Obviously Unuseful Predictions. We discard predictions that offer little or no additional information. Specifically, we remove predictions whose lemma is already present in the RS; such predictions provide no new hints about potentially missing terminology. We further remove predictions that either (1) are among the top 250 most common words in English, or (2) belong to the union of Berry et al.'s [20] and Arora et al.'s [21] [22] sets of vague words and stopwords in requirements. The output of this step is a list of predictions cleared of obviously unuseful terms.

Step 4) Generating Domain-specific Corpus. Using WikiDoMiner [19] (introduced in Sect. 2), we automatically extract from Wikipedia a domain-specific corpus for the input RS. WikiDoMiner has a depth parameter that controls the expansion of the corpus. When this parameter is set to zero, we obtain a corpus of articles containing a direct match to the key phrases in the RS. Increasing the depth generates larger corpora, with each level further expanding the sub-categories of Wikipedia articles included. In our work, we restrict our search to direct article matches (i.e., *depth = 0*). This enables quick corpus generation and further scopes terminology expansion to what is most immediately pertinent to the RS at hand. In our exploratory investigation, we observed that larger depth values significantly increase corpus size, diluting its domain-specificity and in turn reducing the effectiveness of filtering in Step 6 of the approach.

Step 5) Building Feature Matrix for Filtering. For each prediction from Step 3, we compute a feature vector as input for a ML-based classifier that decides whether the prediction is "relevant" or "non-relevant" to the input RS. Our features are listed and explained in Table 1. The main principle behind our feature design has been to keep the features generic and in a normalized form. Being generic is important because we do not want the features to rely on any particular domain or terminology. Having the features in a normalized form is important for allowing labelled data from multiple documents to be combined for training, and for the resulting ML models to be applicable to unseen documents. The output of this step is a feature matrix where each row represents a prediction (from Step 3) and each column represents a feature as defined in Table 1.

Step 6) Filtering Noise from Predictions The predictions from Step 3 are noisy (i.e., have many false positives). To reduce the noise, we subject the predictions to a pre-trained ML-based filter. The most accurate ML algorithm for this purpose is selected empirically (see RQ2 in Sect. 4). The selected algorithm is trained on the development and training portion of our dataset (P_1 in Table 2, as we discuss in Sect. 4). Due to our features in Table 1 being generic and normalized, the resulting ML model can be used as-is over unseen documents without re-training (see RQ3 in Sect. 4 for evaluation of effectiveness). The output of this step is the list of BERT predictions that are classified as "relevant" by our filter; duplicates are excluded from the final results.

4   Evaluation

In this section, we empirically evaluate our approach. During the process, we also build the pre-trained ML model required by Step 6 of the approach (Fig 2).

Table 1. Features for Learning Relevance and Non-relevance of Predictions by BERT.

ID	Type (T), Definition (D) and Intuition (I)
MF1	**(T)** Nominal **(D)** POS tag of the masked word (noun or verb). **(I)** This feature is helpful if nouns and verbs happen to influence relevance in different ways
F2	**(T)** Nominal **(D)** POS tag of the prediction; this is obtained by replacing the masked word with the predicted word and running the NLP pipeline on the resulting sentence. **(I)** The intuition is similar to F1, except that predictions are not necessarily nouns or verbs and can, e.g., be adjectives or adverbs
F3	**(T)** Nominal (Boolean) **(D)** True if F1 and F2 match; otherwise, False. **(I)** A mismatch between F1 and F2 could be an indication that the prediction is non-relevant
F4	**(T)** Numeric **(D)** Length (in characters) of the masked word. **(I)** Words that are too short may give little information. As such, predictions resulting from masking short words could be non-relevant
F5	**(T)** Numeric **(D)** Length (in characters) of the prediction. **(I)** Predictions that are too short could be non-relevant
F6	**(T)** Numeric **(D)** $\min(F4, F5)/\max(F4, F5)$. **(I)** A small ratio (i.e., a large difference in length between the prediction and the masked word) could indicate non-relevance
F7	**(T)** Numeric **(D)** The confidence score that BERT provides alongside the prediction. **(I)** A prediction with a high confidence score could have an increased likelihood of being relevant
F8	**(T)** Numeric **(D)** Levenshtein distance between the prediction and the masked word. **(I)** A small Levenshtein distance between the prediction and the masked word could indicate relevance
F9	**(T)** Numeric **(D)** Semantic similarity computed as cosine similarity over word embeddings. **(I)** A prediction that is close in meaning to the masked word could have a higher likelihood of being relevant
F10*	**(T)** Ordinal **(D)** A value between zero and nine, indicating how frequently the prediction (in lemmatized form) appears across *all BERT-generated predictions* over a given RS. **(I)** A smaller value could indicate a higher likelihood of relevance
F11*[†]	**(T)** Ordinal **(D)** A value between zero and nine, indicating how frequently the prediction (in lemmatized form) appears in the *domain-specific corpus*. **(I)** A smaller value could indicate a higher likelihood of relevance
F12[†][‡]	**(T)** Numeric **(D)** Average TF-IDF rank of the prediction across all articles in the domain-specific corpus. **(I)** A higher rank could indicate a higher likelihood of relevance
F13[†][‡]	**(T)** Numeric **(D)** Maximum TF-IDF rank of the prediction across all articles in the domain-specific corpus. **(I)** Same intuition as that for F12

Zero is most frequent (top ten percentile) and nine is least frequent (bottom ten percentile). [†] Feature uses domain-specific corpus. [‡] TF-IDF values are normalized by Euclidean norm.

4.1   Research Questions (RQs)

Our evaluation answers the following RQs using part of the PURE dataset [1]. In lieu of expert input about incompleteness for the documents in this dataset, we apply the withholding strategy discussed in Sect. 1 to simulate incompleteness.

RQ1. How accurately can BERT predict relevant but missing terminology for an input RS? The number of predictions that BERT generates per mask is a configurable parameter. RQ1 examines what value for this parameter offers the best trade-off for producing useful recommendations.

RQ2. Which ML classification algorithm most accurately filters unuseful predictions made by BERT? Useful recommendations from BERT come alongside a considerable amount of noise. In RQ2, we examine different ML algorithms to filter noise. We further study the impact of data balancing and cost-sensitive learning to prevent over-filtering.

RQ3. How accurate are the recommendations generated by our approach over unseen documents? In RQ3, we combine the best BERT configuration from RQ1 with the filter models built in RQ2, and measure the accuracy of this combination over unseen data.

4.2   Implementation and Availability

We have implemented our approach in Python. The NLP pipeline is implemented using SpaCy 3.2.2. For extracting word embeddings, we use GloVe [13]. To obtain masked language model predictions from BERT, we use the Transformers 4.16.2 library by Hugging Face (https://huggingface.co/) and operated in PyTorch 1.10.2+cu113. Our ML-based filters are implemented in WEKA 3-8-5 [23]. To implement the ML features listed in Table 1, we use standard implementations of cosine similarity (over word embeddings) and Levenshtein distance [24]. The TFIDF-based features in this table (F12-13) use TfidfVectorizer from scikit-learn 1.0.2. Our implementation and evaluation artifacts are publicly available [8].

4.3   Dataset

Our evaluation is based on 40 documents from the PURE dataset [1] – a collection of public-domain requirements specifications. Many of the documents in PURE require manual cleanup (e.g., removal of table of contents, headers, section markers, etc.) We found 40 to be a good compromise between the effort that we needed to spend on cleanup and having a dataset large enough for statistical significance testing, mitigating the effects of random variation, and training ML-based filters. The selected documents, listed in Table 2, cover 15 domains. We partition the documents into two (disjoint) subsets P_1 and P_2. P_1 is used for approach development and tuning, i.e., for answering RQ1 and RQ2. P_2, i.e., the documents unseen during development and tuning, is used for answering RQ3. Our procedure for assigning documents to P_1 or P_2 was as follows: We first randomly selected one document per domain and put it into P_2; this is to maximize domain representation in RQ3. From the rest, we randomly selected 20 documents for inclusion in P_1, while attempting to have P_1 represent half of the data in terms of token count. Any remaining document after this process was assigned to P_2, thus giving us 20 documents in P_2 as well. Table 2 provides domain information and summary statistics for documents in P_1 and P_2 after cleanup.

Table 2. Our Dataset (Subset of PURE [1]). P_1 is for development and training and P_2 for testing.

| | Domain | | | | | | | | | | | | | | | Statistics | |
	A	B	C	D	E	F	G	H	I	J	K	L	M	N	O	# of sentences	# of tokens
P_1 Dev & Training		gamma, jse	tachonet, nasa x38, nenios, libra	evla back, gemini	pnnl	elsfork, ctc network	beyond	-	-	space fractions, multi-mahjong	-	clarus low, grid bgc	-	-	watcom gui, sce api, hats, watcom	12137	192761
P_2 Testing	sprat, cctns, dii	e-procurement	inventory	esa, telescope	themas, elsfork	agentmom, tcs	evla corr	micro care	npac	qheadache	colorcast	clarus high	ijis	rlcs	grid 3D	10958	192403

A : Security, *B* : Finance, *C* : Administration, *D* : Astronomy, *E* : Energy, *F* : Communications, *G* : Hardware Design, *H* : Medicine, *I* : Databases, *J* : Games, *K* : Art, *L* : Weather, *M* : Legal, *N* : Transport, *O* : UX/Visualization.

4.4   Analysis Procedure

EXPI. This experiment answers RQ1. For every document $p \in P_1$, we randomly partition the set of sentences in p into two subsets of (almost) equal sizes. In line with our arguments in Sect. 1, we *disclose* one of these subsets to BERT and *withhold* the other. We apply Steps 1, 2 and 3 of our approach (Fig. 2) to the disclosed half, *as if this half were the entire input document.* We run Step 2 of our approach with four different numbers of predictions per mask: 5, 10, 15, and 20. For every document $p \in P_1$, we compute two metrics, *Accuracy* and *Coverage*, defined in Sect. 4.5. As the number of predictions per mask increases from 5 to 20, the predictions made by BERT reveal more terms that are relevant to the withheld half. Nevertheless, as we will see in Sect. 4.6, the benefits diminish beyond 15 predictions per mask.

To ensure that the trends we see as we increase the number of predictions per mask are not due to random variation, we pick different shuffles of each document p across different numbers of predictions per mask. For example, the disclosed and withheld portions for a given document p when experimenting with 5 predictions per mask are different random subsets than when experimenting with 10 predictions per mask.

EXPII. This experiment answers RQ2 and further constructs the training set for the ML classifier in Step 6 of our approach (Fig. 2). We recall the disclosed and withheld halves as defined in EXPI. For every document $p \in P_1$, we label predictions as "relevant" or "non-relevant" using the following procedure: Any prediction matching some term in the withheld half is labelled "relevant". The criterion for deciding whether two terms match is a cosine similarity of $\geq 85\%$ over GloVe word embeddings (introduced in Sect. 2). All other predictions are labelled "non-relevant". The conservative threshold of 85% ensures that only terms with the same lemma or with very high semantic similarity are matched. For each prediction, a set of features is calculated as detailed in Step 5 of our approach. It is paramount to note that Step 4, which is a prerequisite to Step 5, *exclusively* uses the content of the disclosed half without any knowledge of the withheld half. The above process produces labelled data for each $p \in P_1$. We aggregate all the labelled data into a single *training set*. This is possible because our features (listed in Table 1) are generic and normalized.

Equipped with a training set, we compare five widely used ML algorithms: Feed Forward Neural Network (NN), Decision Tree (DT), Logistic Regression (LR), Random Forest (RF) and Support Vector Machine (SVM). All algorithms are tuned with optimal hyperparameters that maximize classification accuracy over the training set. For tuning, we apply multisearch hyperparameter optimization using random search [25]. The basis for tuning and comparing algorithms is ten-fold cross validation. We experiment with under-sampling the "non-relevant" class with and without cost-sensitive learning (CSL); the motivation is reducing false negatives (i.e., relevant terms incorrectly classified as "non-relevant"). For CSL, we assign double the cost (penalty) to false negatives compared to false positives (i.e., noise). We further assess the importance of our features using information gain (IG) [14]. In our context, IG measures how efficient a given feature is in discriminating "non-relevant" from "relevant" predictions. A higher IG value implies a higher discriminative power.

EXPIII. This experiment answers RQ3 by applying our end-to-end approach to unseen requirements documents, i.e., P_2. To conduct EXPIII, we need a pre-trained classifier for Step 6 of our approach (Fig. 2). This classifier needs to be completely independent of P_2. We build this classifier using the training set derived from P_1, as discussed in EXPII. EXPIII follows the same strategy as in EXPI, which is to randomly withhold half of each document p (now in P_2 rather than in P_1) and attempting to predict the novel terms of the withheld half. In contrast to EXPI, in EXPIII, predictions made by BERT are post-processed by a filter aimed at reducing noise. We repeat EXPIII *five times* for each $p \in P_2$. This mitigates random variation resulting from the random selection of the disclosed and withheld halves, thus yielding more realistic ranges for performance. In EXPIII, we study three levels of filtering. Noting that there are 20 documents in P_2, the results reported for EXPIII use $20 * 5 * 3 = 300$ runs of our approach.

4.5 Metrics

We define separate metrics for measuring (1) the quality of term predictions and (2) the performance of filtering. The first set of metrics is used in RQ1 and RQ3 and the second set is used in RQ2 and RQ3. To define our metrics, we need to introduce some notation. Let Lem : bag → bag be a function that takes a bag of words and returns another bag of words by *lemmatizing* every element in the input bag. Let U : bag → set be a function that removes duplicates from a bag and returns a set. Let C denote the set of common words and stopwords as explained under Step 3 in Sect. 3. Given a document p treated as a bag of words, the terminological content of p's disclosed half, denoted h_1, is given by set $X = U(Lem(h_1))$. In a similar vein, the terminological content of p's withheld half, denoted h_2, is given by set $Y = U(Lem(h_2))$. What we would like to achieve through BERT is to predict as much of the *novel* terminology in the withheld half as possible. This novel terminology can be defined as set $N = (Y - X) - C$. Let bag V be the output of Step 3 (Fig. 2) when the approach is applied *exclusively* to the disclosed half of a given document (i.e., h_1). Note that V is already free of any terminology that appears in the disclosed half, as well as of all common words and stopwords.

Quality of Term Predictions. Let set D denote the (duplicate-free) lemmatized predictions that have the potential to hint at novel terminology in the withheld half of a given document. Formally, let $D = \mathsf{U}(\mathsf{Lem}(V))$. We define two metrics, *Accuracy* and *Coverage* to measure the quality of D. *Accuracy* is the ratio of terms in D matching some term in N, to the total number of terms in D. That is, $Accuracy = |\{t \in D \mid t \text{ matches some } t' \in N\}|/|D|$. A term t matches another term t' if the word embeddings have a cosine similarity of $\geq 85\%$ (already discussed under EXPII in Sect. 4.4). The second metric, *Coverage*, is defined as the ratio of terms in N matching some term in D, to the total number of terms in N. That is, $Coverage = |\{t \in N \mid t \text{ matches some } t' \in D\}|/|N|$. The intuition for Accuracy and Coverage is the same as that for the standard Precision and Recall metrics, respectively. Nevertheless, since our matching is inexact and based on a similarity threshold, it is possible for more than one term in D to match an individual term in N. Coverage, as we define it, excludes multiple matches, providing a measure of how much of the novel terminology in the withheld half is hinted at by BERT.

Quality of Filtering. As explained earlier, our filter is a binary classifier to distinguish relevance and non-relevance for the outputs from BERT. To measure filtering performance, we use the standard metrics of *Classification Accuracy*, *Precision* and *Recall*. True positive (TP), false positive (FP), true negative (TN) and false negative (FN) are defined as follows: A TP is a classification of "relevant" for a term that has a match in set N (defined earlier). A FP is a classification of "relevant" for a term that does *not* have a match in N. A TN is a classification of "non-relevant" for a term that does not have a match in N. A FN is a classification of "non-relevant" for a term that does have a match in N. *Classification Accuracy* is calculated as $(TP + TN)/(TP + TN + FP + FN)$. *Precision* is calculated as $TP/(TP + FP)$ and *Recall* as $TP/(TP + FN)$.

4.6   Results

RQ1. Figure 3 provides box plots for Accuracy and Coverage with the number of predictions by BERT ranging from 5 to 20 in increments of 5. Each box plot is based on 20 datapoints; each datapoint represents one document in P_1. We perform statistical significance tests on the obtained metrics using the nonparametric pairwise Wilcoxon's rank sum test [26] and Vargha-Delaney's effect size [27]. Table 3 shows the results of the statistical tests. Each column in the table compares Accuracy and Coverage across two levels of predictions per mask. For example, the *5 vs. 10* column compares the metrics for when BERT generates 5 predictions per mask versus when it generates 10.

For Accuracy, Fig. 3 shows a downward trend as the number of predictions per mask increases. Based on Table 3, the decline in Accuracy is statistically significant with each increase in the number of predictions, the exception being the increase from 15 to 20, where the decline is not statistically significant.

For Coverage, Fig. 3 shows an upward but saturating trend. Five predictions per mask is too few: all other levels are significantly better. Twenty is too many, notably because of the lack of a significant difference for Coverage in the *10 vs. 20* column of Table 3. The choice is thus between 10 and 15. We select 15 as this yields an average increase of 3.2% in Coverage compared to 10 predictions per mask. This increase is not statistically significant. Nevertheless, the price to pay is an average decrease of (14.12 − 11.97 =) 2.15% in Accuracy. Given the importance of Coverage, we deem 15 to be a better compromise than 10.

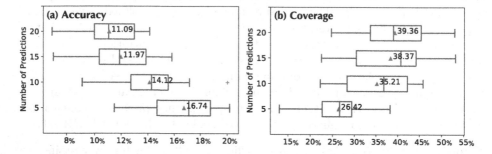

Fig. 3. (a) Accuracy and (b) Coverage for Different Numbers of Predictions per Mask. Each box plot represents 20 datapoints (one datapoint per $p \in P_1$) as computed by EXPI in Sect. 4.4.

Table 3. Statistical Significance Testing for the Results of Fig. 3.

	5 vs. 10		5 vs. 15		5 vs. 20		10 vs. 15		10 vs. 20		15 vs. 20	
	p-value	Â12	p-value	Â12	p-value	Â12	p-value	Â12	p-value	Â12	p-value	Â12
Accuracy	0.0051	0.245 (L)	5E-06	0.1075 (L)	2E-08	0.045 (L)	0.0143	0.725 (M)	0.0002	0.8275 (L)	0.242	0.61 (S)
Coverage	0.0015	0.795(L)	5E-05	0.855(L)	2E-05	0.87(L)	0.2184	0.385(S)	0.2084	0.3825(S)	0.883	0.485 (N)

Effect size: Large (L), Medium (M), Small (S), Negligible (N)

> The answer to **RQ1** is: When requirements omissions are simulated by with-holding, having BERT make 15 predictions per mask is the best trade-off for detecting missing terminology. BERT predicts terms that, on average, hint at ≈4 out of 10 omissions (Coverage ≈38%). On average, ≈1 in 8 predictions is relevant (Accuracy ≈12%).

RQ2. Table 4 shows the results for ML-algorithm selection using the full (P_1) training set (61,996 datapoints), the under-sampled training set (36,842 data-points), and the under-sampled training set alongside CSL. Classification Accuracy, Precision and Recall are calculated using ten-fold cross validation. In the table, we highlight the best result for each metric in bold. When one uses the full training set (*option 1*) or the under-sampled training set without CSL (*option 2*), Random Forest (RF) turns out to be the best alternative. When the under-sampled training set is combined with CSL (*option 3*), RF still has the best Accuracy and Precision. However, Support Vector Machine (SVM) presents a moderate advantage in terms of Recall. Since option 3 is meant at further improving the filter's Recall, we pick SVM as the best alternative for this particular option. Figure 4 lists the features of Table 1 in descending order of information gain (IG), averaged across options 1, 2 and 3. We observe that our corpus-based features (F11–F13) are among the most influential, thus justifying the use of a domain-specific corpus extractor in our approach.

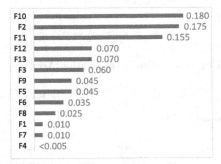

Compared to option 1, options 2 and 3 get progressively more lax by filtering *less*. We answer RQ3 using RF for options 1 and 2 and SVM for option 3. For better intuition, we

Fig. 4. Feature Importance (Avg).

refer to option 1 as *strict*, option 2 as *moderate* and option 3 as *lenient*.

> The answer to **RQ2** is: RF and SVM yield the most accurate filter for unuseful predictions. RF is a better alternative for more aggressive filtering, whereas SVM is a better alternative for more lax filtering (thus better preserving Recall).

RQ3. Without filters and over our test set (P_2 in Table 2), the 15 predictions per mask made by BERT have an average Accuracy of 12.11% and average Coverage of 40.04%. Box plots are provided in Fig. 5. We recall from Sect. 4.4 (EXPIII) that five different random shuffles are performed for each $p \in P_2$. The plots in Fig. 5 are based on $5 * 20 = 100$ runs.

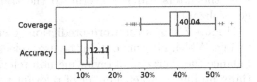

Fig. 5. Accuracy and Coverage over Test Set (P_2) without Filtering.

Table 4. ML Algorithm Selection (RQ2). All algorithms have tuned hyperparameters.

	Neural Network			Decision Tree			Logistic Regression			Random Forest			Support Vector Machine		
	A(%)	P(%)	R(%)	A(%)	P(%)	R(%)	A(%)	P(%)	R(%)	A(%)	P(%)	R(%)	A(%)	P(%)	R(%)
Full Training Set (P_1)*	81.1	69.5	65.1	82.3	74.6	61.5	81.3	73.5	58.1	**84.1**	**76.4**	**67.0**	81.4	74.1	57.4
Under-sampled †	78.9	77.6	81.2	78.9	78.3	79.9	79.0	**79.1**	78.7	**80.3**	78.8	**83.0**	79.2	77.9	81.7
Under-sampled + CSL ‡	77.6	72	90.4	77.3	71.7	90.3	77.1	71.7	91.7	**79.1**	**74.5**	88.6	76.8	70	**92.4**

A(%) = Classification Accuracy, P(%) = Precision of the "relevant" class, R(%) = Recall of the "relevant" class; all values are percentages.
*Strict filtering (option 1), †Moderate filtering (option 2), ‡Lenient filtering (option 3)

Figure 6 shows the performance of our three filters, namely strict, moderate, and lenient, over P_2. We observe that for all three filters, Precision levels over unseen data are lower than the cross-validation results in RQ2 (Table 4). This discrepancy was particularly expected for the moderate and lenient filters, noting that, for these two filters, Table 4 reports performance over an under-sampled

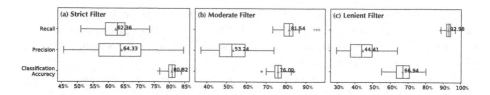

Fig. 6. Filtering Classification Accuracy, Precision and Recall over Test Set (P_2).

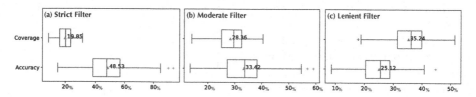

Fig. 7. Accuracy and Coverage over Test Set with (a) Strict, (b) Moderate, and (c) Lenient Filter.

dataset. As for Recall, the results of Fig. 6 indicate that the filters behave consistently with trends seen in cross-validation. This consistency provides evidence that the filters did not overfit to the training data and are thus sufficiently generalizable.

Figure 7 shows the word-prediction Accuracy and Coverage results *after* filtering. Which filtering option the user selects depends on how the user wishes to balance the overhead of reviewing non-relevant recommendations against potentially finding a larger number of relevant terms missing from requirements.

The lenient filter increases Accuracy by an average ≈13% while decreasing Coverage by ≈5%. The strict filter increases Accuracy by an average ≈36% while decreasing Coverage by ≈20%. All filters impact both Accuracy and Coverage in a statistically significantly manner with medium to large effect sizes.

> *The answer to* **RQ3** *is: Depending on how aggressively one chooses to filter noise from BERT's masked-word predictions, the average Accuracy of our approach ranges between ≈49% and ≈25%. With a strict filter, approximately one in two recommendations made by our approach is relevant, whereas with a lenient filter, approximately one in four is. With a lenient filter, the recommendations hint at ≈35% of the (simulated) missing terminology. With a strict filter, this number decreases to ≈20%.*

4.7 Limitations and Validity Considerations

Limitations. We did not have access to domain experts for identifying genuine cases of incompleteness. Our evaluation therefore simulates incompleteness by withholding content from existing requirements. Future user studies remain necessary for developing a qualitative interpretation of our results and drawing more definitive conclusions about the usefulness of our approach.

Validity Considerations. The validity considerations most pertinent to our evaluation are internal, construct and conclusion validity. With regard to *internal validity*, we note that the disclosed and withheld portions were chosen randomly. To mitigate random variation, we used a sizable dataset (40 documents) and further employed repeated experimentation (RQ3). With regard to *construct validity*, we note that our metrics for term predictions discard any terms already seen in the disclosed portion as well as any duplicates, common words and stopwords. This helps ensure that our metrics provide a legitimate assessment of the quality of the predictions made by BERT. That being said, our current quality assessment is not based on human judgement, and instead hinges on an automatically calculated similarity measure. A human-based validation of the predictions by BERT is therefore essential for better gauging the practical value of our approach. With regard to *conclusion validity*, we note that, we chose a 50-50 split between the disclosed and withheld portions. Assuming that the useful predictions by BERT are evenly distributed across the withheld portion, similar benefits should be seen with different split ratios, as long as the withheld portion is not excessively small. We anticipate that there will be a limit to how small the withheld portion can be, before it becomes too difficult to make relevant predictions. This limit determines the sensitivity of our approach to incompleteness. Further research is required to establish this limit.

5   Related Work

Ferrari et al. [5] propose an NLP-based approach for automatically extracting relevant terms and relations from descriptions such as client-meeting transcripts. Based on the extracted results, they make recommendations for improving the completeness of requirements. Dalpiaz et al. [6] develop a technique based on NLP and visualization to explore commonalities and differences between multiple viewpoints and thereby help stakeholders pinpoint occurrences of ambiguity and incompleteness. In the above works, the sources of knowledge used for completeness checking (descriptions or alternative viewpoints) are existing development artifacts. Our work uses a generative language model, BERT, for completeness checking. In contrast to the above works, our approach does not assume the existence of any user-provided artifacts against which to compare the completeness of requirements.

Arora et al. [4] use domain models for detecting incompleteness in requirements. The authors simulate requirements omissions and demonstrate that domain models can signal the presence of these omissions. Again, there is an assumption about the existence of an additional development artifact – in this case, a domain model. This limits the applicability of the approach to when a sufficiently detailed domain model exists.

Bhatia et al. [28] address incompleteness in privacy policies by representing data actions as semantic frames. They identify the expected semantic roles for a given frame, and consequently determine incompleteness by identifying missing role values. Cejas et al. [29] use NLP and ML for completeness checking of privacy policies. Their approach identifies instances of pre-defined concepts such as "controller" and "legal basis" in a given policy. It then verifies through rules whether all applicable concepts are covered. The above works deal with privacy policies only and have a predefined conceptual model for textual content. Our BERT-based approach is not restricted to a particular application domain and does not have a fixed conceptualization of the textual content under analysis. Instead, we utilize BERT's pre-training and attention mechanism to make contextualized recommendations for improving completeness.

Shen and Breaux [30] propose an NLP-based approach for extracting domain knowledge from user-authored scenarios and word embeddings. While this approach is not concerned with checking the completeness of requirements, it uses BERT's MLM for generating alternatives by masking words in requirements statements. Our approach uses BERT's MLM in a similar manner. In contrast to this earlier work, we take steps to address the challenge arising from such use of BERT over requirements, namely the large number of non-relevant alternatives (false positives) generated. We propose a ML-based filter that uses a combination of NLP and statistics extracted from a domain-specific corpus to reduce the incidence of false positives.

6   Conclusion

Our results indicate that masked-word predictions by BERT, when complemented with a mechanism to filter noise, have potential for detecting incompleteness in requirements. In future work, we plan to conduct user studies to more conclusively assess this potential. Other directions for future work include experimentation with BERT variants and improving accuracy via fine-tuning.

Acknowledgements. This work was funded by the Natural Sciences and Engineering Research Council of Canada (NSERC) under the Discovery and Discovery Accelerator programs. We are grateful to Shiva Nejati, Sallam Abualhaija and Jia Li for helpful discussions. We thank the anonymous reviewers of REFSQ 2023 for their constructive comments.

References

1. Ferrari, A., Spagnolo, G.O., Gnesi, S.: PURE: a dataset of public requirements documents. In: RE (2017)
2. Zowghi, D., Gervasi, V.: The three Cs of requirements: consistency, completeness, and correctness. In: REFSQ (2003)
3. Zowghi, D., Gervasi, V.: On the interplay between consistency, completeness, and correctness in requirements evolution. IST **45**(14), 993–1009 (2003)
4. Arora, C., Sabetzadeh, M., Briand, L.C.: An empirical study on the potential usefulness of domain models for completeness checking of requirements. Empir. Softw. Eng. **24**(4), 2509–2539 (2019). https://doi.org/10.1007/s10664-019-09693-x

5. Ferrari, A., dell'Orletta, F., Spagnolo, G.O., Gnesi, S.: Measuring and improving the completeness of natural language requirements. In: Salinesi, C., van de Weerd, I. (eds.) REFSQ 2014. LNCS, vol. 8396, pp. 23–38. Springer, Cham (2014). https://doi.org/10.1007/978-3-319-05843-6_3

6. Dalpiaz, F., van der Schalk, I., Lucassen, G.: Pinpointing ambiguity and incompleteness in requirements engineering via information visualization and NLP. In: Kamsties, E., Horkoff, J., Dalpiaz, F. (eds.) REFSQ 2018. LNCS, vol. 10753, pp. 119–135. Springer, Cham (2018). https://doi.org/10.1007/978-3-319-77243-1_8

7. Devlin, J., Chang, M.-W., Lee, K., Toutanova, K.: BERT: pre-training of deep bidirectional transformers for language understanding. In: NAACL-HLT (2019)

8. Luitel, D., Hassani, S., Sabetzadeh, M.: Replication package (2023). https://doi.org/10.6084/m9.figshare.22041341

9. Jurafsky, D., Martin, J.: Speech and Language Processing, 2nd edn. Prentice Hall, Upper Saddle River (2009)

10. Hey, T., Keim, J., Koziolek, A., Tichy, W.F.: NoRBERT: transfer learning for requirements classification. In: RE (2020)

11. Ezzini, S., Abualhaija, S., Arora, C., Sabetzadeh, M.: Automated handling of anaphoric ambiguity in requirements: a multi-solution study. In: ICSE (2022)

12. Mikolov, T., Yih, W., Zweig, G.: Linguistic regularities in continuous space word representations. In: NAACL-HLT (2013)

13. Pennington, J., Socher, R., Manning, C.: GloVe: global vectors for word representation. In: EMNLP (2014)

14. Witten, I.H., Frank, E., Hall, M.A.: Data Mining: Practical Machine Learning Tools and Techniques, 4th edn. Morgan Kaufmann, Boston (2017)

15. Berry, D.M., Cleland-Huang, J., Ferrari, A., Maalej, W., Mylopoulos, J., Zowghi, D.: Panel: context-dependent evaluation of tools for NL RE tasks: recall vs. precision, and beyond. In: RE (2017)

16. Ezzini, S., Abualhaija, S., Arora, C., Sabetzadeh, M., Briand, L.: Using domain-specific corpora for improved handling of ambiguity in requirements. In: ICSE (2021)

17. Cui, G., Lu, Q., Li, W., Chen, Y.R.: Corpus exploitation from Wikipedia for ontology construction. In: LREC (2008)

18. Ferrari, A., Donati, B., Gnesi, S.: Detecting domain-specific ambiguities: an NLP approach based on Wikipedia crawling and word embeddings. In: AIRE (2017)

19. Ezzini, S., Abualhaija, S., Sabetzadeh, M.: WikiDoMiner: wikipedia domain-specific miner. In: ESEC/FSE (2022)

20. Daniel, M., Berry, E.K., Krieger, M.: From contract drafting to software specification: linguistic sources of ambiguity, a handbook (2003)

21. Arora, C., Sabetzadeh, M., Briand, L., Zimmer, F.: Automated checking of conformance to requirements templates using natural language processing. IEEE TSE 41(10), 944–968 (2015)

22. Arora, C., Sabetzadeh, M., Briand, L., Zimmer, F.: Automated extraction and clustering of requirements glossary terms. IEEE TSE 43(10), 918–945 (2017)

23. Witten, I.H., Frank, E., Hall, M.A., Pal, C.J.: The WEKA Workbench: Online Appendix for "Data Mining: Practical Machine Learning Tools and Techniques", 4th edn. Morgan Kaufmann Publishers Inc., Boston (2016)

24. Manning, C., Raghavan, P., Schütze, H.: Introduction to Information Retrieval. In: Syngress (2008)

25. Bergstra, J., Bengio, Y.: Random search for hyper-parameter optimization. JMLR 13(2), 1–25 (2012)

26. Capon, J.A.: Elementary Statistics for the Social Sciences: Study Guide. In: Wadsworth (1991)
27. Vargha, A., Delaney, H.: A critique and improvement of the CL common language effect size statistics of McGraw and Wong. J. Educ. Behav. Stat. **25**(2), 101–132 (2000)
28. Bhatia, J., Breaux, T.: Semantic incompleteness in privacy policy goals. In: RE (2018)
29. Cejas, O.A., Abualhaija, S., Torre, D., Sabetzadeh, M., Briand, L.: AI-enabled automation for completeness checking of privacy policies. IEEE TSE **48**(11), 4647–4674 (2022)
30. Shen, Y., Breaux, T.: Domain model extraction from user-authored scenarios and word embeddings. In: AIRE (2022)

Requirement or Not, That is the Question: A Case from the Railway Industry

Sarmad Bashir[1,2], Muhammad Abbas[1,2(✉)], Mehrdad Saadatmand[1], Eduard Paul Enoiu[2], Markus Bohlin[2], and Pernilla Lindberg[3]

[1] RISE Research Institutes of Sweden, Västerås, Sweden
{sarmad.bashir,mehrdad.saadatmand}@ri.se
[2] Mälardalen University, Västerås, Sweden
muhammad.abbas@ri.se, {eduard.paul.enoiu,markus.bohlin}@mdu.se
[3] Alstom, Västerås, Sweden
pernilla.lindberg@alstomgroup.com

Abstract. [**Context and Motivation**] Requirements in tender documents are often mixed with other supporting information. Identifying requirements in large tender documents could aid the bidding process and help estimate the risk associated with the project. [**Question/problem**] Manual identification of requirements in large documents is a resource-intensive activity that is prone to human error and limits scalability. This study compares various state-of-the-art approaches for requirements identification in an industrial context. For generalizability, we also present an evaluation on a real-world public dataset. [**Principal ideas/results**] We formulate the requirement identification problem as a binary text classification problem. Various state-of-the-art classifiers based on traditional machine learning, deep learning, and few-shot learning are evaluated for requirements identification based on accuracy, precision, recall, and F1 score. Results from the evaluation show that the transformer-based BERT classifier performs the best, with an average F1 score of 0.82 and 0.87 on industrial and public datasets, respectively. Our results also confirm that few-shot classifiers can achieve comparable results with an average F1 score of 0.76 on significantly lower samples, i.e., only 20% of the data. [**Contribution**] There is little empirical evidence on the use of large language models and few-shots classifiers for requirements identification. This paper fills this gap by presenting an industrial empirical evaluation of the state-of-the-art approaches for requirements identification in large tender documents. We also provide a running tool and a replication package for further experimentation to support future research in this area.

Keywords: Requirements identification · Requirements classification · tender documents · NLP

1 Introduction

Like many other industries, the project acquisition in the railway industry also starts with a call for tender. A tender document is a formal request calling for competing offers from different potential suppliers or contractors. A tender document

© The Author(s), under exclusive license to Springer Nature Switzerland AG 2023
A. Ferrari and B. Penzenstadler (Eds.): REFSQ 2023, LNCS 13975, pp. 105–121, 2023.
https://doi.org/10.1007/978-3-031-29786-1_8

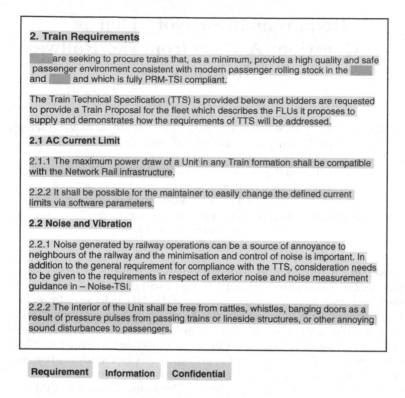

Fig. 1. Motivating example of requirements mixed with supporting text

typically consists of chunks of English text in the form of high-level technical specifications, supporting information, and contractual obligations. The very competitive market of railway vehicle manufacturing necessitates a quick response to such a call for tender. This requires estimating the risk associated with the call and the time required to deliver the end product. Extracting high-level technical specifications (requirements) from the tender document becomes crucial to estimate the risk and time required for a call. The high-level requirements are used to derive low-level requirements to be agreed upon. Furthermore, the requirements in the tender document are compared with already delivered products to estimate risk based on the novelty of the requirements. Therefore, identifying technical specifications from tender documents becomes a pre-requisite to enable project acquisition and later Requirement Engineering (RE) tasks. Moreover, errors and inaccuracies in this phase can have cascading effects on the rest of the development process.

Figure 1 shows a motivating example from a real tender document showing requirements mixed with technical specifications. Manually identifying the requirements in large tender documents could be time-consuming and prone to human error. Requirement identification in a large document can be automated by formulating it as a binary text classification problem. According to Berry [7], the automated solutions for RE tasks should ideally have a 100% recall rate;

however, this is not often achieved in Machine learning (ML)-based solutions. Despite this, utilizing automated solutions could still accelerate the process since human input would only be required to sanitize the final output. The ultimate goal must be to optimize the performance of these automated classification solutions to alleviate the workload in practical settings. In RE, requirement classification is one of the most prominent activities, as reported in literature [39]. Related work on distinguishing requirements from other information often experimented with traditional ML-based approaches for classification [4,13,35]. Furthermore, the work of Abualhaija *et al.* [3,4] considers single sentences as a unit of classification. However, requirements and information could range over multiple sentences in our case. Nevertheless, the same approaches could be modified to take multisentence input. However, the performance of large transformer-based language models and few-shot classifiers in the task is still unclear. On the other hand, work on distinguishing functional and non-functional requirements [5,15,20,29] is a different use case, and studies in the domain often use public datasets with some exceptions.

This study is conducted in close collaboration with Alstom, Sweden (Alstom), a world-leading railway vehicle manufacturing company. The main objective of this study is to find a practical solution to the requirements identification problem at Alstom. Therefore, this study reports an empirical evaluation of 20+ different classification pipelines for distinguishing requirements from supporting text in large documents. The selected seminal pipelines include approaches from traditional ML, deep learning, and transformer-based classifiers. In addition, we leverage new approaches based on few-shot learning to address the common challenge of data scarcity in the RE domain. Furthermore, to support further research on the topic, we evaluated the same pipelines on a public dataset and provided a replication package with a running tool[1]. This paper may also refer to requirements identification as distinguishing requirements or classification.

The rest of the paper is structured as follows. Section 2 provides a brief overview of the related work and background. Section 3 presents the study design and the selected classification pipelines for requirements identification. Section 4 presents and discusses the results. Section 5 presents potential validity threats and limitations. Finally, Sect. 6 concludes the paper with future directions.

2   Related Work and Background

Related Work. This paper focuses on identifying requirements through automated classification. Binkhonain and Zhao [8] performed a systematic literature review on one aspect of the RE process, i.e., automated requirements classification, specifically providing solutions to distinguish between functional (FRs) and non-functional requirements (NFRs). Often FRs are related to the core functionality, and NFRs describe the properties and constraints of the system. The distinction between FRs and NFRs impacts the handling of requirements

---

[1] Replication package and Tool: https://github.com/a66as/REFSQ2023-ReqORNot.

elicitation, documentation, and validation process [12]. Therefore, the task of automatic extraction and classification of requirements has been the focus of RE researchers. Within this group of studies, Jindal *et al.* [20] employs an automated approach to extract and classify security requirements. They use term-frequency inverse document frequency (tfidf) weight vectors to analyze the security requirements with the goal of further classifying into sub-categories of security based on the Decision Tree (DT) algorithm. Moreover, Varenov *et al.* [32] proposes a sentence-level classifier based on fine-tuned DistilBERT [29] to allocate security requirements into predefined groups. Recently, Alhoshan *et al.* [5] leverages a Zero-Shot Learning (ZSL) technique on a subset of the PROMISE dataset to classify NFRs into two categories, i.e., Usability and Security. Furthermore, Herwanto *et al.* [15] propose an automated approach to identify privacy requirements in user stories based on the Named Entity Recognition (NER) model, trained on Bi-directional Long Short Term Memory Networks (BI-LSTM) with conditional random field [18].

The other more related thread of work is distinguishing requirements from other information. Similar to our use case, Winkler *et al.* [35] propose a deep learning (DL) classifier based on Convolution Neural Networks (CNNs) to identify requirements from additional material stored in IBM DOORS. Falkner *et al.* [13] propose a Naive Bayes (NB) classifier—trained on unique words—to identify requirements from Request of Proposal (RFP) documents within the railway safety domain. Furthermore, Abualhaija *et al.* [4] proposes an automated ML-based approach to demarcate requirements in textual specifications by considering one sentence as a unit of classification. They empirically evaluate ML classifiers on the industrial dataset consisting of 12 documents. In addition, Sainani *et al.* [28] defines a two-step methodology to first extract requirements from 20 Software Engineering (SE) contracts and then allocate them to their specific types. For identification and extraction of requirements, Bi-LSTM yields the best results compared to ML algorithms. To allocate identified requirements in sub-classes, BERT (Bi-directional Encoder Representations from Transformers) performed better in terms of F-1 score.

While our work shares the same general objective as the above-mentioned approaches, we address the need for extensive empirical evaluation in automated requirements identification and classification on industrial and public datasets. Furthermore, we evaluated a new approach, namely a few-shot classifier, to identify requirements based on a limited dataset, a well-known problem in the RE domain.

Background. Most of the ML or DL algorithms for classification do not work with raw text but instead require transformed data as feature vectors. The feature vectors can be generated with information retrieval (IR)-based lexical approaches or with semantic approaches. In our work, we utilize tfidf vectors—a lexical approach—to represent and train data on classical supervised ML algorithms. We further apply the dimensionality reduction technique, i.e., principal component analysis (PCA), on tfidf vectors to increase interpretability through the creation of newly uncorrelated features with maximum variance.

Traditionally, Language Models (LMs) capture regularities, morphological and distributional properties of a language. For DL algorithms, we consider state-of-the-art semantic strategies based on LMs and neural networks. The semantic-based LMs are coupled with a statistical classifier to perform classification. We use FastText (FT) [9] and GloVe (GLV) [24] semantic representations to train LSTM neural network for the identification of requirements in large documents. Furthermore, we fine-tuned multiple token-based BERT LM variations based on transformer architecture [33]. Originally, token-based BERT LM comes in two variants for language representation, BERT base and large, pre-trained on 16 GB data from Toronto BookCorpus and English Wikipedia dataset. With the advent of transfer learning, token-based BERT LMs have been widely used for different downstream tasks—in our case, classification to distinguish requirements.

Additionally, we perform few-shot fine-tuning on different variations of Sentence Transformers [25] (ST)—a modified version of the pre-trained BERT LM based on the siamese network. Specifically, we fine-tuned Sentence-BERT [25] (S-BERT) and MiniLM-L12-v2 [34] (Mini-LM) on our datasets. Originally, ST LMs are pre-trained for tasks like clustering and semantic search. However, we can fine-tune ST LMs through Sentence Transformer Fine-Tuning [31] (SETFIT) framework with a small number of examples for our requirements classification task. Few-shot methods are an attractive solution and can address the long-standing problem of data shortage in the RE domain.

3   Study Design

This work can be regarded as an *exploratory* case study oriented towards improving the project acquisition process at Alstom. Following the guidelines of Runeson and Höst [26], this section outlines the context, objectives, data collection, and analysis procedure.

3.1   Case Context

Rail vehicle manufacturing is a globally competitive market. Like many other industries, customers in the railway industry also publish a call for tender to which companies respond. The tender document often contains contractual obligations, supporting information, and technical specifications of the required product. In response to the call for tender, in addition to understanding the contractual obligations, companies must also identify potential requirements from the documents to achieve the following objectives.

a) The extracted technical specifications must be reflected in deriving the customer requirements to be agreed upon. This can aid the project acquisition process.
b) The risk associated with the new project must be estimated to enable the project resource and time management. This is done by comparing the extracted technical specification to the already delivered projects, currently based on experience.

Alstom is continuously looking for ways to improve the process of project acquisition with tool support. As a first step, automated approaches for distinguishing
requirements from other supporting information are investigated in this study.
In this regard, for this paper, the *case* under study is the performance of various
classification pipelines in requirements identification. The *units* under analysis
are five tender documents from Alstom and the public dronology dataset [10].

3.2   Objective and Research Questions

Our main goal is to improve the project acquisition phase in the studied context.
As an initial step (this study), we first need to identify the requirements within
the tender documents. Requirements identification problem can be formulated
as a binary text classification problem. There have been a number of approaches
proposed for binary classification over the years. Therefore, this work is not
"reinventing the wheel" but instead aims to find an already existing practical—
in terms of execution time—solution for the problem in the studied context. As
discussed in the following sections, we consider seminal state-of-the-art classifiers for this study. In addition, since the considered approaches for classification might react differently to text pre-processing, we also study the impact of
pre-processing on classification performance. To this end, we pose the following
research questions (RQs):

- *RQ1: What is the performance of different classification pipelines in require-
  ments identification?*
- *RQ2: What is the impact of pre-processing on classification performance?*
- *RQ3: What is the execution time of each classification pipeline?*

Table 1. Datasets

Dataset	Reqs.	Info.	Sent.	AW	pAW	TRD	TSD
Industrial	1680	1293	8332	39	20	2378	595
Public	99	280	533	25	13	303	76

* AW = Avg. words, pAW = Avg. words when
pre-processed, TRD = Avg. training dataset rows,
TSD = Avg. test dataset rows

3.3   Data Collection

Industrial Case. We had access to five already annotated tender documents from
our industrial partner. The tender documents contain multi-sentence chunks of
text explaining the technical specifications, contractual obligations, and supporting information. The requirements among the documents were already tagged,
and the projects were already delivered to customers. Therefore, the ground truth
on whether a chunk of text is a requirement or not is already available in the
dataset. Note that the selected pipelines (see coming sections) for distinguishing requirements require annotated input for training only. We selected all the

requirements and non-requirements among these five documents using the following steps. First, we removed all the duplicates across the five files and considered unique chunks of text. To avoid selecting potential non-requirements as requirements, we selected only the requirements that were also allocated to a team for development. A total set of 1680 requirements and 1293 non-requirements from the industrial documents was reached, as shown in the first row of Table 1.

Public Dronology Dataset. The dronology public dataset consists of 398 entries of various types such as "components", "requirements", "design definitions", and "sub-task". Among these entries, 99 entries are tagged as requirements. We prepared the dronology dataset for this study as follows. First, we considered all the requirements as requirements and components, design definitions, and sub-tasks as non-requirements. Then, we dropped (19 entries) entries with no text. A total set of 99 requirements and 280 non-requirements was reached, as shown in the last row of Table 1. Note that this dataset does not directly represent the studied context; however, we argue that evaluating the pipelines on this similar dataset would support replication and reproducibility of our results.

All the considered classification pipelines (see the coming section) are fed the data with and without pre-processing. As shown in the Sent. column, the total number of sentences in the considered datasets are 8,332 and 533, respectively. On average, each of the entries consists of 39 and 25 words. After pre-processing and stop word removal, the average words across all entries drops to 20 and 13 for the industrial and public datasets, respectively. Due to the uneven distribution of data over the labels, the *p-fold* cross-validation method is not employed for evaluation [11]. As typical, we used stratified five-fold cross-validation to evaluate the selected pipelines. The average number of entries per fold in the training set and the test are 2378 and 595, respectively.

3.4   Pipelines for Distinguishing Requirements

For this study, we considered the most seminal text classification approaches for evaluation in distinguishing requirements from ordinary text. As typical in the NLP domain, pre-processing of the input text might impact classification performance. Therefore, we also consider the datasets both with and without pre-processing. In addition, we also consider a baseline random pipeline (W. Rand.) that classifies input as a requirement or not based on their frequency distribution in the dataset.

Pre-processing: Our pre-processing pipeline consists of tokenization, stop-words removal, part-of-speech (POS) tagging, and lemmatization of the input text using spaCy [17]. An output of the pre-processing pipeline for the requirement '6.3.1' from Fig. 1 is presented as follows. "maximum power draw unit train formation compatible network rail infrastructure".

Traditional ML-Based classifiers: For lexical classifiers, we considered widely used and recommended ML algorithms, e.g., Support Vector Machines (SVM),

Logistic Regression (LR), DT, Random Forest (RF), and NB. For a fair comparison and tuning, we applied random multi-search optimization [6] to select the optimal hyperparameters. SVM and LR achieved better results on evaluation metrics when trained with normalized and reduced tfidf vectors using PCA. However, the rest of the ML pipelines—RF, DT, and NB—performed better with normalized TF-IDF vectors without PCA-based dimensionality reduction.

Deep Semantic Representation Based Classifiers: For the training and evaluation of DL-based LSTM networks, we use FT and GLV LMs—pre-trained and custom (self-trained)—embeddings for semantic representation. To generate the custom embeddings, we train the FT LM on 20 epochs, with word embeddings (WE) dimension size set to 100 and window size set to three. For custom GLV embeddings, we get the best results when the window size is set to 10, the learning rate is set to 0.05, the WE dimension size is 100, and the epochs are set to 30. We defined a two-layer LSTM network to train on custom and pre-trained WEs. To minimize the training loss function, we used Adam [21] optimizer with a learning rate of 0.001. Furthermore, we prevent the over-fitting of the network by appending a dropout layer—randomly dropping units with their connections—with a rate of 0.1 after every LSTM layer. The batch, epochs, and maximum sequence sizes are set as 32, 10, and 128, respectively.

We selected widely used BERT variants, i.e., SciBERT, RoBERTa, BERT base, XLMRoBERTa (XRBERT), DistilBERT (DisBERT), and XLNet. To fine-tune different variations of the token-based BERT family, we employ a BERT WordPiece [37] tokenizer to prepare the datasets. The WordPiece tokenizer splits the words of a text into one word per token or into word pieces—where one word is tokenized into multiple words. We use the AdamW-optimizer, an adoption of Adam with a weight decay of 0.01, to optimize the weights while fine-tuning the token-based BERT network [23]. Furthermore, we select a maximal learning rate of 2e-05 instead of aggressive learning rates with the purpose of avoiding catastrophic forgetting of BERT pre-trained knowledge [30]. We set a practical batch and maximum sequence size as 16 and 128 across all the token-based BERT pipelines. We set the epoch size as 10 to iterate the datasets over the BERT's network. The reason behind selecting a higher number of epochs on relatively smaller datasets is that BERT's common one-size-fits-all is sub-optimal and needs more training time to stabilize the network [38]. However, some studies in the literature have set an even higher number of epochs (e.g. [16]), but we argue it may lead to over-fitting.

Few-Shot Learning Based on Sentence Transformers: To fine-tune different variations of pre-trained ST, we utilize the SETFIT framework for our downstream requirements classification task. SETFIT consists of a two-step training approach. In the first step, we fine-tuned ST on a limited dataset—few shots—with a contrastive training approach—frequently used for image similarity [22]. In a few-shot scenario, contrastive fine-tuning enlarges the training dataset by creating positive and negative pairs through in-class and out-class selection. In the second step, we train an LR (Logistic Regression) model as a classification head on the embeddings—encoded through fine-tuned ST—with original labeled training data. For evaluation, fine-tuned ST generates the sentence embeddings of unseen examples, and then the LR model predicts the class label of the input sentence embeddings. To fine-tune the ST model, we use the cosine-similarity loss function with a learning rate of 2e-5 and a batch size of 16. We set the number of iterations for the generation of text pairs for contrastive learning to 20 with one epoch.

3.5   Metrics for Evaluation

We use the standard evaluation metrics for text classification, as follows. *Accuracy (A)* is the ratio of the number of correct predictions and the total predictions. *Precision (Prec.)* is the ratio of correct positive predictions and the total number of positive predictions. *Recall (Rec.)* quantifies the number of correct positive predictions from all possible positive predictions. *F1 score (F1)* is the harmonic mean of precision and recall. We report the macro and weighted average across the fold for all our evaluation metrics. However, to answer our research questions, we use weighted averages of the metrics for simplicity.

3.6   Execution Procedure

Both datasets' tagged requirements and information are moved to two separate files. Using random stratified five-fold sampling [19], we created five folds from each dataset for cross-validation. As mentioned, each fold consists of 80% of the randomly sampled data in the training set and 20% of the data in the holdout set. All the selected pipelines were fed with the five folds for training the models, and the holdout sets were used to compute the evaluation metrics. In the case of the few-shot classification pipelines, we only selected 10% and 20% of the training set as folds to train the model and evaluated it using the entire holdout set. We executed all the experiments on a local server using parallel computing. The server is configured with four Nvidia Tesla M10 graphics processing units, an Intel Xeon Gold 5122 processor @ 3.60GHz, and primary memory of 256 GB.

4   Results and Discussion

Table 2 and Table 3 show the experiment's results. The names of Pipelines starting with a 'p' indicate that our pre-processing pipeline was coupled with the classification pipeline. The `Weighted Average` and `Macro Average` columns show the weighted and macro averages of our evaluation metrics across the five folds. `Avg. A.` shows the average accuracy of the pipelines.

RQ1: Performance. As shown in Table 2, among the traditional machine learning-based approaches, SVM slightly outperformed the others in terms of F1 score. Regarding accuracy, RF and LR slightly outperformed all other requirement identification pipelines based on ML. A similar trend can also be observed in the public dataset. This is in line with the results in the literature. Interestingly, the deep-learning-based LSTM model combined with the word embeddings model does not exhibit a significant improvement compared to traditional

Table 2. Performance and execution time of the pipelines on industrial case

Pipeline	Setup	Weighted Average			Macro Average			Avg. A.	Time (mins)	
		Prec.	Rec.	F1	Prec.	Rec.	F1	A	Tr	Ts
W. Rand.	Freq. based	.49	.49	.49	.49	.49	.48	.49	–	–
SVM	Norm., PCA	.79	.79	.79	.80	.78	.78	.78	.70	.02
pSVM	Norm., PCA	.78	.78	.78	.79	.77	.77	.78	.74	.09
NB	Norm	.74	.69	.69	.73	.71	.69	.69	<.01	<.01
pNB	Norm	.74	.68	.67	.72	.70	.68	.68	.29	.07
DT	Norm	.72	.72	.72	.71	.71	.71	.71	<.01	<.01
pDT	Norm	.71	.71	.71	.71	.71	.71	.71	.29	.07
LR	Norm., PCA	.79	.79	.78	.79	.77	.78	.79	.30	<.01
pLR	Norm., PCA	.78	.78	.78	.79	.76	.77	.78	.41	.07
RF	Norm	.79	.79	.79	.79	.78	.78	.79	<.01	<.01
pRF	Norm.	.79	.78	.78	.79	.77	.77	.78	.38	.07
LSTM	FT custom	.77	.77	.77	.76	.76	.76	.77	1	.02
pLSTM	FT custom	.75	.75	.75	.75	.75	.74	.75	1	.08
LSTM	FT pre-train	.75	.75	.75	.74	.74	.74	.75	2	.02
pLSTM	FT pre-train	.72	.72	.72	.72	.72	.72	.72	1.2	.08
LSTM	GLV custom	.77	.77	.77	.77	.76	.76	.77	2	.02
pLSTM	GLV custom	.76	.76	.76	.76	.75	.76	.76	1.2	.09
LSTM	GLV pre-train	.78	.78	.78	.78	.77	.78	.78	2	.02
pLSTM	GLV pre-train	.78	.78	.78	.78	.77	.78	.78	1.3	.08
SciBERT	uncased	**.82**	.81	.81	**.82**	.80	.80	.81	34	.25
pSciBERT	uncased	.80	.78	.76	.81	.75	.75	.78	32	.30
RoBERTa	base	.81	.81	.81	**.82**	.80	.80	.81	39	.27
pRoBERTa	base	.80	.79	.79	.81	.78	.78	.79	37	.32
BERT	base, cased	**.82**	**.82**	.81	**.82**	**.81**	**.81**	**.82**	35	.29
pBERT	base, cased	.79	.79	.79	.79	.79	.79	.80	32	.32
BERT	base, uncased	**.82**	**.82**	**.82**	**.82**	**.81**	**.81**	**.82**	34	.29
pBERT	base, uncased	.80	.80	.80	.80	.79	.79	.80	32	.33
XRBERT	base	**.82**	.81	.81	**.82**	.80	**.81**	.81	57	.29
pXRBERT	base	.78	.77	.77	.78	.76	.76	.77	41	.25
DisBERT	base, cased	.81	.81	.81	.81	.80	.80	.81	31	.13
pDisBERT	base, cased	.80	.80	.80	.80	.79	.79	.80	25	.18
DisBERT	base, uncased	.81	.81	.81	.81	**.81**	.80	.81	31	.15
pDisBERT	base, uncased	.80	.80	.70	.81	.78	.79	.80	29	.21
XLNet	base	.81	.81	.80	.81	.80	.80	.81	47	.36
pXLNet	base	.81	.80	.80	.81	.79	.79	.80	47	.42
S-BERT	10% train	.75	.75	.75	.75	.74	.75	.75	24	.14
pS-BERT	10% train	.73	.73	.73	.72	.72	.72	.73	18	.20
Mini-LM	10% train	.74	.74	.74	.74	.74	.74	.74	7	.04
pMini-LM	10% train	.72	.72	.72	.72	.72	.71	.72	6	.10
S-BERT	20% train	.77	.77	.76	.76	.76	.76	.77	45	.17
pS-BERT	20% train	.74	.74	.74	.74	.74	.74	.74	37	.20
Mini-LM	20% train	.75	.75	.75	.75	.74	.74	.75	14	.03
pMini-LM	20% train	.72	.72	.72	.72	.72	.72	.72	11	.10

Table 3. Performance and execution time of the pipelines on the Dronology public dataset

Pipeline	Setup	Weighted Average			Macro Average			Avg. A.	Time (mins)	
		Prec.	Rec.	F1	Prec.	Rec.	F1	A	Tr	Ts
W. Rand.	Freq. based	.60	.58	.59	.48	.48	.48	.58	–	–
SVM	Norm., PCA	.78	.79	.75	.76	.63	.64	.78	.18	< .01
pSVM	Norm., PCA	.78	.77	.70	.80	.57	.55	.77	.03	<.01
NB	Norm	.70	.55	.58	.58	.61	.54	.55	<.01	<.01
pNB	Norm	.71	.56	.58	.60	.62	.55	.56	.03	<.01
DT	Norm	.74	.74	.74	.67	.66	.66	.74	<.01	<.01
pDT	Norm	.72	.72	.72	.64	.63	.63	.72	.03	<.01
LR	Norm., PCA	.74	.75	.67	.73	.54	.51	.75	.14	.02
pLR	Norm., PCA	.70	.75	.64	.67	.52	.46	.74	.03	< .01
RF	Norm	.76	.78	.75	.72	.64	.65	.78	.01	<.01
pRF	Norm.	.77	.78	.75	.74	.63	.65	.78	.04	<.01
LSTM	FT custom	.75	.78	.73	.72	.61	.61	.78	.24	.01
pLSTM	FT custom	.76	.77	.75	.71	.66	.67	.77	.18	.02
LSTM	FT pre-train	.74	.76	.75	.67	.66	.66	.76	.23	.01
pLSTM	FT pre-train	.67	.68	.68	.58	.57	.58	.68	.17	.02
LSTM	GLV custom	.73	.75	.73	.65	.66	.63	.75	.23	.01
pLSTM	GLV custom	.77	.78	.77	.72	.69	.69	.78	.17	.02
LSTM	GLV pre-train	.80	.80	.79	.74	.73	.73	.80	.26	.01
pLSTM	GLV pre-train	.74	.75	.74	.68	.65	.66	.75	.18	.02
SciBERT	uncased	.84	.84	.83	.79	.78	.78	.83	5	.03
pSciBERT	uncased	.87	.87	.86	.84	.80	.82	.87	5	.03
RoBERTa	base	.82	.86	.84	.76	.78	.77	.86	5	.03
pRoBERTa	base	.80	.81	.79	.77	.69	.70	.81	5	.04
BERT	base, cased	**.88**	**.88**	**.87**	**.85**	.83	**.83**	**.88**	3	.01
pBERT	base, cased	.83	.84	.82	.81	.74	.76	.84	3	.03
BERT	base, uncased	**.88**	**.88**	**.87**	.84	**.84**	**.83**	**.88**	3.5	.02
pBERT	base, uncased	.83	.84	.83	.80	.75	.77	.84	3	.03
XRBERT	base	.86	.86	.86	.82	.83	.82	.86	7	.03
pXRBERT	base	.86	.86	.86	.82	.83	.82	.81	7	.04
DisBERT	base, cased	.85	.85	.85	.80	.81	.80	.85	3	.01
pDisBERT	base, cased	.83	.83	.82	.79	.75	.76	.83	3	.02
DisBERT	base, uncased	.85	.86	.85	.81	.81	.80	.86	3	.01
pDisBERT	base, uncased	.82	.83	.82	.79	.74	.75	.83	3	.02
XLNet	base	**.88**	.87	**.87**	**.85**	.83	**.83**	.87	6	.04
pXLNet	base	.82	.83	.83	.78	.76	.77	.83	6.5	.05
S-BERT	10% train	.75	.65	.66	.64	.67	.62	.65	3	.04
pS-BERT	10% train	.68	.59	.61	.57	.58	.55	.59	2	.04
Mini-LM	10% train	.76	.67	.69	.66	.70	.64	.67	1	<.01
pMini-LM	10% train	.70	.56	.58	.58	.60	.54	.56	.40	.01
S-BERT	20% train	.75	.65	.66	.64	.67	.62	.67	4	.03
pS-BERT	20% train	.74	.65	.67	.63	.66	.62	.65	4	.04
Mini-LM	20% train	.79	.68	.70	.68	.73	.66	.68	2	<.01
pMini-LM	20% train	.77	.71	.72	.67	.72	.67	.71	1	.01

ML algorithms for classification in both public and industrial cases. Our results indicated that SVM is closely followed by LSTM coupled with the pre-trained GLV WE model. We argue that this could be because of the impact of feature engineering in SVM. Generally, SVM's performance is similar or better compared to artificial neural networks (ANN) when there is less training dataset. On the other hand, on average, the BERT family slightly outperformed all other pipelines with traditional fine-tuning. The BERT base uncased-based pipeline for requirements identification slightly outperformed all other pipelines across all evaluation metrics, with an average F1 score of 0.82 for the industrial dataset and 0.87 on the public dataset. This could be explained by BERTs' ability to capture the long-range dependencies in sequential data through its so-called self-attention mechanism. Additionally, the capability of fine-tuning BERT pre-trained parameters on task-specific datasets allows the model to better incorporate domain-specific knowledge than other traditional approaches. All other sub-families of BERT—SciBERT, RoBERTA, XRBERT, and DisBERT—closely followed the BERT base uncased-based pipeline. In addition, the performance of XLNet is also close to the performance of the BERT family. The architecture of XLNet and BERT family models are different, but they share a similar pre-training objective to capture the contextual relationships in natural language data. Therefore, when fine-tuned on a similar dataset for a classification task, there is not a significant difference in terms of evaluation metrics.

The lack of larger datasets in the RE domain is a commonly highlighted problem in the literature [5,14,39]. Therefore, evaluating few-shot learning approaches for requirements identification is equally important. Our results show that our selected sentence transformer-based few-shot classification pipelines for requirements identification achieved comparable results with as little data as 20% of the training set used for training the models. The few-shots classification pipelines also performed very well when only 10% of training data was used for training. In our industrial case, the best-performing few-shot classification pipeline is the pipeline based on the S-BERT model. For comparison, fine-tuned S-BERT achieves an F1 performance score of 0.76, which is only 0.06 less than the best-performing fined-tuned BERT uncased pipeline on a complete dataset. This is a significant step in the RE domain because S-BERT only requires a few samples to fine-tune the model, and it can address the standing challenge of insufficient annotated RE datasets. Furthermore, this can help RE researchers to completely exploit DL classifiers' usage in various phases of the RE.

Based on the presented results, we summarize an answer to RQ1.

> **Answer to RQ1.** The BERT base uncased-based pipeline for distinguishing requirements from general text slightly outperformed all other pipelines across the two datasets with an average F1 score of 0.85. However, no significant difference in the performance of the BERT family is observed. Results further indicate that few-shot classification pipelines for distinguishing requirements perform well (with an F1 score of 0.76) on significantly fewer samples.

Note that the current performance evaluation of the pipelines is based solely on the results obtained from the already annotated datasets. However, to further assess and validate the effectiveness of the pipelines in practice, it is necessary to conduct a controlled experiment in an industrial setting. This is because the real-world efficiency of the automated pipelines for such tasks is also dependent on the environment in which it will be used [7]. Additionally, the evaluation of automated pipelines must consider the context for its performance evaluation relative to the task manually performed by humans. Therefore, following the study of Winkler *et al.* [36], in the future, we plan to perform an empirical study to further validate our solution in practice. The findings of such an experiment would lead to further improvements in the pipelines and would highlight the avenues for future research in the studied context.

RQ2: Impact of pre-processing. Table 2 and 3 also contains the evaluation results of the pipelines with pre-processing. Based on our experience and literature in NLP, traditional ML-based approaches typically improve performance when pre-processing is applied [1,2]. However, we found a general trend in the task of distinguishing requirements; on average, pre-processing has a negative impact on classification performance. Particularly among the traditional approaches, all pipelines (except LR and NB) show a decrement of up to .02 in the F1 score when pre-processing is applied. The LSTM family also shows a negative relationship between pre-processing and F1 score. However, the LSTM pipeline based on the GLV word embedding model for distinguishing requirements shows no impact on performance when pre-processing is applied. Finally, as expected for all the transformer-based models, pre-processing has a negative impact on the model performance. Based on these results, we summarized the answer to RQ2 as follows.

> *Answer to RQ2.* Generally, we observed a negative impact of pre-processing (with stop words removal and lemmatization) on model performance in the task of distinguishing requirements from general text.

RQ3: Execution Time. The `Time (mins)` column in Table 2 and Table 3 also shows the average execution time of the pipelines per fold both in training (`Tr`) and in inference mode (`Ts`). Pipelines with pre-processing—both in training and inference mode—also report an average pre-processing time added to the overall time. In other words, the `Time (mins)` column shows the pipeline's average end-to-end execution time per fold.

As expected, the traditional ML-based and LSTM-based approaches converge faster, with an average end-to-end execution time of under two minutes in training. Likewise, the average end-to-end execution time in inference mode across the folds—with 595 and 76 entries per fold in the industrial and public dataset—is also under a minute. In addition, for fine-tuning the BERT family, the end-to-end execution time is under an hour in the worst case. Note that the pre-processing for LSTM and BERT family does add an overhead. However, in some cases, the same pipeline with pre-processing takes even less time than the one without pre-processing. This could be explained by the fact that the same model has to train on less vocabulary than when the data was not pre-processed. Nevertheless, the reported execution time in training still shows a trend of increase as the size of the model increases. BERT family averaged an inference time of under a minute per fold. As fine-tuning is done only once per task and can be done at night, the engineers do not have to wait for more than a minute in inference mode—which is how end-users use these models. Based on the results, we summarised an answer to RQ3 as follows.

> **Answer to RQ3.** Pipeline for distinguishing requirements produces results for input (500+ entries as input) in under two minutes in the worst case. Fine-tuning large language models for classification tasks could take hours on high-compute units when training on a dataset with 2300 entries. However, fine-tuning is often done once per task. Therefore, the approaches could still be practical in a real-world context and can aid the project acquisition process.

5   Threats to Validity

This section presents validity threats according to Runeson et al. [26].

Construct Validity. As typical, we cast the requirement identification as a binary text classification problem. Our unit of classification ranges over multiple sentences. However, in some cases, the input might contain some sentences that are requirements and others that are not. We do not tackle such cases. We argue that considering already delivered projects' tender documents where experts tagged requirements and allocated them to teams for implementation resolves such issues.

Internal Validity. Internal validity threats affect the validity and credibility of our results. We based our implementation on open-source libraries and publicly available language models to address potential internal validity threats. Furthermore, we shared the replication package and a running tool to support future research.

External Validity. Our results are obtained from five representative documents from one company that might not represent the whole railway domain. Therefore, for the generalizability of our results, we also include a public dataset for evaluation. However, as typical for case studies, we do not claim the generalizability of our results beyond the studied context.

6   Conclusion and Future Work

Requirements identification in larger documents enables a quick response to the call for tenders and could help later RE tasks such as retrieval for reuse and deriving low-level requirements. This study is oriented toward finding a practical solution to the requirements identification problem in a large railway company using classification. Therefore, the study evaluates a variety of classification approaches in the requirements identification contexts. Our results show that the transformer-based approaches slightly outperform all other approaches in the requirements identification task. Particularly, the BERT base uncased-based pipeline performs the best in terms of F1 score and produces results in practical time. Finally, results also indicate that few-shot classifiers can achieve comparable performance with as little as 20% of the training data. We argue that the use of few-shot learning in RE tasks should be investigated further.

In the future, we plan to conduct a controlled experiment in the studied settings to evaluate the effectiveness of the developed solution for comparison to manual requirements identification. Additionally, we aim to pre-train large language models on railway industry-specific documents and compare the results in two classification tasks, i.e., requirements identification and allocation of requirements to different teams. Extending the current tool to estimate the risk associated with a tender call is also planned for future work.

Acknowledgement. This work is partially funded by the AIDOaRt (KDT) and SmartDelta [27] (ITEA) projects.

References

1. Abbas, M., Ferrari, A., Shatnawi, A., Enoiu, E., Saadatmand, M., Sundmark, D.: On the relationship between similar requirements and similar software. Requir. Eng. **28**, 1–25 (2022)
2. Abbas, M., Saadatmand, M., Enoiu, E., Sundamark, D., Lindskog, C.: Automated reuse recommendation of product line assets based on natural language requirements. In: Ben Sassi, S., Ducasse, S., Mili, H. (eds.) ICSR 2020. LNCS, vol. 12541, pp. 173–189. Springer, Cham (2020). https://doi.org/10.1007/978-3-030-64694-3_11
3. Abualhaija, S., Arora, C., Sabetzadeh, M., Briand, L.C., Traynor, M.: Automated demarcation of requirements in textual specifications: a machine learning-based approach. Empir. Softw. Eng. **25**(6), 5454–5497 (2020). https://doi.org/10.1007/s10664-020-09864-1
4. Abualhaija, S., Arora, C., Sabetzadeh, M., Briand, L.C., Vaz, E.: A machine learning-based approach for demarcating requirements in textual specifications. In: 2019 IEEE 27th International Requirements Engineering Conference (RE), pp. 51–62. IEEE (2019)
5. Alhoshan, W., Zhao, L., Ferrari, A., Letsholo, K.J.: A zero-shot learning approach to classifying requirements: a preliminary study. In: Gervasi, V., Vogelsang, A. (eds.) REFSQ 2022. LNCS, vol. 13216, pp. 52–59. Springer, Cham (2022). https://doi.org/10.1007/978-3-030-98464-9_5
6. Bergstra, J., Bengio, Y.: Random search for hyper-parameter optimization. J. Mach. Learn. Res. **13**(2), 281–305 (2012)
7. Berry, D.M.: Empirical evaluation of tools for hairy requirements engineering tasks. Empir. Softw. Eng. **26**(6), 1–77 (2021). https://doi.org/10.1007/s10664-021-09986-0

8. Binkhonain, M., Zhao, L.: A review of machine learning algorithms for identification and classification of non-functional requirements. Expert Syst. Appl. X **1**, 100001 (2019)
9. Bojanowski, P., Grave, E., Joulin, A., Mikolov, T.: Enriching word vectors with subword information. Trans. Assoc. Comput. Linguist. **5**, 135–146 (2017)
10. Cleland-Huang, J., Vierhauser, M., Bayley, S.: Dronology: an incubator for cyber-physical systems research. In: 2018 IEEE/ACM 40th International Conference on Software Engineering: New Ideas and Emerging Technologies Results (ICSE-NIER), pp. 109–112 (2018)
11. Dell'Anna, D., Aydemir, F.B., Dalpiaz, F.: Evaluating classifiers in se research: the ecser pipeline and two replication studies. Empir. Softw. Eng. **28**(1), 1–40 (2023)
12. Eckhardt, J., Vogelsang, A., Fernández, D.M.: Are "non-functional" requirements really non-functional? an investigation of non-functional requirements in practice. In: 38th International Conference on Software Engineering, pp. 832–842 (2016)
13. Falkner, A., Palomares, C., Franch, X., Schenner, G., Aznar, P., Schoerghuber, A.: Identifying requirements in requests for proposal: a research preview. In: Knauss, E., Goedicke, M. (eds.) REFSQ 2019. LNCS, vol. 11412, pp. 176–182. Springer, Cham (2019). https://doi.org/10.1007/978-3-030-15538-4_13
14. Ferrari, A., Dell'Orletta, F., Esuli, A., Gervasi, V., Gnesi, S.: Natural language requirements processing: a 4D vision. IEEE Softw. **34**(6), 28–35 (2017)
15. Herwanto, G.B., Quirchmayr, G., Tjoa, A.M.: A named entity recognition based approach for privacy requirements engineering. In: 2021 IEEE 29th International Requirements Engineering Conference Workshops (REW). IEEE (2021)
16. Hey, T., Keim, J., Koziolek, A., Tichy, W.F.: Norbert: transfer learning for requirements classification. In: 2020 IEEE 28th International Requirements Engineering Conference (RE), pp. 169–179. IEEE (2020)
17. Honnibal, M., Montani, I.: spacy 2: natural language understanding with bloom embeddings, convolutional neural networks and incremental parsing. To Appear **7**(1), 411–420 (2017)
18. Huang, Z., Xu, W., Yu, K.: Bidirectional lstm-crf models for sequence tagging. arXiv:1508.01991 (2015)
19. Hubert, M., Rousseeuw, P.: International encyclopedia of statistical science (2010)
20. Jindal, R., Malhotra, R., Jain, A.: Automated classification of security requirements. In: 2016 International Conference on Advances in Computing, Communications and Informatics (ICACCI), pp. 2027–2033. IEEE (2016)
21. Kingma, D.P., Ba, J.: Adam: a method for stochastic optimization. arXiv preprint arXiv:1412.6980 (2014)
22. Koch, G., Zemel, R., Salakhutdinov, R., et al.: Siamese neural networks for one-shot image recognition. In: ICML Deep Learning Workshop, Lille, vol. 2 (2015)
23. Loshchilov, I., Hutter, F.: Fixing weight decay regularization in adam (2018)
24. Pennington, J., Socher, R., Manning, C.D.: Glove: Global vectors for word representation. In: Proceedings of the 2014 Conference on Empirical Methods in Natural Language Processing (EMNLP), pp. 1532–1543 (2014)
25. Reimers, N., Gurevych, I.: Sentence-bert: sentence embeddings using siamese bert-networks. arXiv preprint arXiv:1908.10084 (2019)
26. Runeson, P., Höst, M.: Guidelines for conducting and reporting case study research in software engineering. Empir. Softw. Eng. **14**(2), 131–164 (2009)
27. Saadatmand, M., Enoiu, E.P., Schlingloff, H., Felderer, M., Afzal, W.: Smartdelta: automated quality assurance and optimization in incremental industrial software systems development. In: 25th Euromicro Conference on Digital System Design (DSD) (2022)

28. Sainani, A., Anish, P.R., Joshi, V., Ghaisas, S.: Extracting and classifying requirements from software engineering contracts. In: 2020 IEEE 28th International Requirements Engineering Conference (RE), pp. 147–157. IEEE (2020)

29. Sanh, V., Debut, L., Chaumond, J., Wolf, T.: Distilbert, a distilled version of bert: smaller, faster, cheaper and lighter. arXiv:1910.01108 (2019)

30. Sun, C., Qiu, X., Xu, Y., Huang, X.: How to fine-tune BERT for text classification? In: Sun, M., Huang, X., Ji, H., Liu, Z., Liu, Y. (eds.) CCL 2019. LNCS (LNAI), vol. 11856, pp. 194–206. Springer, Cham (2019). https://doi.org/10.1007/978-3-030-32381-3_16

31. Tunstall, L., et al.: Efficient few-shot learning without prompts. arXiv:2209.11055 (2022)

32. Varenov, V., Gabdrahmanov, A.: Security requirements classification into groups using nlp transformers. In: 2021 IEEE 29th International Requirements Engineering Conference Workshops (REW), pp. 444–450. IEEE (2021)

33. Vaswani, A., et al.: Attention is all you need. Adv. Neural Inf. Process. Syst. **30**, 1–11 (2017)

34. Wang, W., Wei, F., Dong, L., Bao, H., Yang, N., Zhou, M.: Minilm: deep self-attention distillation for task-agnostic compression of pre-trained transformers. Adv. Neural Inf. Process. Syst. **33**, 5776–5788 (2020)

35. Winkler, J., Vogelsang, A.: Automatic classification of requirements based on convolutional neural networks. In: 2016 IEEE 24th International Requirements Engineering Conference Workshops (REW), pp. 39–45. IEEE (2016)

36. Winkler, J.P., Grönberg, J., Vogelsang, A.: Optimizing for recall in automatic requirements classification: An empirical study. In: 2019 IEEE 27th International Requirements Engineering Conference (RE), pp. 40–50. IEEE (2019)

37. Wu, Y., et al.: Google's neural machine translation system: bridging the gap between human and machine translation. arXiv preprint arXiv:1609.08144 (2016)

38. Zhang, T., Wu, F., Katiyar, A., Weinberger, K.Q., Artzi, Y.: Revisiting few-sample bert fine-tuning. arXiv preprint arXiv:2006.05987 (2020)

39. Zhao, L., et al.: Natural language processing for requirements engineering: a systematic mapping study. ACM Comput. Surv. (CSUR) **54**(3), 1–41 (2021)

Summarization of Elicitation Conversations to Locate Requirements-Relevant Information

Tjerk Spijkman[1,2](✉) ⓘ, Xavier de Bondt[1,2], Fabiano Dalpiaz[1](✉) ⓘ,
and Sjaak Brinkkemper[1] ⓘ

[1] Department of Information and Computing Sciences, Utrecht University, Utrecht,
The Netherlands
{f.dalpiaz,s.brinkkemper}@uu.nl
[2] fizor., Utrecht, The Netherlands
{tjerk.spijkman,xavier.de.bondt}@fizor.com

Abstract. [**Context and motivation**] Conversations around requirements, such as interviews and workshops, are a key activity of requirements elicitation, and play a significant role in the creation of requirements specifications. [**Question/problem**] While these conversations contain a wealth of knowledge, requirements engineers use them mainly through note-taking during the conversation and by recalling the information from their memory. There is potential for supporting practitioners by retrieving important information from the recordings of these conversations. [**Principal ideas/results**] Although transcriptions can be automatically generated with good accuracy, they often contain excessive text to be efficiently used for processing requirements elicitation sessions. Thus, we observed a need to transform these datasets into a useful format for requirements engineers to analyze. [**Contribution**] We present RECONSUM, a prototype that utilizes Natural Language Processing (NLP) to summarize requirements conversations. RECONSUM takes as input a transcribed conversation, and it filters the speaker turns by keeping only those that include a question and that are expected to contain, or to be answered with, requirements-relevant information. In addition to presenting RECONSUM, we experiment with different algorithms to assess the most effective combination.

Keywords: Requirements Elicitation · Natural Language Processing · Conversational RE · Requirements-Relevant Information

1 Introduction

Requirements elicitation concerns the activities of seeking, uncovering, acquiring, and elaborating requirements [38]. This information is often gathered through conversational activities in which a requirements engineer (or analyst) works with system stakeholders to get an understanding of the goals and design of the system [9]. According to the NaPiRE survey [35], interviews and facilitated sessions such as workshops are the most frequently used elicitation techniques: 73% and 67% of the respondents state to be using them, respectively.

© The Author(s), under exclusive license to Springer Nature Switzerland AG 2023
A. Ferrari and B. Penzenstadler (Eds.): REFSQ 2023, LNCS 13975, pp. 122–139, 2023.
https://doi.org/10.1007/978-3-031-29786-1_9

Researchers have studied requirements conversations, notably interviews, and found that note-taking is a useful activity [16] for the early detection of common problems such as ambiguity [33]. However, these conversations can range from a few hours to multiple days [2,28], thereby making it not only likely for the analyst to miss out on certain information, but also cognitively demanding as they would need to focus both on the note-taking and on keeping a natural flow.

The recordings of requirements conversations contain valuable information that can easily be lost in the overall picture of the elicitation. While creating and investigating transcriptions can be time consuming, the increasing remote work – including the online conduction of interviews and workshops – offers the opportunity to use the capability of modern online meeting tools like Microsoft Teams and Zoom Meetings to generate transcriptions that consistently improve their precision through neural network approaches [3].

Although manual reviews are possible for short conversations and they are a useful educational tool [30], we argue that analysts need to be supported in the analysis of longer real-life conversations, and that Natural Language Processing (NLP) can be fruitfully used to such an extent.

In this paper, we propose RECoNSUM (Requirements Elicitation Conversations Summarizer), a NLP prototype tool that can assist practitioners and researchers in processing elicitation conversations by summarizing the transcriptions and extracting requirements-relevant information. We utilize the *Question & Answer (Q&A)* structure prevalent in conversations, as discussed in Sect. 2. RECoNSUM retains only relevant questions and their answers through *extractive summarization* [12], thereby making transcripts easier to review, as practitioners would see a short version of the original conversation transcript. The outputs of RECoNSUM are meant to be used in a front-end to enable exploration of such conversations, as per the mockup of Fig. 1.

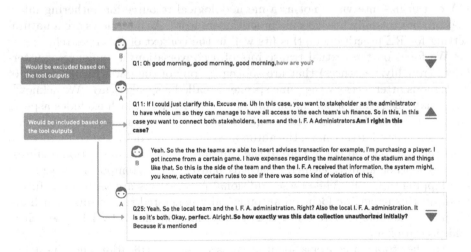

Fig. 1. Mockup visualization of the outputs of RECoNSUM. In this example, questions Q1–Q10 and Q12–Q24 are hidden as they are expected to be irrelevant.

To effectively summarize a document, one needs to gain a deep understanding of the document to gather the relevant information. In our context, this amounts to identifying and extracting *requirements-relevant information* from the transcript of a requirements conversation. We build on previous research [28,29]: through data investigation and experimentation with a focus group of RE students and practitioners, we gained a first understanding of what requirements-relevant information exists in requirements conversations. Based on these premises, we define the following research questions:

MRQ. How can we identify requirements-relevant information in a transcript of a requirements elicitation conversation?

RQ1. How can we define requirements relevance in a transcript?

RQ2. How to design an automated approach for locating requirements-relevant information?

The rest of this paper is structured as follows. In Sect. 2, we discuss background and related work. Section 3 outlines the research method and describes RECONSUM. We report on a validation of the approach in Sect. 4, and finally, we present a discussion and future works in Sect. 5.

2   Background

Conversation Structures. There are many different types of conversations, including small talk, troubles telling, and elicitation conversations [18]. Conversation analysis (CA) is the systematic analysis of the talk produced in everyday situations of human interaction [20]. CA goes beyond the scope of the spoken words and it includes video recordings of the workplace, or the onscreen activities for a conversation between people playing a game. Mondada [23] states that CA can utilize interviews not as a methodological resource for gathering information, but to study how specific practitioners work. As interviews are a natural setting for RE practitioners, this fits well in the context of this research.

We focus on the textual transcripts of a conversation: a sequence of utterances (roughly, sentences) that are spoken by one of the participants. A set of contiguous utterances by the same speaker is called a speaker turn. We acknowledge that this is only a partial picture of a conversation, which excludes aspects such as the use of artifacts (e.g., whiteboards), the analysis of intonation cues and visual cues. Furthermore, the automated transcripts delivered through video conferencing tools do not adhere to the standards used in conversation analysis, as they do not identify elements such as pauses, speed/tempo of speech, and overlapping talk [19]. While these additional perspectives are part of our future work, in this paper, we focus on automatically generated transcripts, which are already a valuable resource that is generally not considered by RE researchers and practitioners.

Another topic of conversational analysis is the identification and characterization of recurrent interaction practices [26]. These consist of a set of actions: asking, telling, requesting, inviting, complaining, etc. One of the key concepts in interaction practices is the *adjacency pair* [31], resulting from the turn-taking format that is common in conversations [34]. Adjacency pairs are based on the understanding that an utterance is related to what comes before, and what comes next. An adjacency pair is composed of two speaker turns uttered by different speakers and placed adjacently [24]. For instance, a typical type of pair is "Request for information" followed by "Informative answer".

Stolcke *et al.* [32] discuss the lack of consensus on describing discourse structure; however, they argue that dialogue acts (DAs) are a useful first level of analysis. A DA is roughly equivalent [32] to speech acts and adjacency pairs. There are, however, differences between these theories. Take two questions such as: "Did you do it?" and "What did you wear today?". While speech acts [25] consider both questions as illocutionary acts, DAs classify these into a Yes-No question and Wh-question, respectively. DAs are a method for classifying discourse data using 42 different labels (see Table 1 for some examples). In our research, we utilize libraries that enable the automated identification of DAs to perform extractive summarization.

Related Works. In our previous research, we performed an empirical study to determine the contents of one particular RE conversation common in practice: fit-gap analysis [28], which aims to distinguish between those parts of a software product that already fit the client from those gaps that needs to be addressed via configuration or customization. The understanding gained from this work was used in designing the TRACE2CONV prototype tool that assists in establishing automated pre-requirements specification traceability [29].

While focused on a different type of artifact, i.e., requirements specifications, Abualhaija *et al.* [1] apply supervised machine learning to recognize and demarcate requirements in a free-form requirements specification. In their work, modal verbs are used to determine important segments of the text, and parts of their NLP pipeline inspired this work.

Another adjacent area of research is that of automated requirements classification [6], which led to the development of NLP tools that organize requirements into categories. A typical classification distinction is between functional and non-functional requirements. To achieve such a classification, Kurtanović and Maalej [22] apply supervised machine learning through support vector machines. Their research shows that POS tags, word n-grams, modal verbs and the POS tag 'cardinal number' were the most informative. Similarly, higher-level linguistic dependencies can be useful in classification, as shown by Dalpiaz *et al.* [7]. This work also observes that the performance of classifiers degrades when used on other datasets; this led to the birth of the ECSER pipeline for a rigorous evaluation of classifiers in software engineering [10].

Only a few scholars have studied requirements conversations in depth. Alvarez and Urla [2] performed a manual analysis of interview transcripts concerning the construction of an ERP system; they studied the role of stakeholders and of client stories. Ferrari *et al.* [16] conducted and analyzed 34 simulated interviews and identified four facets of ambiguity (unclarity, multiple understanding, incorrect disambiguation and correct disambiguation); moreover, they found that ambiguity can be a cue for the elicitation of tacit knowledge. Their follow-up work [14] explores the use of voice and biofeedback to identify emotions that may represent engagement. Bano *et al.* [4] analyze interviews by novices and they build an educational framework for teaching how to conduct requirements interviews. The same authors extend their approach by including (reverse) role-playing elements [15].

3   RECONSUM: A Tool for Summarizing RE Conversations

We designed RECONSUM according to the phases of the engineering cycle by Wieringa [36], focusing on the first three phases (the design cycle) as the current prototype has not yet been applied to practical cases. The code can be found in GitHub[1] and a persistent copy is in the online appendix [27]. In this section, we discuss the problem investigation and solution design steps of Wieringa's design cycle. Sect. 4 reports on the validation step.

3.1   Problem Investigation and Solution Design Iterations

The problem addressed by this research is based on observations of the artifacts from our previous research [28,29]. We found that RE conversations contain useful information, but that their manual analysis is time consuming. This indicates an opportunity for designing intelligent tools that can reduce the necessary effort through automation.

To further explore the problem domain, we analyzed nine recordings of requirements interviews conducted by master's students at Utrecht University in a simulated setting. These recordings were split across three different cases (see [8] for details): IFA – the international football association portal, UMS – an urban mobility simulator, and HMS – a hospital management system. All interviews had a similar time frame (max. 60 min) and structure, thereby allowing us to consider multiple cases per domain as well as different domains.

Given the limited existing literature, the treatment design is approached in an exploratory way. We went through multiple design iterations, which had the definition of the term *requirements-relevant information* as a recurring theme. Although a seemingly simple term, we found that answering the question "what does it mean to be relevant" is hard. There is no single recipe to define if something is relevant to an analyst, as this may depend on the specific use case, e.g., authoring requirements, searching for missing requirements, tracing requirements backwards, and implementing the requirements.

Discarded Designs. We initially expected to utilize the categories of requirements-relevant information that we defined in previous research [28,29]. However, we found out that these categories were too context reliant. For example, the *as-is process* is fundamental when replacing a legacy system, while the *to-be process* is more important for a new system design [28]. Additionally, our

---

[1] https://github.com/RELabUU/REConSum.

collection of transcripts was heterogeneous regarding the design activity (green-field vs. brownfield), and several sentences were hard to classify as they described the as-is by indicating something about the to-be.

We also attempted the design of a machine learning-based automated binary classification of the transcript, with *speaker turns* being either relevant or irrel-evant. The main challenges with this approach, however, were the heterogeneity of the transcripts and the too limited amount of labeled data.

Extractive Question-Based Summarization. Through further exploration of the artifacts, we found that the Q&A structure could effectively be used to produce a condensed version of the conversation while still covering most of the content. The idea was that of retaining only those questions that are (poten-tially) relevant, and the analyst could then further navigate the conversation by zooming in on the answers of those questions as shown in Fig. 1. This led to the design detailed in this section; an approach that first classifies the text based on its structure, focusing on the identification of a special kind of adjacency pairs: questions and answers to those questions.

3.2   Question Identification and Relevance Detection

We aim to obtain an extractive summarization of a requirements conversation consisting of only those speaker turns that include questions that are potentially relevant for requirements engineers.

Formally, let a conversation $C = (T_1, T_2, \ldots, T_n)$ be a sequence of speaker turns, where $n \in \mathcal{N}^+$ is a positive natural number. A speaker turn T_i is a sequence of utterances (roughly, sentences) that are spoken by the same speaker: given $i, m \in \mathcal{N}^+$, $T_i = (U_1, U_2, \ldots, U_m)$. An utterance U_j is a sequence of words by the same speaker: given $j, p \in \mathcal{N}^+$, $U_j = (w_j^1, \ldots, w_j^p)$. We define two func-tions. *IsQuestion* $: \mathcal{U} \to \{0, 1\}$ is a Boolean function that returns 1 if and only if $U \in \mathcal{U}$ is a question. *IsRelevant* $: \mathcal{U} \to \{0, 1\}$ is a Boolean function that returns 1 if and only if $U \in \mathcal{U}$ is a relevant utterance for a requirements engineer. We can now formally define our summarization function $Summ : \mathcal{C} \to \mathcal{C}$, where $\mathcal{C}$ is the domain of conversations, as follows:

$$Summ(C) = S.\ S \text{ is a sub-sequence of } C \text{ and } \forall T = (U_1, \ldots, U_m) \in S,$$
$$\exists k \in [1, m].\ IsQuestion(U_k) \wedge IsRelevant(U_k)$$

REConSum Process. Figure 2 provides an overview of how REConSum implements the summarization function *Summ*. It takes a requirements inter-view transcript as an input, and first determines which speaker turns contain a question. This is done utilizing Part-of-Speech tagging and/or Dialogue Act classification. Then, REConSum determines if these questions are relevant by assessing whether they contain domain-specific terms; our assumption is that the presence of those terms is an indicator of relevance.

We determine whether a speaker turn includes domain-specific terms by cal-culating Term Frequency–Inverse Document Frequency technique (TF–IDF). We

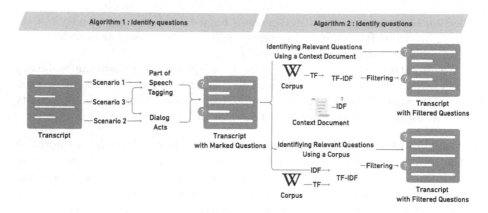

Fig. 2. A process flow overview of RECONSUM.

first compute the Inverse Document Frequency (IDF) either of the transcript itself (bottom-right scenario in Fig. 2) or of a context document (top-right scenario in Fig. 2). We then calculate Term Frequency (TF) of a Wikipedia corpus, which we take as a general-purpose corpus where the distribution of the terms is not expected to reflect the specificity of the conversation domain. Then, RECONSUM retains only those speaker turns that both include a question as well as words with a high TF–IDF score, indicating that these terms are much more frequent in the domain than in Wikipedia.

Algorithm 1: Identify Questions. The first stage of RECONSUM is to identify the questions in a transcript through (i) a deep learning classifier based on dialogue acts, (ii) the occurrence of sequences of Part-of-Speech (POS) tags, or (iii) either of the previous. The dialogue acts approach assigns a dialogue act to each sentence in the speaker turn, while the approach based on POS tags assigns these tags to parts of each sentence. These POS tags were taken from the Penn Treebank POS Tagset[2], which contains two clause-level tags that can indicate questions [5]: SBARQ and SQ. These tags indicate four types of questions: *wh*-questions, *yes-no*-questions, *tag*-questions, and *choice*-questions.

Our dialogue act-based approach relies on the off-the-shelf classifier DialogTag[3], which uses a neural architecture based on BERT [11] to assign a dialogue act to a sentence. DialogTag uses a subset of the Switchboard-1 corpus; the latter was created using 2,400 telephone conversations, with conversations among 543 speakers on 70 topics [21]. Our implementation labels as question those speaker turns that include a sentence that denotes one of the question types that DialogTag identifies; frequent examples are in Table 1.

We also propose a third approach that aims at supporting those scenarios where recall is more important than precision: we execute both of the previous approaches and retain the speaker turns if they are included in either approach.

---

[2] Our implementation is inspired by that of https://github.com/garcia2015/NLP_QuestionDetector.

[3] https://github.com/bhavitvyamalik/DialogTag.

Table 1. Examples of question types, based on dialogue act classification, that are used by function DIALOGUEACTS in Algorithm 1.

Tag	Example
Yes-No-Question	*Is there already some data that can be gathered from the existing systems that can already be put in the new one or not?*
Wh-Question	*I'm gonna ask you, how long does it take for that person to analyze the situation and uh monitor a certain road or urban traffic situations?*
Declarative Yes-No-Question	*So it would be a manual change, not a new iteration of the automated schedule.*
Backchannel in question form	*Um, this should also be made available I imagine, during a match for instance, the score of the match should be updated immediately once it's changed. Right?*
Open-Question	*What do you mean with 'local' I.F.A.?*
Rhetorical-Questions	*(...) We have done it in the in the other city the other year. So why shouldn't it work now?*
Or-Clause	*So you think there should be the same rights for every system user? Or do you think that one user should have less rights capability?*
Tag-Question	*Right?*

Algorithm 1. Identify Questions

Input: C a set of speaker turns,
Output: T the set of speaker turns, with the questions marked

```
1: function DIALOGUEACTS(C)
2:     for all sent ∈ C do
3:         T[sent] ← sent
4:         T[sent]_question ← False
5:         for all tag ∈ DIALOGTAG.DIALOGUE_ACTSTOKENIZEsent do
6:             if ANY({-Question, Or-Clause}) ∈ tag then
7:                 T[sent]_question ← True
8:     return T
```

```
1: function PART-OF-SPEECH-TAGS(C)
2:     for all sent ∈ C do
3:         T[sent] ← sent
4:         T[sent]_question ← False
5:         for all subtree ∈ NLP_ANNOTATEsent do
6:             if subtree.POS_tag ∈ {SBARQ, SQ} then
7:                 T[sent]_question ← True
8:     return T
```

```
1: function COMBINED(C)
2:     T_1 ← DIALOGUEACTSC
3:     T_2 ← PART-OF-SPEECH-TAGSC
4:     for i = 0; i ¡ |C|; i++ do
5:         T[sent] ← C[i]
6:         T[sent]_question ← T_1[i]_question ∨ T_2[i]_question
7:     return T
```

The first step of RECONSUM is detailed in Algorithm 1. An input set C of speaker turns is turned into a version T where the speaker turns with a question are marked. The DIALOGUEACTS function loops through each sentence in C, retrieves the dialogue acts that apply to that sentence through the `DialogTag` BERT-based classifier, and determines if the sentence contains one of the dialogue acts that indicate a question. Similarly, the PART-OF-SPEECH-TAGS function generates the POS trees of a sentence, and explores each of them to see if it contains a question indication (`SBARQ` or `SQ`). The combined approach (function COMBINED) returns those speaker turns that are identified as questions by at least one of the other two functions.

Algorithm 2. Categorize Relevant Questions

Input: T a set of speaker turns, with the questions marked,
    F a file to compare the relevance to; either a context document or the conversation transcript,
    num_words the number of unfiltered words you would like to categorize the questions on
    (set to 60 in our experiments based on empirical testing),
Output: the set of speaker turns, with the questions and their relevance marked

```
1: function CREATEWORDLIST(F, num_words)
2:     IDF ← LOAD_WIKI_TF
3:     File ← PREPROCESS_FILE F
4:     Words ← CALCULATE_TF-IDF IDF, File
5:     Word_List ← TAKEFIRSTNSORT Words, num_words
6:     Word_List ← STEM w ∈ Word_List | w ∉ stop_words
7:     return Word_List
```

```
1: function FILTERQUESTIONS(T, F)
2:     Word_List ← CREATEWORDLIST F, 60
3:     for all sent ∈ T do
4:         T[sent]_relevant ← False
5:         if T[sent]_question then
6:             for all word ∈ SPLIT_SENTENCE sent do
7:                 if STEM word ∈ Word_List then
8:                     T[sent]_relevant ← False
9:     return T
```

Algorithm 2: Categorize Relevant Questions. In the second stage (right-hand side of Fig. 2), RECONSUM looks for *relevant* questions based on the outputs of the first stage. RECONSUM implements two approaches, both reliant on TF–IDF. In both cases, TF is calculated on the basis of a Wikipedia dataset [17]. In the first approach, we calculate IDF on a document that describes the application context (a *context document*), which provides us with words that can indicate the relevance of a question. Based on TF–IDF, we retain only questions that include words with a high TF–IDF value. In our second approach, we follow the same process, but IDF is calculated on the transcript without the need to use a context document.

This functionality is described in Algorithm 2, which includes two functions. The first one (CREATEWORDLIST) takes as input (i) the file F we use to calculate IDF: either the transcript or the context document, and (ii) an integer that indicates the number of unfiltered words to categorize the outputs. After loading the Wikipedia Term-Frequency and processing the file (F), the TF–IDF scores can be calculated. After that, we sort on these scores and take the number of unfiltered words that we specified. Finally, we remove stopwords, and we stem the remaining words. The second function (FILTERQUESTIONS) takes as inputs the set of speaker turns, with the questions marked, from Algorithm 1 and the same file F used by CREATEWORDLIST. It first creates a list of domain-specific words by calling CREATEWORDLIST, and then marks the questions that contain one of these words as relevant, thus returning a set of speaker turns where the questions are marked and have their relevance indicated.

4   Evaluation

After explaining the design of our evaluation and golden standard in Sect. 4.1, we present qualitative and quantitative results in Sect. 4.2.

4.1   Designing the Golden Standard

With the aim of measuring the performance of RECONSUM in identifying only the relevant questions in a conversation, we set off to design a golden standard. To do so, we performed a number of design iterations that were meant to define the instrument through which the golden standard could be created. The goal was to have this standard created by people who are not the authors of this paper. To this end, the tagging was facilitated through a survey.

A key decision was establishing what context (how many speaker turns) should be shown for each question, as we expected it would be difficult to rate relevance without that information. We eventually decided to include the speaker turn that includes the question, the previous turn, and the next one.

Table 2. Categorization of requirements-relevant information in the tagging

Functional requirement	The speaker turn refers to functionality that the software system has to exhibit. For example, register users, schedule events, calculate something or allow messaging
Non functional requirement	Software quality or non-functional requirement. The speaker turn refers to qualities that the system should provide while delivering its functionalities, e.g., speed, security, capacity, compatibility, reliability, usability, portability
System users	The speaker turn mentions the users of the system, or other stakeholders that do not use the system
Current process understanding	The speaker turn contains information about the current process or system as-is, including current problems that the interviewee is facing
Within or outside of the scope	The speaker turn explicitly contains a discussion of elements that should be in the system to-be or not. These define the boundaries of the system's scope
No requirements-relevant information	The speaker turn does not contain any relevant information.

Another challenge concerned deciding whether a segment was relevant. To facilitate this, we defined a categorization of relevance inspired by our earlier work [28], as shown in Table 2. Sometimes the question itself was not relevant, yet the surrounding text was. Therefore, the taggers were asked whether requirements-relevant information: (a) could be expected in the answer to the question; or (b) could be found in the speaker turn shown after the question.

The taggers were first asked to decide whether the segment included one of the relevant categories in Table 2, and, if so, they could answer the questions regarding where the relevant information was located. All taggers were provided with a tagging guide (in our online appendix alongside the source code and the results [27]), and an overview of the case being discussed.

Execution of Tagging. We created the golden standard for the nine datasets shown in Table 3 by recruiting 18 taggers: either students familiar with requirements engineering or practitioners. Two of them tagged each dataset using a Qualtrics survey: each participant was assigned a case, and they would see all of the questions in the conversation in chronological order, as in Fig. 3. They would go through the conversation one question at a time, with the option to return to the previous question. As per Table 3, the participants saw on average circa 71% of the conversation, and they could tag for relevance 58.8% of the conversation. This difference arises because the participants could not tag the speaker turn before the one where the question is located.

Table 3. Evaluation datasets. The UMS/IFA/HMS identifier refers to the case name as per Sect. 3.1. The table also shows recording length, number of speaker turns, then number (#) and percentage (%) of speaker turns (a) shown to the taggers, (b) that could be tagged, and (c) that include questions. The 'Relevant' columns characterize the gold standard defined by the taggers, and 'Agreement' shows inter-rater agreement in percentage and using Cohen's kappa.

Set	Length mm:ss	Speaker Turns												
		Total	Shown		Taggable		Relevant		Questions		Relevant		Agreement	
			#	%	#	%	#	%	#	%	#	%	%	k
1-UMS	50:23	167	117	70.1%	95	56.9%	61	64.2%	49	29.3%	31	63.3%	54.7%	0.20
2-IFA	49:15	148	107	72.3%	85	57.4%	67	78.8%	46	31.1%	34	73.9%	58.8%	0.21
3-UMS	41:29	98	69	70.4%	56	57.1%	36	64.3%	30	30.6%	14	46.7%	60.7%	0.16
4-HMS	23:05	69	50	72.5%	41	59.4%	31	75.6%	21	30.4%	15	71.4%	90.2%	0.75
5-IFA	58:06	179	132	73.7%	105	58.7%	51	48.6%	56	31.3%	20	35.7%	76.2%	0.53
6-HMS	38:25	116	77	66.4%	64	55.2%	44	68.8%	34	29.3%	17	50.0%	60.1%	0.22
7-IFA	47:12	162	109	67.3%	91	56.2%	79	86.8%	46	28.4%	41	89.1%	59.3%	0.07
8-HMS	39:24	155	115	74.2%	98	63.2%	68	69.4%	54	34.8%	38	70.4%	89.8%	0.76
9-HMS	30:31	80	65	81.3%	55	68.8%	39	70.9%	28	35.0%	20	71.4%	92.7%	0.80
Average	49:15	130	93	71.6%	77	58.8%	53	69.0%	40	31.0%	26	63.2%	71.6%	0.41
Total		1174	841		690		476		364		230			

Although the design was meant to ensure a common understanding, the inter-rater agreement is low (the macro-average of 0.41 is at the boundary between *fair* and *moderate*). We mainly ascribe this to the fact that we did not define a clear use case, and the notion of relevance may depend on the task at hand and domain experience. In general, we identified three common types of disagreements: (i) the statement 'Do you expect the question to be answered with requirements-relevant information' was often misread as 'Does the question include ...'; (ii) whether yes-no answers should be considered relevant; and (iii) if the summary of a previous answer, made by the analyst, should be considered relevant. The disagreements were first manually validated by the second author to identify obvious sloppiness, and those cases were discarded. When this analysis did not resolve the disagreements, we took an inclusive approach in which a speaker turn was considered relevant if one of the taggers tagged it as such.

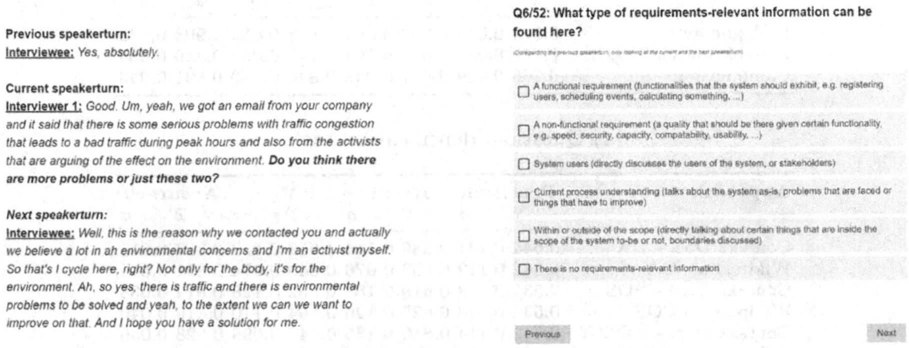

Fig. 3. Illustration of the tagging tool. On the left, the speaker turn including a question, together with the adjacent ones, are shown. On the right, the tagger selects whether and what kind of requirements-relevant information exists.

4.2   RECONSUM Results

To determine the effectiveness of RECONSUM (its implementation of the extractive summarization function *Summ* in Sect. 3.2), NLP summarization task metrics could be applied, e.g., coherence, consistency, percentage of text shown [13]. Other metrics such as BLEU and ROUGE assume the existence of a reference summary, which we do not possess at this stage of our research. In this paper, we assess RECONSUM's ability of filtering, and therefore utilize standard information retrieval metrics: precision, recall, F_1-score, and accuracy. As a unit of analysis, we take speaker turns; in other words, we measure (i) if the speaker turn contains a question, and (ii) if a speaker turn with a question is relevant and should therefore be retained in the summary.

Question Detection. For the first part of RECONSUM, we compare the three variants described in Algorithm 1: POS tagging, dialogue acts, and their combination. Table 4a presents a summary of the results, while the results for each dataset can be found online. The approach based on a deep learning classifier for dialogue acts leads to higher precision and higher recall than the approach based on POS tagging, perhaps thanks to the higher number of question types that it recognizes. Combining both approaches increases the number of true positives, but at the cost of increasing the false positives too. Using the combined approach reduces the summarization rate, but at the same time it decreases the likeliness of missing requirements-relevant information.

Table 4. Performance metrics, showing macro-average and standard deviation across the nine datasets of Table 3. The best results are highlighted in green.

Approach	Precision		Recall		F_1		Accuracy	
	$\bar{x}$	σ	$\bar{x}$	σ	$\bar{x}$	σ	$\bar{x}$	σ
Dialogue acts	0.820	0.077	0.906	0.066	0.858	0.054	0.908	0.035
Part of Speech Tags	0.778	0.096	0.696	0.101	0.727	0.061	0.839	0.033
Combination	0.766	0.089	0.951	0.048	0.846	0.060	0.891	0.046

(a) Question identification task

Approach	Precision		Recall		F_1		Accuracy	
	$\bar{x}$	σ	$\bar{x}$	σ	$\bar{x}$	σ	$\bar{x}$	σ
Context Doc. - DA	0.644	0.131	0.689	0.144	0.653	0.117	0.867	0.015
Wikipedia - DA	0.642	0.112	0.653	0.076	0.640	0.074	0.861	0.011
Context Doc. - POS	0.535	0.158	0.619	0.153	0.562	0.128	0.819	0.038
Wikipedia - POS	0.532	0.148	0.625	0.126	0.564	0.110	0.816	0.040
Context Doc. - COMB	0.542	0.114	0.810	0.135	0.641	0.099	0.828	0.036
Wikipedia - COMB	0.548	0.119	0.808	0.106	0.646	0.097	0.831	0.041

(b) Question relevance task

Categorization of Relevant Questions. Once the questions are found in the transcript, Algorithm 2 determines if they are requirements relevant. The algorithm includes two approaches for calculating term frequency, either from the conversation itself, or from a contextual document. These can then be applied to all three approaches for question detection, leading to six combinations. The results in Table 4b show that the highest precision is obtained by combining (a) dialogue act tagging for question identification with (b) TF–IDF using a context document for relevance detection. The highest recall is obtained through the COMBINED algorithm for question detection. The latter result is not surprising, as Algorithm 2 takes as input the outputs of Algorithm 1, and the combined approach had by far the highest recall (over 95%, see Table 4a). Based on these results, we cannot identify a clear winner, as the decision depends on the relevant metrics for the use case. Precision is more important for a specification task as enrichment to note-taking, while recall is more important when searching for missed requirements.

Locating Requirements-Relevant Information in Questions. The results show that RECONSUM is able to effectively extract the questions from the conversations. In Table 3, circa 63% of these questions were tagged as relevant by the taggers. The tagging results, however, indicate a higher relevance for all taggable items (questions, plus the following speaker turn) than just the questions: 69%. Thus, the answers contain more requirements-relevant information than the questions, thereby indicating their importance for users when exploring a conversation. If we would include those answers in the summary, we would increase relevance by reducing the summarization rate.

5 Conclusions, Limitations, and Future Work

In this paper, we presented RECONSUM as a step towards the summarization of requirements conversation transcripts, building on and extending the knowledge in the field of Conversational RE [29]. RECONSUM employs NLP techniques to extract questions from a transcript and to determine their relevance. This approach was validated against an assembled golden standard, reaching an F_1 score around 65%. While just showing the questions might not contain all the necessary information for a RE practitioner, this is meant to be a starting point for further exploring parts of a transcribed conversation.

Although the definition of *requirements relevance* (**RQ1**) in conversations is not final, the discussions and creation of a golden standard provides further knowledge in this domain. The high disagreement across taggers shows that determining requirements relevance depends on the perspective of the individual tagger and on the use case at hand: why is the transcript being explored? Similarly, relevance cannot be determined in a vacuum (a single speaker turn), and we allowed taggers to read a context that includes the previous and following speaker turns (both adjacency pairs before and after the question). The survey design also defines a minimal categorization of data as presented in Table 2 which is the result of multiple design iterations.

To automate the identification of requirements-relevant information (**RQ2**), the best results (see Table 4) are obtained through a combination of dialogue acts classification and TF–IDF that allows question recognition and to determine their relevance. RECONSUM provides these questions as a summary of the entire transcript that can facilitate exploration of the conversations by third parties or reviewing by part-taking practitioners.

The answers to RQ1 and RQ2 allow us to address **MRQ**: RECONSUM is our initial answer for the automated identification of requirements-relevant information in a requirements conversation.

Threats to Validity. The obtained results should be seen in light of the threats to validity, which we classify according to Wohlin *et al.* [37].

Conclusion Validity. The comparison against the golden standard has some limitations, as it was created by one pair of taggers per conversations. This means we are not only comparing our tool to the golden standard, but also to the human performance in creating this standard. Additionally, we have used classic information retrieval metrics, but we did not employ classic summarization metrics at this stage, which require possessing a reference summary. Additionally, the relevance of the questions and answers were tagged by perceived relevance, but we did not tag the remaining speaker turns (those that do neither include a question or that constitute an answer to a question). It needs to be confirmed which speaker turns include the highest percentage of requirements-relevant information.

Internal Validity. To make the tagging exercise easier for the participants, we used a non-exhaustive list of relevance categories, which might have impacted their perception of relevance for the tagged speaker turns. Additionally, while we utilized generated transcripts, these were post-processed to remove transcription errors; this is likely to have a positive effect on the findings.

Construct Validity. All the cases used in our validation and theory building consisted of interviews focused on the creation of one of three information systems. This homogeneity probably has an impact on the type of information to be found in the transcripts. Our results are based on the golden standard, but requirements relevance remains a term that is up to the interpretation and use case for reviewing the context. This means that the lack of clear boundaries for requirements relevance has an impact on our findings and conclusions.

External Validity. Our validation and design relied on a set of simulated interviews conducted by students. Whether the results generalize to practical settings can only be determined by using interviews from real-world projects.

Future Works. The research leading to RECONSUM is part of *conversational RE*: "the analysis of requirements elicitation conversations aimed at identifying and extracting requirements-relevant information" [29]. We expect to support this goal by building RE tools that can reduce the effort for practitioners to review and explore the conversations for requirements-relevant information. We first discuss direct improvements for RECONSUM, followed by more general research directions concerning conversational RE.

As an additional functionality for RECONSUM, we experimented with applying different learning approaches (Machine Learning, Transfer Learning, and Zero-Shot Learning) to categorize questions similar to our tagging exercise. While these outputs were not significant due to the limited labeled data, we expect that extending the golden standard could enable an effective learning technique to classify data within the categories of requirements relevance. Also, an investigation of the options for user interaction starting from the outputs of RECONSUM is necessary to allow the use of the tool in practice. The interface shown in Fig. 1 is only an initial idea that shall be further developed.

Beyond RECONSUM, we can extend the *conversational RE* field in many ways. For instance, the generation of domain/data models from conversations could speed up development drastically especially in the low-code development domain. The field of conversation analysis offers many avenues for a rich exploration of conversations, e.g., exploiting multi-modal data that includes video footage, screensharing, whiteboard contents, and prototypes. Another open topic is to extend the analysis beyond single conversations into a more extensive approach that can be utilized throughout a project linking all conversations together and keeping track of changes in requirements over time.

Acknowledgements. We thank all the participants who acted as taggers. The use of the recorded and transcribed dataset is made possible thanks to the ethical Science-Geosciences Ethics Review Board of Utrecht University (case S-20339).

References

1. Abualhaija, S., Arora, C., Sabetzadeh, M., Briand, L.C., Traynor, M.: Automated demarcation of requirements in textual specifications: a machine learning-based approach. Empir. Softw. Eng. **25**, 5454–5497 (2020)
2. Alvarez, R., Urla, J.: Tell me a good story: using narrative analysis to examine information requirements interviews during an ERP implementation. ACM SIG-MIS Database **33**(1), 38–52 (2002)
3. Archibald, M.M., Ambagtsheer, R.C., Casey, M.G., Lawless, M.: Using zoom videoconferencing for qualitative data collection: Perceptions and experiences of researchers and participants. Int. J. Qual. Methods **18** (2019)
4. Bano, M., Zowghi, D., Ferrari, A., Spoletini, P., Donati, B.: Teaching requirements elicitation interviews: an empirical study of learning from mistakes. Requir. Eng. **24**(3), 259–289 (2019). https://doi.org/10.1007/s00766-019-00313-0
5. Bies, A., et al.: Bracketing guidelines for Treebank II style Penn Treebank project. University of Pennsylvania, Technical report (1995)
6. Cleland-Huang, J., Settimi, R., Zou, X., Solc, P.: Automated classification of non-functional requirements. Requir. Eng. **12**(2), 103–120 (2007)
7. Dalpiaz, F., Dell'Anna, D., Aydemir, F.B., Çevikol, S.: Requirements classification with interpretable machine learning and dependency parsing. In: IEEE International Requirements Engineering Conference, pp. 142–152 (2019)
8. Dalpiaz, F., Gieske, P., Sturm, A.: On deriving conceptual models from user requirements: an empirical study. Inf. Softw. Technol. **131**, 106484 (2021)
9. Davis, A., Dieste, O., Hickey, A., Juristo, N., Moreno, A.M.: Effectiveness of requirements elicitation techniques: empirical results derived from a systematic review. In: IEEE International Requirements Engineering Conference, pp. 179–188 (2006)
10. Dell'Anna, D., Aydemir, F.B., Dalpiaz, F.: Evaluating classifiers in SE research: the ECSER pipeline and two replication studies. Empir. Softw. Eng. **28**(1), 1–40 (2023)
11. Devlin, J., Chang, M.W., Lee, K., Toutanova, K.: BERT: pre-training of deep bidirectional transformers for language understanding (2018), https://arxiv.org/abs/1810.04805
12. El-Kassas, W.S., Salama, C.R., Rafea, A.A., Mohamed, H.K.: Automatic text summarization: a comprehensive survey. Expert Syst. Appl. **165**, 113679 (2021)
13. Fabbri, A.R., Kryściński, W., McCann, B., Xiong, C., Socher, R., Radev, D.: Summeval: re-evaluating summarization evaluation. Trans. Assoc. Comput. Linguist. **9**, 391–409 (2021)
14. Ferrari, A., Huichapa, T., Spoletini, P., Novielli, N., Fucci, D., Girardi, D.: Using voice and biofeedback to predict user engagement during requirements interviews. arXiv:2104.02410 (2021)

15. Ferrari, A., Spoletini, P., Bano, M., Zowghi, D.: SaPeer and ReverseSaPeer: teaching requirements elicitation interviews with role-playing and role reversal. Requir. Engi. **25**(4), 417–438 (2020)
16. Ferrari, A., Spoletini, P., Gnesi, S.: Ambiguity and tacit knowledge in requirements elicitation interviews. Requir. Eng. **21**(3), 333–355 (2016). https://doi.org/10.1007/s00766-016-0249-3
17. Galkin, M., Malykh, V.: Wikipedia TF-IDF Dataset release (2020). https://doi.org/10.5281/zenodo.3631674
18. Hakulinen, A.: Conversation types. In: D'hondt, S., Verschueren, J., Östman, J.O. (eds.) The Pragmatics of Interaction, pp. 55–65 (2009)
19. Hepburn, A., Bolden, G.B.: The conversation analytic approach to transcription. In: Stivers, T., Sidnell, J. (eds.) The Handbook of Conversation Analysis, pp. 57–76 (2013)
20. Hutchby, I., Wooffitt, R.: Conversation Analysis: Principles, Practices and Applications. Wiley, Hoboken (1998)
21. John, J., Godfrey, E.H.: Switchboard-1 release 2 (1993). https://doi.org/10.35111/sw3h-rw02
22. Kurtanović, Z., Maalej, W.: Automatically classifying functional and non-functional requirements using supervised machine learning. In: IEEE International Requirements Engineering Conference, pp. 490–495 (2017)
23. Mondada, L.: The conversation analytic approach to data collection. In: Stivers, T., Sidnell, J. (eds.) The Handbook of Conversation Analysis, pp. 32–56 (2013)
24. Schegloff, E.A., Sacks, H.: Opening up closings. Semiotica **8**(4), 289–327 (1973)
25. Searle, J.R., Searle, J.R.: Speech Acts: An Essay in the Philosophy of Language. Cambridge University Press, Cambridge (1969)
26. Sidnell, J.: Basic conversation analytic methods. In: Stivers, T., Sidnell, J. (eds.) The Handbook of Conversation Analysis, pp. 77–99. Wiley Online Library (2013)
27. Spijkman, T., de Bondt, X., Dalpiaz, F., Brinkkemper, S.: Online appendix to Summarization of Elicitation Conversations to Locate Requirements-Relevant Information (2023). https://doi.org/10.5281/zenodo.7650324
28. Spijkman, T., Dalpiaz, F., Brinkkemper, S.: Requirements elicitation via fit-gap analysis: a view through the grounded theory lens. In: International Conference on Advanced Information Systems Engineering, pp. 363–380 (2021)
29. Spijkman, T., Dalpiaz, F., Brinkkemper, S.: Back to the roots: linking user stories to requirements elicitation conversations. In: IEEE International Requirements Engineering Conference (RE@Next! track) (2022)
30. Spoletini, P., Ferrari, A., Bano, M., Zowghi, D., Gnesi, S.: Interview review: an empirical study on detecting ambiguities in requirements elicitation interviews. In: International Working Conference on Requirement Engineering: Foundation for Software Quality, pp. 101–118 (2018)
31. Stivers, T.: Sequence organization. In: Stivers, T., Sidnell, J. (eds.) The Handbook of Conversation Analysis, pp. 191–209 (2013)
32. Stolcke, A., et al.: Dialogue act modeling for automatic tagging and recognition of conversational speech. Comput. Linguisti. **26**(3), 339–373 (2000)
33. Sutcliffe, A., Sawyer, P.: Requirements elicitation: towards the unknown unknowns. In: IEEE International Requirements Engineering Conference, pp. 92–104 (2013)
34. Traum, D.R., Hinkelman, E.A.: Conversation acts in task-oriented spoken dialogue. Comput. Intell. **8**(3), 575–599 (1992)
35. Wagner, S., et al.: Status quo in requirements engineering: a theory and a global family of surveys. ACM Trans. Softw. Eng. Methodol. **28**, 1–48 (2019)

36. Wieringa, R.J.: Design Science Methodology for Information Systems and Software Engineering. Springer, Heidelberg (2014). https://doi.org/10.1007/978-3-662-43839-8
37. Wohlin, C., Runeson, P., Höst, M., Ohlsson, M.C., Regnell, B., Wesslén, A.: Experimentation in Software Engineering. Springer, Heidelberg (2012). https://doi.org/10.1007/978-3-642-29044-2
38. Zowghi, D., Coulin, C.: Requirements elicitation: a survey of techniques, approaches, and tools. In: Engineering and Managing Software Requirements, pp. 19–46. Springer, Heidelberg (2005). https://doi.org/10.1007/3-540-28244-0_2

Ontology-Based Automatic Reasoning and NLP for Tracing Software Requirements into Models with the OntoTrace Tool

David Mosquera[1]([🖂]) [iD], Marcela Ruiz[1] [iD], Oscar Pastor[2] [iD],
and Jürgen Spielberger[1] [iD]

[1] Zürich University of Applied Sciences, Gertrudstrasse 15, 8400 Winterthur, Switzerland
{mosq,ruiz,spij}@zhaw.ch
[2] PROS-VRAIN: Valencian Research Institute for Artificial Intelligence, Universitat Politècnica de València, València, Spain
opastor@dsic.upv.es

Abstract. Context and motivation. Traceability is an essential part of quality assurance tasks for software maintainability, validation, and verification. However, the effort required to create and maintain traces is still high compared to their benefits. **Problem.** Some authors have proposed traceability tools to address this challenge, yet some of those tools require historical traceability data to generate traces, representing an entry barrier to software development teams that do not do traceability. Another common requirement of existing traceability tools is the scope of artefacts to be traced, hindering the adaptability of traceability tools in practice. **Principal ideas.** Motivated by the mentioned challenges, in this paper we propose OntoTraceV2.0: a tool for supporting trace generation of arbitrary software artefacts without depending on historical traceability data. The architecture of OntoTraceV2.0 integrates ontology-based automatic reasoning to facilitate adaptability for tracing arbitrary artefacts and natural language processing for discovering traces based on text-based similarity between artefacts. We conducted a quasi-experiment with 36 subjects to validate OntoTraceV2.0 in terms of efficiency, effectiveness, and satisfaction. **Contribution.** We found that OntoTraceV2.0 positively affects the subjects' efficiency and satisfaction during trace generation compared to a manual approach. Although the subjects' average effectiveness is higher using OntoTraceV2.0, we observe no statistical difference with the manual trace generation approach. Even though such results are promising, further replications are needed to avoid certain threats to validity. We conclude the paper by analysing the experimental results and limitations we found, drawing on future challenges, and proposing the next research endeavours.

Keywords: Traceability · Ontology · NLP · Automatic reasoning · OntoTrace

1 Introduction

Traceability in software development refers to generating, maintaining, and using traces between software artefacts [1, 2]. A trace is a triplet of elements composed of a source

© The Author(s), under exclusive license to Springer Nature Switzerland AG 2023
A. Ferrari and B. Penzenstadler (Eds.): REFSQ 2023, LNCS 13975, pp. 140–158, 2023.
https://doi.org/10.1007/978-3-031-29786-1_10

artefact, a target artefact, and a trace link [2]. Software artefacts vary depending on the software development context and can be of different formats such as: textual requirements, source code, mock-ups, test cases, graphical software models, among others. Keeping such artefacts traced is essential to quality assurance tasks such as software maintainability, validation, and verification [3, 4]. However, in practice, the effort required to trace artefacts outweighs traceability benefits [5]. Thus, some authors have proposed novel approaches, especially for generating traces between software artefacts [5–15]. These proposals have attempted to decrease the effort required for generating traces between artefacts. Yet, some of them depend on historical traceability data—a.k.a. training traceability data—[5, 7, 9, 14], are fixed to specific artefact types [8, 10–13], and lack decision-making support techniques for trace generation [6, 15].

We had previously conceived OntoTrace: a tool for supporting trace generation of arbitrary software artefacts using ontology-based automatic reasoning [15] (see Onto-TraceV2.0 research timeline in Fig. 1). In this paper, we evolve OntoTrace into Onto-TraceV2.0 providing it with a Natural Language Processing (NLP) layer that support decision-making on generating traces between artefacts. Although OntoTrace supports trace generation of arbitrary software artefacts, we scope our research to trace software requirements—i.e., user stories—into software models—i.e., Existence Dependency Graph (EDG) models. Thus, OntoTrace users can use an automatic reasoner together with NLP to infer traceability-related information such as: i) which artefacts are not yet traced; ii) which are the traceable source/target artefacts; and iii) given a specific artefact, which are the possible recommended traces between it and other artefacts based on text-based similarity.

We conducted a quasi-experiment with 36 subjects to validate OntoTraceV2.0 in terms of subjects' efficiency, effectiveness, and satisfaction in the context of the rapid software prototyping course at the Zürich University of Applied Sciences (ZHAW). Experimental results show how OntoTraceV2.0 positively affects the subjects' efficiency and satisfaction during trace generation compared to a manual approach. Although the subjects' average effectiveness is higher using OntoTraceV2.0, we observed no statistical difference with the manual trace generation approach in terms of effectiveness. Even though such results are promising, we identified some validity threats such as maturity, low statistical power, and generality threats that requires further replications to validate our results. Finally, we discuss our conclusions and the subsequent challenges to a complete technology transference.

Fig. 1. OntoTraceV2.0 research timeline and overview.

The paper is structured as follows: in Sect. 2, we review the related works; in Sect. 3, we exhibit the problem scope, main definitions, and exemplify how to configure Onto-Trace for tracing user stories and EDG models [16]; in Sect. 4, we present all new features included in OntoTraceV2.0; in Sect. 5, we show the OntoTraceV2.0 validation results; and, finally, in Sect. 6, we discuss conclusions and future work.

2 Related Works

Automating totally or partially the trace generation in software development has gained researchers' attention. Thus, they have proposed novel traceability tools. Some authors propose tools for generating traces between artefacts based on historical traceability data—a.k.a. training traceability data—such as: artificial neural networks [5, 14], historical-similarity-based algorithms [9], and Bayes classifiers [7]. Although these tools are helpful, they depend on extensive and well-labelled training data sets based on historical traceability data. This represents an entry barrier for software development teams that currently do not trace their artefacts. Other authors propose tools that do not rely on historical traceability data, such as ontology-based recommendation systems [12, 13], expert systems [8], pattern languages [11], and metamodel-based ontologies [10]. However, these tools are limited to generating traces between specific artefacts. Thus, software development teams cannot adapt such tools to their software development traceability needs. For instance, some tools [8, 12, 14] limit their source/target artefacts to text-based artefacts—e.g., source code, standards, and textual requirements. Therefore, non-textual artefacts such as models, UIs, and mock-ups are beyond their scope. Having that in mind, in previous work we have proposed OntoTrace as a tool for generating traces between arbitrary artefacts without the need to rely on historical traceability data [15]. Nevertheless, like the Capra tool proposed in [6], they both lack decision-making support for analysts to decide on which traces need to be generated—i.e., both lack support for recommending which artefacts should be traced.

To address such gaps, we propose to evolve OntoTrace to OntoTraceV2.0: an ontology-based automatic reasoning NLP (Natural Language Processing) tool for generating traces between software artefacts. Like its predecessor, OntoTraceV2.0 does not rely on historical traceability data and is not restricted to a specific set of traceable artefacts. In addition, we combine automatic reasoning with NLP to support decision-making on which traces should be generated between artefacts. Thus, OntoTraceV2.0 is a step forward in improving software trace generation, having such combination as the main technical novelty.

3 Problem Scope

The goal of this paper is: to **analyse** *the OntoTrace tool* **for the purpose of** *supporting software traceability* **with respect to** *effectiveness, efficiency, and satisfaction of Onto-Trace users* **from the point of view of** *the researchers* **in the context of** *software trace generation tasks.* To address this goal, we have taken the following decisions:

- The OntoTrace tool proposed in [15] is founded on general traceability definitions taken from [1, 2, 17], which supports trace generation in any traceability context. We define *traceability context* as the set of SOURCE and TARGET software artefacts to be connected by means of traces. For instance, the traceability context of this paper and the controlled quasi-experiment presented in Sect. 5 is the generation of traces between User Stories [18] as SOURCE and EDG models—a UML-class-diagram-like model [16]—as TARGET software artefacts. The reason is that User Stories and UML models are widely used by software development teams to document software requirements. Moreover, EDG models are supported by teaching and learning tools like Merlin, which are good fit for teaching and experimental purposes [16].

- We define *traceability activity* as any activity involved in the traceability process such as generating, using, and maintaining traces [2].
- We define *OntoTrace user* as any software development team role carrying out a traceability activity using OntoTrace [15].
- The traceability activity that we select for this paper is *trace generation*. Other traceability activities are out of this paper's scope.

Based on these decisions, we propose the following research questions:

RQ1: How to improve OntoTrace to allow for automatic trace recommendations? We consider different NLP techniques [19–22] to provide trace recommendations between artefacts and refactor the OntoTrace architecture [15], reflecting all new features. As a result, we propose OntoTraceV2.0.

RQ2: When the subjects use OntoTraceV2.0, is their effectiveness, efficiency, and satisfaction in establishing traceability links among User Stories and EDG models affected? To answer this question, we conduct a quasi-experiment to compare effectiveness, efficiency, and satisfaction of subjects that did software traceability with OntoTraceV2.0 and the traditional way (without OntoTraceV2.0).

3.1 Traceability Context: Tracing User Stories and EDG Models

In this Section we show the application of the method Ontology101 [23] to establish the traceability context for this paper: *Tracing User Stories [18] as SOURCE and EDG models [16] as TARGET software artefacts*. As a result, we create an ontology based on such traceability context, containing the structure of artefacts and traces. This ontology is the main input for using OntoTrace since it relies on automatic reasoning based on the defined ontological structure. We present a summary with the application of each Ontology101 step, a set of guidelines to specialise each Ontology101 step for establishing the traceability context, and the outcome of applying each guideline (see Table 1).

Having the strategy selected from G11, the context-dependent traceability ontology is ready to be translated into a computational-readable knowledge representation language as OWL (Ontology Web Language [26]) and then used with OntoTrace.

4   Evolving OntoTrace into OntoTraceV2.0[1]

In previous work, we proposed OntoTrace as an ontology-based automatic reasoning trace generation tool [15]. In this Section, we address RQ1 presented in Sect. 3, improving

---

[1] OntoTraceV2.0 code is available here: https://tinyurl.com/4d45utrf.

Table 1. Establishing traceability context

Ontology101 Step (S)	Guidelines (G) for establishing traceability context	Result: traceability context user stories and EDG
S1: Determine the domain and scope of the ontology	**G1.** Specify context-dependent artefacts that require to be traced	User story parts EDG model elements
	G2. Classify the context-dependent artefacts into source and target artefacts	Source: User story parts Target: EDG model elements
S2: Consider reusing existing ontologies	**G3.** Reuse existing metamodels, tools, domain models, syntaxes, documentation, libraries, and vocabulary that describe context-dependent artefacts	We reuse the following ontology and metamodel: - Ontology for User Stories [18] - EDG metamodel [16]
S3: Enumerate important terms in the ontology	**G4.** List terms representing context-dependent source artefacts	User story role, user story action, user story goal, user story object
	G5. List terms representing context-dependent source artefacts	EDG object, EDG attribute, EDG dependency, EDG method
	G6. Specify context-dependent trace properties to link source and target artefacts	We propose the traceability matrix in Table 2 based on literature on transforming user stories into EDG models [24, 25] (EDG can be transformed into UML and vice versa [16]). This traceability matrix represents the context-dependent trace properties based on researchers' [24, 25] and authors' experience
S4: Define the classes and the class hierarchy	**G7.** Use the class hierarchy for defining the *source/target* sub-classes based on the resulting terms from G4 and G5	See Fig. 2

(continued)

Table 1. (*continued*)

Ontology101 Step (S)	Guidelines (G) for establishing traceability context	Result: traceability context user stories and EDG
	G8. Use the class hierarchy from G7 for defining the *trace* sub-classes based on the resulting trace properties from G6. Each sub-class relates to a traceability link defined as follows: ---*Trace hasSource some Source AND Trace hasTarget some Target* Constraint: Define *trace* sub-classes until all possible trace properties resulting from G6 have been covered with at least one trace sub-class	
S5: Define the properties of classes	**G9.** Define context-dependent traceability properties with the following naming: *has + Source/Target + Artefact Name* Constraint: All target/source sub-classes must be related to at least one traceability-related property	Property *hasSourceUserStoryRole*, inheriting from the *hasSource* property Property *hasTargetEDGObject*, inheriting from the *hasTarget* property
S6: Define the facets of the properties	**G10.** Define the range and domain of context-dependent traceability properties as follows: Set the domain as all possible trace sub-classes from G8 that have the source/target artefact as its range Set the range as the source/target artefact of the trace	---*hasSourceUserStoryRole* **property:** this property's domain is a *Trace Between User Story Role and EDG Object* class instance, and its range is *User Story Role* class instances *hasTargetEDGObject* **property:** this property's domain is a *Trace Between User Story Role and EDG Object* class instance, and its range is *EDG Object* class instances

(*continued*)

Table 1. (*continued*)

Ontology101 Step (S)	Guidelines (G) for establishing traceability context	Result: traceability context user stories and EDG
S7: Create instances	**G11.** For each artefact, select one of the following individual instance creation strategies: *Manual:* Artefacts are difficult to access programmatically, such as physical documentation *Automatic*: Artefacts are contained in accessible repositories allowing for programmatic retrieval operations	***Automatic strategy*** for creating source artefacts since User Stories are stored digitally ***Automatic strategy*** for EDG models since they are digital

Table 2. Traceability matrix in our running example: User story parts vs EDG model elements.

Source artefact	EDG: target artefact			
	Object	Attribute	Dependency	Method
User story role	✓			
User story action			✓	✓
User story object	✓	✓		
User story goal		✓	✓	

✓: Traceability link; **EDG**: Existence Dependency Graph.

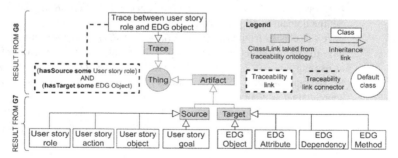

Fig. 2. Excerpt of trace sub-classes, hierarchy, and traceability links of our running example.

OntoTrace to allow for automatic trace recommendation. We show which are the new OntoTraceV2.0 architecture elements in Fig. 3 compared to OntoTrace. Moreover, we describe the new modules in the following paragraphs.

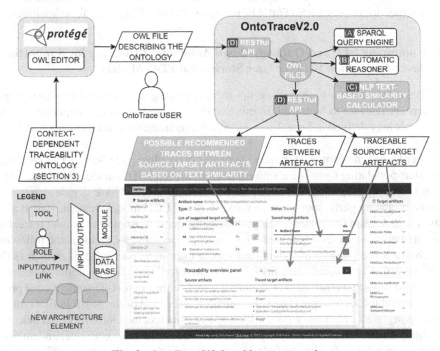

Fig. 3. OntoTraceV2.0 architecture overview

First, OntoTrace user specifies a context-dependent ontology (see Sect. 3.1) and creates an OWL file [26] describing it using an external tool such as Protégé. Then, the OntoTrace user provides this OWL file to OntoTrace to use the following modules:

- **Module A.** OntoTrace provide information about source and target artefacts using a set of SPARQL queries. All the information is retrieved using the context-dependent traceability ontology.
- **Module B.** OntoTrace uses an automatic reasoner together with the SPARQL query engine to answer the following traceability related questions: i) which source/target artefacts are traceable; ii) which are the traces between artefacts; iii) which possible traces exist between source/target artefacts.

Module A and B allow OntoTrace users to store traces between artefacts based on automatic reasoning. Nevertheless, OntoTrace cannot recommend which of the possible source/target artefacts are relevant to be traced. This is problematic, mainly when an artefact can be traced to many different artefacts, motivating us to evolve OntoTrace [15] and propose OntoTraceV2.0. Therefore, we create a new web-based user interface and include the following two modules to OntoTraceV2.0:

- **Module C.** We provide OntoTraceV2.0 with an NLP layer for suggesting traces between artefacts based on their text-based similarity (see Sect. 4.1), addressing the aforementioned gap.
- **Module D.** Now, OntoTraceV2.0 is a web-based tool instead of a standalone tool. OntoTrace users uses a RESTful API to access all OntoTraceV2.0 functionalities.

4.1 Combining NLP and Ontology-Based Automatic Reasoning for Supporting Trace Generation Between User Stories and EDG Models

OntoTrace [15] have a limitation on not recommending which possible source/target artefacts are relevant to be traced. In this Section, we use the context-dependent traceability ontology defined in Sect. 3.1 to exemplify this limitation and show how NLP can solve it.

After having the context-dependent traceability ontology (see Sect. 3.1), OntoTrace users start populating OntoTrace with user story parts and EDG model elements. These artefact instances represent the set of all traceable artefacts A. We divide A into two subsets: user story parts (source artefacts) $S_A \subseteq A$ and EDG model elements (target artefacts) $T_A \subseteq A$. OntoTrace uses ontology-based automatic reasoning to answer traceability related questions, creating subsets of S_A and T_A. Specially, we focus on the following traceability-related question[2]: having selected a user story part $s_a \in S_a$, which is the set of possible EDG model elements $PT_a \subseteq T_a$ to trace? OntoTrace automatic reasoner answers this question creating the PT_a subset based on the context-dependent trace properties (see Table 2). Now, the OntoTrace user can select one of the possible EDG model elements $pt_a \in PT_a$ to create a trace with s_a. . For instance, the OntoTrace user selects a User Story Role Secretary and OntoTrace answers based on the context-dependent trace properties (see Table 2) with the following possible EDG model elements PT_a to trace: EDG Object Aircraft Manager, EDG Object Aircraft, and EDG Object Secretary. Now, the OntoTrace user can select EDG Object Secretary as pt_a to create a trace with the User Story Role Secretary as s_a. We graphically show this example in Fig. 4.

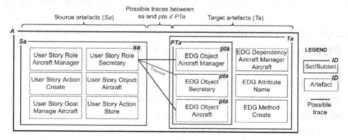

Fig. 4. Previous OntoTrace version automatic reasoning result.

---

[2] Notice that this question can also be written as: having selected a EDG model element $t_a \in T_a$, which is the set of possible user story parts $PS_a \subseteq S_a$ to trace? However, we use the source-to-target variant instead of target-to-source variant for simplicity.

Notice that OntoTrace's automatic reasoner filtered out all target artefacts that the OntoTrace user must not trace to a User Story Role, such as EDG Dependencies, Methods, and Attributes. However, the OntoTrace user still needs to decide which EDG model element pt_a from the PT_a subset is the correct one to trace. Whether all possible EDG model elements $pt_a \in PT_a$ are equally valid is a problem that limits the scope of the automatic reasoner for trace generation. To address such a problem, we provide OntoTraceV2.0 with an NLP layer to recommend which EDG model element pt_a is relevant to be traced to a selected user story part s_a based on text-based similarity. OntoTraceV2.0's NLP layer comprises three sub-layers: extracting artefacts' text data, processing extracted text, and calculating the similarity between artefacts. As a result, OntoTraceV2.0 provide a similarity value with the possible traces between artefacts. We show how OntoTraceV2.0 transforms ontology-based automatic reasoning output using the NLP layer in Fig. 5.

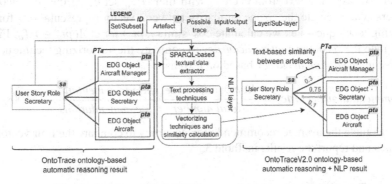

Fig. 5. OntoTraceV2.0 NLP layer and sublayers explained.

In the first sub-layer, we extract the textual data from the selected user story part s_a and all possible EDG model elements $pt_a \in PT_a$. We gather artefacts' relevant data from s_a and each $pt_a \in PT_a$ using SPARQL [27] queries. Then, we transform the information retrieved by the SPARQL queries into a textual description as input for the next sub-layer. In the second sub-layer, we process the textual description resulting from last layer. We apply text processing techniques [22], such as removing punctuation, lowercasing, tokenization, stop word removal, and lemmatization. In the third sub-layer, we receive the processed text and calculate the cosine similarity between the angle of the s_a and pt_a vectors [19], having as a result a text-based similarity value between 0 to 1. Then, we provide the possible traces between the selected user story part s_a and all possible EDG model elements $pt_a \in PT_a$ with the text-based similarity value. Finally, OntoTraceV2.0 recommend tracing s_a to a $pt_a \in PT_a$ if the calculated text-based similarity is higher or equal to a recommendation threshold. We show a detailed example on how this text go through all three NLP sub-layers in Fig. 6.

So far, we briefly discussed the technical details of each NLP sub-layer. In this paper, we specially focus on the third sub-layer's vectorizing techniques [20, 21] and how such techniques affect the OntoTraceV2.0 recommendation accuracy. To do so, we test four vectorizing techniques: Count Vectorizer, TFIDF Vectorizer, Doc2Vec,

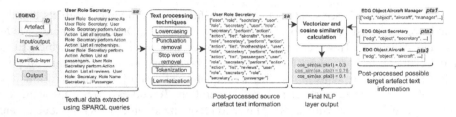

Fig. 6. Detailed NLP layer example.

and Universal Sentence Encoder. Count and TFIDF Vectorizers are mathematical-based techniques for vectorizing text, using word frequency to create a vector representation [21]. The Doc2Vec and the Universal Sentence Encoder are machine-learning-based techniques for vectorizing text using word embeddings [20, 21]. To compare them, we gather and process the text from each s_a with their PT_a set using the first two NLP sub-layers. We calculate the cosine similarity between vector representations for each vectorizing technique—i.e., we calculate cos_sim(s_a, pt_a) for each $pt_a \in PT_a$. Finally, we calculate the *recommendation accuracy* to compare the vectorizing techniques. We define *recommendation accuracy* based on [28] as:

$$Recommendation\ accuracy = \frac{Number\ of\ successful\ recommendations}{Total\ number\ of\ recommendations} * 100\% \quad (1)$$

We calculate the average recommendation accuracy to compare the four vectorizing techniques and report the results in Table 3.

Table 3. Vectorizing techniques and their recommendation accuracy.

Vectorizing technique	Recommendation threshold	Recommendation accuracy (AVG)
Count vectorizer	0.9025	71.80%
TFIDF vectorizer	0.9139	75.89%
Doc2Vec	0.9965	20.21%
Universal Sentence Encoder	0.9625	60.37%

We observe that the TFIDF vectorizer has the highest recommendation accuracy in average among vectorizing techniques. Therefore, we select the TFIDF vectorizer to implement the third sub-layer vectorizing technique. For the sake of space, we include a detailed description on how we designed the SPARQL queries, implemented the text processing techniques, and selected the vectorizing techniques and similarity calculation method in an annexe repository[3]. However, evaluating the effect on recommendation accuracy with different similarity calculation formulas, text processing techniques, SPARQL queries, and other NLP techniques—e.g., using transformer models—is still work in progress.

---

[3] https://doi.org/10.5281/zenodo.7589791.

5   Evaluating OntoTraceV2.0

We have conducted a quasi-experiment to measure the extent OntoTraceV2.0 affects trace generation effectiveness, efficiency, and satisfaction. We design and execute this quasi-experiment based on Wohlin et al. [29] and Moody's [30] Technology Acceptance Model (TAM), addressing RQ2 presented in Sect. 3. Our quasi-experiment is fixed to User Stories and EDG models. However, we consider answering RQ2 as the first step for future experiment replications with other artefacts and as a first step to find general conclusions.

5.1   Experimental Design

The experimental goal according to the Goal/Question/Metric template [31] is to **analyse the use of** *OntoTraceV2.0* **for the purpose of** *trace generation between user stories and EDG models* **with respect to** *effectiveness, efficiency, and satisfaction* **from the point of view of** *engineering bachelor students* in **the context of** *a bachelor course on rapid software prototyping (RASOP) at the ZHAW in Switzerland.*

Experimental Subjects. The quasi-experiment was conducted with 36 subjects, all of them engineering students enrolled in the RASOP course. The subjects are part of diverse engineering programs such as: Information technology (IT; 36.1%), Computer Sciences (30.6%), Industrial engineering (8.3%), Systems engineering (5.6%), Mechanical engineering (5.6%), Business engineering (5.6%), Aviation (2.8%), Electrical engineering (2.8%), and Energy and Environmental Engineering (2.8%). More than a half (58.3%) of the subjects have between 0.5 to 10 years of industry experience (2.8 years average $\pm$ 3.1 years std) in field such as software engineering, data mining, mechanics, and semi-conductor industry, among others. However, only one subject (2.8%) has one year of previous experience on software traceability. The other subjects (97.2%) have no previous experience on software traceability. Subjects were informed about data collection, and they executed the experimental tasks as part of the course graded activities. Nevertheless, we inform them that there are no direct benefits in the grades to let us collect their data.

Variables. We consider *one independent variable*: generating traces with and without OntoTraceV2.0. On the other hand, we consider *three independent variables* grouped by effectiveness, efficiency, and satisfaction based on Moody's evaluation model [30]. For effectiveness, we decide to measure subject's precision during trace generation. For efficiency, we plan to measure subject's number of generated traces per minute. For satisfaction, we propose to measure three qualitative variables based on a 1-to-5 Likert scale: Perceived ease of use (PEU), perceived usefulness (PU), and Intention to Use (ITU).

Hypotheses. We define null hypotheses (represented by a 0 in the subscript) stating that OntoTraceV2.0 do not affect the trace generation effectiveness, efficiency, and satisfaction. The alternative hypotheses (represented by a 1 in the subscript) suppose there is an influence. We show our hypotheses in Table 4, alternative hypotheses are omitted.

Table 4. Null hypothesis (H_0) description.

H_0	**Statement:** The use of OntoTraceV2.0 does not affect the subject's...
$H1_0$	...**effectiveness** when generating traces between user stories and EDG models
$H2_0$	...**efficiency** when generating traces between user stories and EDG models
$H3_0$	... **PEU** when generating traces between user stories and EDG models
$H4_0$	... **PU** when generating traces between user stories and EDG models
$H5_0$	... **ITU** when generating traces between user stories and EDG models
PEU: Perceived Ease of Use; **PU**: Perceived Usefulness; **ITU**: Intention to Use	

5.2 Procedure and Data Analysis[4]

We conducted the quasi-experiment following a blocked subject-object study having one factor with two treatments experimental design [29]. Hence, we propose randomly dividing subjects into two balanced groups: GR1 and GR2. Both groups received training on traceability during the RASOP lectures. We design two experimental objects (O1 and O2) that both GR1 and GR2 will face in two different sessions. Subjects receive a set of user stories as source artefacts and an EDG as target artefact, having as a task generating the traces between artefacts. Moreover, source and target artefacts are previously labelled (see Fig. 7).

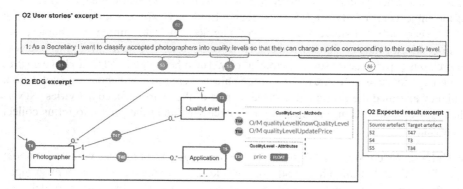

Fig. 7. O2 experimental object excerpt. **S#**: source artefact; **T#**: target artefact.

Before each session, we introduce the experimental task using an experimental training object (O0). Data from O0 is not collected nor evaluated since it is just for training subjects. During the first session, GR1 works on O1 and GR2 works on O2, both groups without using OntoTraceV2.0—i.e., using a manual traceability strategy. During the second session, GR1 works on O2 and GR2 works on O1, but now using OntoTraceV2.0.

---

[4] To facilitate further replications, all material related to the experimental objects, demographics, and results can be found at https://doi.org/10.5281/zenodo.7360221.

We decide to evaluate OntoTraceV2.0 against manual traceability rather than Onto-TraceV1.0 since OntoTraceV2.0 contains all features from OntoTraceV1.0, allowing us to assess not only the new NLP feature but also the ontology-based automatic reasoning feature. At the end of each session, we ask subjects to provide us with the ending time, the generated traces, and a satisfaction questionnaire.

Using the previously discussed configuration, quasi-experiment findings are not entirely dependent on the experimental object since we use two experimental objects. Moreover, we avoid the between-session experimental object learning effect since subjects work on different experimental objects in each session. Furthermore, session 1 and session 2 were performed with a one-week time difference, decreasing the effect on satisfaction variables by the time between sessions. However, we could not prevent a between-task learning effect—i.e., even if we did not reveal the correct results between sessions, subjects learn how to perform the traceability task from session 1 and use that knowledge in session 2—due to time and infrastructure limitations. As a disclaimer, such a learning effect can affect effectiveness and efficiency metrics, requiring further replications to validate our results. We deeply discuss this and other threats to validity in more detail in Sect. 5.3. Finally, all subjects are used in both sessions, avoiding variability among subjects.

Data Analysis. We analyse the descriptive statistics, comparing means of dependent variables (see Fig. 8). Moreover, we run a generalised linear model to test the hypothesis (see Table 5).

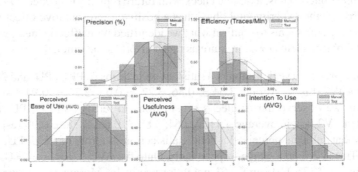

Fig. 8. Quasi-experiment results' distributions, having y-axis as the probability density.

Effectiveness. We observe subject's effectiveness in terms of precision without Onto-TraceV2.0 is in average $73.99\% \pm 15.77\%$ compared to $81.74\% \pm 15.49\%$ with Onto-TraceV2.0. This means that subjects identify 7.75% more correct traces with Onto-TraceV2.0 compared to a manual strategy on average. However, we observe that Onto-TraceV2.0 has no statistical representative effect into subject's precision. Similarly with the interaction between OntoTraceV2.0 and the experimental object. Therefore, we cannot reject $H1_0$. On the other hand, we observe the experimental object has a negative representative effect into subject's precision. This could indicate that one experimental object is more challenging that the other.

Table 5. Statistical generalized linear model test results.

Independent variables	Dependent variables				
	Precision	Efficiency (Traces/min)	PEU (AVG)	PU (AVG)	ITU (AVG)
OntoTraceV2.0	0.0165 (0.0539)	0.670*** (0.230)	0.489* (0.267)	0.649*** (0.231)	0.600* (0.331)
Experimental Object	−0.099* (0.0513)	0.807*** (0.219)	0.256 (0.255)	0.246 (0.221)	−0.167 (0.316)
OntoTraceV2.0 & Experimental Object	0.121 (0.0749)	−0.783** (0.319)	−0.0438 (0.372)	−0.105 (0.322)	0.0569 (0.461)

Standard errors in parentheses; ***: $p < 0.01$; **: $p < 0.05$, *: $p < 0.1$.

Efficiency. We observe subject's efficiency in terms of trace/min without Onto-TraceV2.0 is in average 1.44 traces/min $\pm$ 0.71 traces/min compared to 1.72 traces/min $\pm$ 0.71 traces/min with OntoTraceV2.0. This means that subjects create 0.28 traces/min (16.8 traces/hour) faster compared to a manual strategy on average. Moreover, we observe that OntoTraceV2.0 has a positive statistical representative effect into subject's efficiency. Similarly with interaction between OntoTraceV2.0 and the experimental object. Therefore, we reject $H2_0$ with a 99% of confidence. As a disclaimer, this result could be due to between-task learning validity threat—i.e., a maturity threat—as we previously mentioned. Thus, a double check with future replicas is needed. On the other hand, we observe the experimental object has a positive representative effect into subject's efficiency. This seems to confirm what we identified with effectiveness, where one experimental object seems to not require as much effort as the other.

Satisfaction. We observe that subject's satisfaction in terms of PEU, PU, and ITU without OntoTraceV2.0 is in average 3.64 $\pm$ 0.86, 3.27 $\pm$ 0.63, 2.75 $\pm$ 0.94 respectively. Furthermore, we observe that subject's satisfactions in terms of PEU, PU, and ITU with OntoTraceV2.0 is in average 4.11 $\pm$ 0.64, 3.88 $\pm$ 0.69, 3.38 $\pm$ 0.93 respectively. This means that subjects perceived a better satisfaction in terms of PEU, PU, and ITU using OntoTraceV2.0 on average—specifically, OntoTraceV2.0 increase PEU, PU, and ITU on average 0.47, 0.61, and 0.63 points respectively. Moreover, we observe that OntoTraceV2.0 has a positive statistical representative effect into PEU, PU, and ITU. Therefore, we reject $H3_0$, $H4_0$, and $H5_0$ with a 90%, 99%, and 90% of confidence respectively. On the other hand, we observe that the experimental object nor the interaction between experimental object and OntoTraceV2.0 has statistical representative effect into subject's satisfaction in terms of PEU, PU and ITU.

5.3 Threats to Validity

Internal Validity. GR1 and GR2 group subjects could share information about their experimental objects, materializing a *diffusion* threat. Due to that, we prepared two versions of our experimental objects O1.1, O2.1 and O1.2, O2.2. Thus, we minimize the effect of *diffusion* about experimental objects between sessions since subjects always face new experimental objects. In terms of *maturity*, subjects were able to improve their tracing skills between sessions affecting their efficiency and effectiveness. To minimise this threat, we do not reveal the results of their performance until the end of the second session, avoiding subjects learn from the first session results. However, subjects still could learn how to generate traces between session 1 and session 2—e.g., subjects could learn how to trace more efficiently even if they do it in a wrong way because they do not know the correct result. In our quasi-experiment, we did not assess this *maturity* threat that could affect especially effectiveness and efficiency. We plan to verify such results in further experiment replications.

External Validity. We involved students from different engineering bachelor programs as experimental subjects. This could represent an *interaction of selection and treatment* threat where subjects are not representative of the population we want to generalize. However, all students participating in RASOP lecture are interested on software development and software quality assurance tasks such as traceability based on RASOP syllabus. Thus, we minimize considering RASOP students as potential population to use OntoTraceV2.0. Nevertheless, we acknowledge the limits in the *generalization of the experiment results* since we did not include other subjects of interest such as traceability experts. We plan to replicate this quasi-experiment to generalize our results, including traceability experts and software development teams working in industry.

Construct Validity. Subjects could be afraid of being evaluated affecting their results, materializing an *evaluation apprehension* threat. We minimize this threat letting subjects know that all data is anonymous, and no benefit/penalization is made for letting us collect their data.

Conclusion Validity. Although we conducted a quasi-experiment with 36 subjects, the sample size is still small. This represents a *low statistical power* threat. To mitigate this threat in the future, we plan to replicate this quasi-experiment increasing the sample size. Moreover, there are external *experimental setting* threats we could not mitigate that can affect the experiment results. For instance, RASOP lecture is scheduled from 17:45 to 21:00. During the evening subjects are tired and that can affect their results.

6 Conclusions and Future Work

In this paper, we propose OntoTraceV2.0: an ontology-based automatic reasoning and NLP-based tool for generating traces between software artefacts. OntoTraceV2.0 is built on top of previous work by including an NLP layer for supporting decision-making on generating traces between artefacts—i.e., for recommending traces between artefacts. Then, OntoTraceV2.0 users can use an automatic reasoner together with NLP to infer traceability-related information such as: i) which artefacts are not yet traced; ii) which are the traceable source/target artefacts; and iii) given a specific artefact, which are the possible recommended traces between it and other artefacts based on text-based similarity.

We conducted a quasi-experiment with 36 subjects to analyse OntoTraceV2.0 effect on effectiveness, efficiency, and satisfaction on trace generation. We observed Onto-TraceV2.0 positively affects the subjects' efficiency and satisfaction during trace generation compared to a manual approach. However, although the subjects' average effectiveness is higher using OntoTraceV2.0, we observed no statistical difference with the manual trace generation approach in terms of effectiveness. The lack of significant effect in terms of effectiveness is a limitation. This indicates we still need to improve OntoTraceV2.0 trace recommendation techniques. In the future, we will improve trace recommendations by devising new techniques, combining NLP and machine learning algorithms. In addition, we identified some threats to validity that can affect our results, especially in terms of effectiveness and efficiency. We plan to replicate this quasi-experiment having in mind threats to validity such as *maturity, low statistical power,* and *generalisation of experimental results* to validate our results.

Acknowledgments. This research is fully funded by the ZHAW Institute for Applied Information Technology (InIT), the Innosuisse Flagship SHIFT project, and the ZHAW School of Engineering. Moreover, we would like to thank all RASOP course students for actively participating on the quasi-experiment, allowing us to gather all the data we used to build our research.

References

1. Charalampidou, S., Ampatzoglou, A., Karountzos, E., Avgeriou, P.: Empirical studies on software traceability: a mapping study. J. Softw. Evol. Process **33** (2021)
2. Cleland-Huang, J., Gotel, O., Zisman, A.: Software and Systems Traceability. Springer, London (2012)
3. Antoniol, G., Canfora, G., de Lucia, A.: Maintaining traceability during object-oriented software evolution: a case study. In: IEEE International Conference on Software Maintenance - 1999 (ICSM 1999), pp. 211–219 (1999)
4. Sundaram, S.K., Hayes, J.H., Dekhtyar, A., Holbrook, E.A.: Assessing traceability of software engineering artifacts. Requir Eng. **15**, 313–335 (2010)
5. Lin, J., Liu, Y., Zeng, Q., Jiang, M., Cleland-Huang, J.: Traceability transformed: generating more accurate links with pre-trained BERT models. In: 2021 IEEE/ACM 43rd International Conference on Software Engineering (ICSE), pp. 324–335. IEEE (2021)
6. Maro, S., Steghofer, J.-P.: Capra: a configurable and extendable traceability management tool. In: 2016 IEEE 24th International Requirements Engineering Conference (RE), pp. 407–408. IEEE (2016)
7. Nagano, S., Ichikawa, Y., Kobayashi, T.: Recovering traceability links between code and documentation for enterprise project artifacts. In: 2012 IEEE 36th Annual Computer Software and Applications Conference, pp. 11–18. IEEE (2012)
8. Guo, J., Cleland-Huang, J., Berenbach, B.: Foundations for an expert system in domain-specific traceability. In: 2013 21st IEEE International Requirements Engineering Conference (RE), pp. 42–51. IEEE (2013)
9. Javed, M.A., UL Muram, F., Zdun, U.: On-Demand automated traceability maintenance and evolution. In: Capilla, R., Gallina, B., Cetina, C. (eds.) ICSR 2018. LNCS, vol. 10826, pp. 111–120. Springer, Cham (2018). https://doi.org/10.1007/978-3-319-90421-4_7
10. Narayan, N., Bruegge, B., Delater, A., Paech, B.: Enhanced traceability in model-based CASE tools using ontologies and information retrieval. In: 2011 4th International Workshop on Managing Requirements Knowledge, pp. 24–28. IEEE (2011)

11. Javed, M.A., Stevanetic, S., Zdun, U.: Towards a pattern language for construction and main-tenance of software architecture traceability links. In: Proceedings of the 21st European Conference on Pattern Languages of Programs, pp. 1–20. ACM, New York (2016)

12. Huaqiang, D., Hongxing, L., Songyu, X., Yuqing, F.: The research of domain ontology rec-ommendation method with its applications in requirement traceability. In: 2017 16th Interna-tional Symposium on Distributed Computing and Applications to Business, Engineering and Science (DCABES), pp. 158–161. IEEE (2017)

13. Hayashi, S., Yoshikawa, T., Saeki, M.: Sentence-to-code traceability recovery with domain ontologies. In: 2010 Asia Pacific Software Engineering Conference, pp. 385–394. IEEE (2010)

14. Guo, J., Cheng, J., Cleland-Huang, J.: Semantically enhanced software traceability using deep learning techniques. In: 2017 IEEE/ACM 39th International Conference on Software Engineering (ICSE), pp. 3–14. IEEE (2017)

15. Mosquera, D., Ruiz, M., Pastor, O., Spielberger, J., Fievet, L.: OntoTrace: a tool for supporting trace generation in software development by using ontology-based automatic reasoning. In: De Weerdt, J., Polyvyanyy, A. (eds.) CAiSE 2022, pp. 73–81. Springer, Cham (2022). https://doi.org/10.1007/978-3-031-07481-3_9

16. Snoeck, M.: Enterprise Information Systems Engineering. Springer, Cham (2014)

17. Guo, J., Monaikul, N., Cleland-Huang, J.: Trace links explained: an automated approach for generating rationales. In: 2015 IEEE 23rd International Requirements Engineering Conference (RE), pp. 202–207. IEEE (2015)

18. Thamrongchote, C., Vatanawood, W.: Business process ontology for defining user story. In: 2016 IEEE/ACIS 15th International Conference on Computer and Information Science (ICIS), pp. 1–4. IEEE (2016)

19. Li, B., Han, L.: Distance weighted cosine similarity measure for text classification. In: Yin, H., et al. (eds.) IDEAL 2013, pp. 611–618. Springer, Heidelberg (2013). https://doi.org/10.1007/978-3-642-41278-3_74

20. Cer, D., Yang, Y., et al: Universal Sentence Encoder (2018)

21. Singh, L.: Clustering text: a comparison between available text vectorization techniques. In: Reddy, V.S., Prasad, V.K., Wang, J., Reddy, K.T.V. (eds.) Soft Computing and Signal Processing. AISC, vol. 1340, pp. 21–27. Springer, Singapore (2022). https://doi.org/10.1007/978-981-16-1249-7_3

22. Hickman, L., Thapa, S., Tay, L., Cao, M., Srinivasan, P.: Text preprocessing for text mining in organizational research: review and recommendations. Organ. Res. Methods **25**, 114–146 (2022)

23. Noy, N.F., McFuiness, D.L.: Ontology Development 101: A Guide to Creating Your First Ontology. https://protege.stanford.edu/publications/ontology_development/ontology101.pdf. Accessed 29 Nov 2021

24. Bragilovski, Maxim, Dalpiaz, Fabiano, Sturm, Arnon: Guided derivation of conceptual mod-els from user stories: a controlled experiment. In: Gervasi, Vincenzo, Vogelsang, Andreas (eds.) REFSQ 2022. LNCS, vol. 13216, pp. 131–147. Springer, Cham (2022). https://doi.org/10.1007/978-3-030-98464-9_11

25. Nasiri, S., Rhazali, Y., Lahmer, M., Chenfour, N.: Towards a generation of class diagram from user stories in agile methods. Procedia Comput. Sci. **170**, 831–837 (2020)

26. Web Ontology Language (OWL). https://www.w3.org/OWL/. Accessed 29 Nov 2021

27. SPARQL query language. https://www.w3.org/2001/sw/wiki/SPARQL. Accessed 29 Nov 2021

28. Fayyaz, Z., Ebrahimian, M., Nawara, D., Ibrahim, A., Kashef, R.: Recommendation systems: algorithms, challenges, metrics, and business opportunities. Appl. Sci. **10**, 7748 (2020)

29. Wohlin, C., Runeson, P., Höst, M., Ohlsson, M.C., Regnell, B., Wesslén, A.: Experimentation in Software Engineering. Springer, Heidelberg (2012)

30. Moody, D.L.: The method evaluation model: a theoretical model for validating information systems design methods. In: ECIS 2003 Proceedings, pp. 79–96 (2003)
31. van Solingen, R., Basili, V., Caldiera, G., Rombach, H.D.: Goal Question Metric (GQM) approach. In: Encyclopedia of Software Engineering. Wiley, Hoboken (2002)

Requirements Classification Using FastText and BETO in Spanish Documents

María-Isabel Limaylla-Lunarejo[1]📷, Nelly Condori-Fernandez[2,3(✉)]📷, and Miguel R. Luaces[1]📷

[1] Fac. Informática, Database Lab., Universidade da Coruña, CITIC, A Coruña, Spain
{maria.limaylla,miguel.luaces}@udc.es
[2] CITIUS, Universidad de Santiago de Compostela, Santiago, Spain
n.condori.fernandez@usc.es
[3] Vrije Universiteit Amsterdam, Amsterdam, The Netherlands
n.condori-fernandez@vu.nl

Abstract. *Context and motivation*: Machine Learning (ML) algorithms and Natural Language Processing (NLP) techniques have effectively supported the automatic software requirements classification. The emergence of pre-trained language models, like BERT, provides promising results in several downstream NLP tasks, such as text classification. *Question/problem*: Most ML/DL approaches on requirements classification show a lack of analysis for requirements written in the Spanish language. Moreover, there has not been much research on pre-trained language models, like fastText and BETO (BERT for the Spanish language), neither in the validation of the generalization of the models. *Principal ideas/results*: We aim to investigate the classification performance and generalization of fastText and BETO classifiers in comparison with other ML/DL algorithms. The findings show that Shallow ML algorithms outperformed fastText and BETO when training and testing in the same dataset, but BETO outperformed other classifiers on prediction performance in a dataset with different origins. *Contribution*: Our evaluation provides a quantitative analysis of the classification performance of fastTest and BETO in comparison with ML/DL algorithms, the external validity of trained models on another Spanish dataset, and the translation of the PROMISE NFR dataset in Spanish.

Keywords: Spanish requirements · Automatic classification requirements · fastText · BETO

1 Introduction

Requirements specifications are significant activities that contribute to the success of a software project [16]. Manual classification of requirements is a complex and time-consuming task due to a large number of requirements, the unavailability of experts, or the lack of specification documents [3]. This complexity can

© The Author(s), under exclusive license to Springer Nature Switzerland AG 2023
A. Ferrari and B. Penzenstadler (Eds.): REFSQ 2023, LNCS 13975, pp. 159–176, 2023.
https://doi.org/10.1007/978-3-031-29786-1_11

be reduced using automatic text classification techniques, whose application has increased rapidly in recent years due to the emergence of ML algorithms [19] and NLP techniques [46]. However, it is the recent expansion of the Deep Neural Networks and the introduction of the Transfer Learning concept that has allowed the creation of pre-trained models, i.e. models with a pre-training phase to capture knowledge from several tasks that can be used later for target task with limited data [13]. Pre-trained language models, pre-trained models for NLP, have become beneficial for downstream NLP tasks, e.g. question answering, sentiment analysis, and summarization [31].

Several studies have used and analyzed various ML techniques for classifying software requirements. For instance, Abad et al. [1] proposed a pre-processing approach for improving the performance of some existing ML algorithms. Kurtanovic and Maalej [22] investigated how accurately they can automatically classify requirements employing lexical and syntactical features, and how to handle the imbalanced classes employing sampling strategies. fastText and BERT, pre-trained language models, have been used also in requirements classification [15,23,39], with optimistic results. Despite this interest in the use of ML techniques for classifying requirements, there is still a lack of analysis for the requirements written in Spanish. Spanish is currently the second native language of the world by the number of speakers, the second language by published scientific documents, and the third language by Internet usage after English and Chinese [17]. In the literature, we have found scarce research exploiting requirements written in Spanish. For example, Apaza et al., [4] propose a hybrid model for requirements elicitation, and De Arriba et al., [6] created a Spanish Twitter dataset for applying sentiment analysis). Moreover, standards like ISO/IEC 25010 are also available in Spanish. These initiatives suggest the increasing importance of the Spanish language in RE.

In this paper, we investigate the performance and generalization of two pre-trained language models, fastText and BETO, in comparison with ML and Deep Learning (DL) algorithms for classifying Spanish requirements. We also analyze the impact of the use of fastText and BETO embeddings in combination with the Convolutional Neural Networks (CNN) algorithm. Two Spanish datasets were used in the experiments: first, the PROMISE NFR dataset [38] translated into Spanish, and second, a Spanish requirements dataset presented in [27]. We used this last dataset to confirm the external validity of the models trained on translated PROMISE NFR. The main contributions of this paper are:

- A quantitative analysis of performance metrics to compare the fastText and BETO classifiers with ML/DL algorithms for requirements classification.
- Analyzing the external validity of trained models with a translated version of PROMISE NFR on another Spanish dataset.
- The translation of the PROMISE NFR dataset in Spanish.
- The use of fastText and BETO embeddings for the CNN algorithm.

The paper has been organized in the following way. Section 2 presents the related works on automatic requirements classification. Section 3 presents the research questions and the corresponding methodology used in this research.

Section 4 reports the experiments carried out and their results. Section 5 discusses the results from the previous section. The most relevant threats to validity are presented in Sect. 6. Finally, we conclude the paper in Sect. 7.

2 Related Work

Previous studies have described the use of ML algorithms to classify requirements. Many ML algorithms have been used for the classification of functional and non-functional requirements (FR/NFR) [12,15,22]. Table 1 shows the result of a review that we have performed of several studies related to FR/NFR classification. We found that some studies reported the use of the Shallow ML algorithms[1], like Naive Bayes (NB), Decision Tree (DT), and Logistic Regression (LR); and the use of DL algorithms. Dias and Cordeiro (2020) [12] present a comparison to determine the best combination of some Shallow ML algorithms with text vectorization techniques like Bag of Words (BoW), Chi-Squared (χ^2)), and Term Frequency and Inverse Document Frequency (TF-IDF), to classify requirements. For FR/NFR classification, the combination of TF-IDF and LR gave the best results. Support Vector Machine (SVM) gets an f1-score of 0.93 and 0.92 for FR/NFR classification in [22]. A decision tree algorithm was used for FR/NFR classification in [1], with a processed dataset (i.e. using Part of Speech and Regular Expressions for identifying features instead of using of text vectorization). A comparison between the Random Forest algorithm and the gradient boosting algorithm was performed in [24].

Algorithms based on Neural-networks also have been used recently in the Requirements Engineering (RE) process. Some of them are Recurrent Neural Networks (RNN), CNN, and pre-trained models, e.g. Transformers. CNN achieves an f1-score of 0.77 [30] for FR/NFR classification. Hey et al. (2020) [15] introduce NoRBERT, a language model based on Transformers, which fine-tunes BERT (a pre-trained model used for transfer learning). NoRBERT achieves an f1-score of 0.90 and 0.93 for FR/NFR classification. Tiun et al.(2020) investigate the performance using a complex neural classifier, finding an f1-score value of 0.928 using fastText [39]. The DBGAT model, a combination of BERT and graph attention network, was presented in [23]. In comparison with several algorithms, DBGAT achieved f1-scores of up to 0.91. A BiGru model obtained an f1-score of 0.94 in [2]. Finally, an ensemble of some DL algorithms was presented in [32].

In this review, we found that there is a lack of automatic classification and labeled datasets in other languages different from English. Even though some studies have presented methods to generate classification models trained in English for use in a target language, (e.g. Cross-Language Text Classification [45]), some studies show that models trained and tested in native language outperform better in text classifications [5,44]. However, no relevant research on the automatic F/NF classification of requirements written in Spanish has been found. Another finding is that considering pre-trained models are still new, their

---

[1] defined by [18]: The *Shallow ML* contains simple artificial neural networks and other ML algorithms such as Support Vector Machine, Logistic Regression.

Table 1. Summary of studies related to FR/NFR classification

Ref.	Algorithms	D.	NLP techniques	Best results
[12]	SVM, KNN, MNB, LR	B	BoW/TF-IDF χ^2)	LR with TF-IDF achieve an f1-score of 0.91
[1]	C4.5 DT (J48 in Weka)	A	POS and entity tagging	Processed dataset: average precision of 0.95 and recall of 0.94
[22]	SVM	A	Use of lexical features	SVM achieve an f1-score of 0.93 (FR) and 0.92 (NFR)
[32]	SVM, KNN	B	BoW	SVM achieve an f1-score of 0.90 and KNN an f1-score of 0.82
[24]	Random Forest, Gradient Boosting	C	–	RF achieves an accuracy of 0.826
[30]	CNN	A	Word Embedding	CNN achive an average f1-score of 0.77, precision 0.81 and recall 0.785 for all classes (12)
[15]	NoRBERT (Transfer Learning)	A	BERT (Transformers)	BERT achive an f1-score of 0.90 (FR) and 0.93 (NFR)
[33]	Ensemble DL (LSTM, BiLSTM, GRU, and CNN)	A	Word Embedding	CNN achieves an f1-score of 0.93. The ensemble model achieves an f1-score of 0.96 per class as a weight ensemble, and 0.94 as mean ensemble
[39]	NB, LG, SVM, CNN, fastText	A	Doc2Vec and Word2Vec	fastText achieves an f1-score of 0.928
[23]	BERT and GAT	A, C	–	DBGAT achieve an f1-score of 0.92 (FR) and 0.96 (NFR)
[2]	BiGRU	A	Word Embedding	BiGRU with word model achieve an f1-score of 0.94

Legend: Ref: Reference, D: Datasets, A: PROMISE, B: PROMISE_exp [26], C: Others.

use is increasing in RE due to their promising results. Moreover, we found that most of the studies use the PROMISE NFR dataset, or a newer version, for training and testing, which does not allow us to determine the level of generalization that these approaches have.

3   Research Design

This section explains the research design used in this work. The goal of this paper is to compare the classification performance and generalization of fastText and BETO models with conventional algorithms using two Spanish datasets.

3.1   Research Questions and Metrics

- **RQ1**: How do the fastText and BETO classifiers with other ML/DL algorithms compare in performance using the PROMISE NFR dataset translated into Spanish?
- **RQ2**: How does the use of fastText and BETO embeddings in combination with the CNN algorithm affect the classification results?
- **RQ3**: To which extent the performance of our classification models trained with a Spanish version of PROMISE NFR differ from those trained with the original PROMISE NFR dataset?

- **RQ4**: Is the classification performance of the models trained using the translated PROMISE NFR preserved when tested on another dataset?

From our research questions (RQs), we identified the *classification performance* as our dependent variable and the requirements in the datasets as the independent variable. The most commonly used performance metrics for classification problems are accuracy (A), precision (P), recall (R), and f1-score (F_1).

3.2 Datasets

Datasets are an essential component of any ML experiment. In our research, we used two datasets: the PROMISE NFR dataset translated into Spanish, and a Spanish requirements dataset (called the second dataset).

PROMISE [38] is a repository used in most requirements classification research. It has 84 datasets, and the one used for requirement classification is NFR, with 625 requirement sentences, 255 requirements identified as functional and 370 as non-functional. For this research, we downloaded the *PROMISE NFR dataset* from [8] and translated it into Spanish. We decided to perform a translation of this dataset for two main reasons. First, this dataset has been used as a benchmark by many classification approaches. And second, even if there is a certain imbalance of sub-classes at the level of NFR, the classes are better balanced only considering a binary classification (FR/NFR). The translation of the PROMISE NFR dataset into Spanish was performed with the help of the Google Translate tool[2], a state of art machine translation software, through the Python library deep-translator[3]. Google Translate was chosen as a translation tool because it is free, one of the most popular, and is more accurate when both languages are similar [43]. Then the translation was reviewed, line by line, by the first author, performing some adjustments. In this translation process, the appearance of the English language in some requirements was changed to the Spanish language. However, the proper names and acronyms were preserved. For example, "MDI form" was translated as "Formulario MDI", and "Dr Susan Poslusny" as "La Dra. Susan Poslusny". Punctuation was also added to some requirements where it was missing. Finally, the labels were preserved, like some names and versions of some browsers, identified as old (e.g. "Internet Explorer 5"). Of the 625 records, one was removed due to duplication, resulting in 254 functional requirements and 370 non-functional. The translated dataset and the code used in this research are published in Zenodo[4].

The *second dataset* is the one presented in [27], and consists of 389 requirements, collected from final degree projects from the University of A Coruña, and has 300 requirements labeled as functional and 89 requirements labeled as non-functional.

---

[2] https://translate.google.es/.
[3] https://github.com/prataffel/deep_translator.
[4] https://doi.org/10.5281/zenodo.7311148.

3.3   fastText and BETO Models

The pre-trained language models have presented a promising performance in the Text Classification domain. The BERT model [11], provided by Google, is a pre-trained model and one of the state-of-art NLP tasks. It has been previously used for requirements classification (NoRBERT [15,36]) with a good classification performance and generalization. We selected BETO [7], a BERT-based model, to take advantage that it was pre-trained on a big Spanish corpus (i.e., Wikipedia, Wikinews, ParaCrawl, EUBookshop, and OpenSubtitles) and in several NLP tasks. Moreover, BETO is one of the oldest monolingual models and therefore most used. BETO has been utilized in several Spanish text classifications (e.g., in the classification of radiological reports [41] and for detecting anorexia in social media [29]), in sentiment analysis [6,10] as a support for several RE tasks such as requirements elicitation or prioritization, and recently for Spanish requirements classification [27]. Two models (cased and uncased) of BETO are available. Since the requirements to be analyzed are written in formal language, case model[5] was used for this analysis.

fastText [20] is an open-source library developed by Facebook Research for text classification and word embedding. fasText has been successfully used in various text classification tasks. Recently, it has been also applied in FR/NFR classification [39]. Umer et al. [42] reported good results with the combination of fastText and CNN on the automatic text classification process. fastText has the advantage of handling out-of-vocabulary words through the use of n-gram characters. It also allows the training of a supervised classifier with labeled data and provides automatic hyperparameter optimization. Besides that, fastText provides pre-trained word embedding libraries for 157 languages, including Spanish. The pre-trained model for Spanish used in this paper is cc.es.300.bin[6], trained using CBOW with dimension 300 and character n-grams of length 5.

3.4   Research Method

In order to answer our research questions, two different experiments were performed. In the following, we explain the two experiments conducted: In order to answer our two research questions, two different experiments were performed. In the following, we explain the two experiments conducted:

Experiment 1. We used a similar procedure to the one proposed by Dalal and Zaveri (2011) [9]. The procedure consists of four steps: a) document pre-processing, b) feature extraction, c) ML selection, and d) training and testing the classifier. As a first step, *a pre-processing* of the requirements was performed. For text management, some Natural Language Processing (NLP) techniques were used. A tokenization task, a stopword removal, and a stemming task were carried out for each requirement, only for the Shallow ML algorithms. The NLTK library[7] was used

---

[5] https://huggingface.co/dccuchile/bert-base-spanish-wwm-cased.
[6] https://fasttext.cc/docs/en/crawl-vectors.html.
[7] https://www.nltk.org/.

for these three tasks. The second step is *feature extraction*, which was used to process the text on quantities and frequencies. Two groups of techniques were implemented: frequency-based and embedding-based. The frequency-based one includes the BoW and TF-IDF techniques. The scikit-learn library was used for text vectorization (i.e., the CountVectorizer tool for BoW and TfidfVectorizer for TF-IDF). Both techniques were tested with different values of n-grams: Unigram, Bigram, and Trigram, for every Shallow ML algorithm. The embedding-based one includes the fastText and BETO embeddings, transforming words into their corresponding word embeddings, using with the CNN algorithm. Regarding the third step, we *selected four Shallow ML algorithms*: Naive Bayes (NB), Gaussian Naive Bayes (GNB), Logistic Regression (LR), Support Vector Machine (SVM), and *three DL algorithms*: CNN, BETO, and fastText. The CNN architecture of our model is based on [21], consisting of a Word embedding Layer as an input layer, a One-dimensional Convolutional Layer, a Dropout Layer, a Polling Layer, a Dense Layer with a ReLU action function, and a Dense Layer with a sigmoid activation function as an output layer. The Transformer library[8] was used to obtain the BETO Tokenizer and the BETO pre-trained Model. We used (i) an architecture of four layers: an Input Layer, a BERT Encoder Layer, a Dropout Layer, and a Dense Layer for classification; and (ii) a set of hyperparameters. The library fastText[9] allows the training of a supervised classifier and provides an automatic hyperparameter optimization.

Finally, as a last step, *the classifiers were trained and tested.* This experiment was performed with the translated PROMISE NFR dataset, using only the binary classification (FR/NFR). We used the default hyperparameters for the Shallow ML algorithms, but use a set of several combinations for the DL algorithms (epochs, dropout, learning rate). Google Colaboratory[10], a platform for programming and executing Python and with free access for GPUs, was used to train BETO, due to the necessity of major capacity. A GPU with 12.68 GB of RAM was used. fastText was also trained in Colaboratory, even not needing the GPU, but for the facility in installation. The other algorithms were performed on a laptop computer with Intel(R) Core(TM) i5-1135G7 with 2.40 GHz and 8 GB of RAM. We saved the models (i.e. the outputs of the ML algorithms run on datasets) for later use. We calculated an average accuracy, precision, recall, and f1-score for each class (FR and NFR) using 10-fold cross-validation to evaluate the models' performance.

Experiment 2. In the second experiment, the models with the best f1-score values for each algorithm on experiment 1 were selected and used for testing on the second dataset. This provided a new set of metric values to evaluate whether these models can be generalized. To ensure the performance comparison among ML/DL models has statistical significance, we conducted a multiple hypotheses testing [34], performing first a Cochran's Q test (omnibus test) and later a McNemar test statistics (post hoc tests). We considered a significance level (α)

---

[8] https://github.com/huggingface/transformers.

[9] https://fasttext.cc/.

[10] https://colab.research.google.com/.

of 0.05 for Cochran's Q test and applied a Bonferroni's correction for the post hoc test. We used the library MLxtend that implements both tests in Python.

4 Experiments and Results

4.1 FR/NFR Classification Using PROMISE NFR Dataset (RQ1/RQ2)

Our analysis focuses on a comparison of fastText and BETO classifiers with a set of Shallow ML algorithms (i.e. NB, GNB, LR, and SVM) and one DL algorithm (CNN) to validate the classification performance in functional (F) and non-functional (NF) requirements. Regarding the Shallow ML algorithms in combination with the two text vectorization techniques (BoW, TF-IDF) and n-gram, 24 metrics set were obtained. All classification performance metrics are published in Zenodo[11] and the highest values for each algorithm are presented in Table 2 Table 3 shows the same results for the CNN, fastText, and BETO classifiers, and CNN in combination with fastText and BETO embedding.

When analyzing the results we found that NB with BoW, LR with BoW, and SVM with TF-IDF combinations gave the highest f1-score for the non-functional classification (0.93), and the combination of NB with BoW achieved a better f1-score value for functional classification (0.9). Regarding the use of n-gram with the frequency-based text vectorization, almost all models obtained the best f1-score values using Bigram over Unigram and Trigram, except the NB algorithm that improves using Trigram. The use of n-gram with the NB algorithm has reported an increase in the classification accuracy for text classification in other domains [40]. The use of fastText embeddings with the CNN algorithms outperforms the other DL algorithms with an f1-score of 0.87 for functional classifications and 0.92 for non-functional classification, followed by the fastTest classifier with a slight difference. The BETO classifier and the CNN with BETO were outperformed by the fastText classifier and the Shallow ML algorithms. In terms of time processing in training, the Shallow ML algorithms took less than 2 s for each one, the lowest time for all algorithms, but they need a longer pre-processing step. The CNN and the BETO algorithms took approx. 2 and 35 min respectively, but considering the combination of several hyperparameter values that were tested, the time was greater. The time processing of fastText to perform the automatic hyperparameters optimization was approx. 7 min.

Table 4 shows the hyperparameters that provided the best results in performance for the DL algorithms, after performing several tests with various combinations. CNN algorithm reduces its number of epochs by combining with fastText embedding but increases with BETO embedding. These values suggest that classification models based on Shallow ML algorithms have a good capacity for classifying Spanish requirements (especially the non-functional label). However, it is also necessary to validate whether these models can be generalized.

---

[11] https://doi.org/10.5281/zenodo.7602116.

Table 2. Results of Shallow ML algorithms in combination with text vectorization techniques for FR/NFR classification using translated PROMISE NFR dataset (RQ1)

10-fold													
NB		GNB		GNB		LR		LG		SVM		SVM	
BoW		BoW		TFIDF		BoW		TFIDF		BoW		TFIDF	
Trigram		Bigram		Bigram		Bigram		Bigram		Bigram		Bigram	
F	NF	F	NF	F	NF	F	NF	F	NF	F	NF	F	NF
A	0.92		0.90		0.89		0.91		0.88		0.89		0.91	
P	0.90	0.93	0.87	0.93	0.86	0.92	0.90	0.92	0.94	0.86	0.86	0.90	0.91	0.91
R	0.91	0.92	0.91	0.90	0.89	0.90	0.88	0.94	0.77	0.97	0.86	0.91	0.87	0.94
F_1	**0.90**	**0.93**	0.88	0.91	0.87	0.91	**0.89**	**0.93**	0.84	0.91	0.85	0.91	**0.89**	**0.93**

Table 3. Results of DL algorithms for FR/NFR classification using translated PROMISE NFR dataset (RQ1/RQ2)

10-fold									
CNN+fastText		CNN+BETO		CNN		fastText		BETO	
F	NF	F	NF	F	NF	F	NF	F	NF
A	0.90		0.83		0.88		0.9		0.86	
P	0.90	0.91	0.81	0.85	0.88	0.88	0.89	0.90	0.84	0.89
R	0.86	0.93	0.78	0.87	0.82	0.92	0.85	0.93	0.80	0.88
F_1	**0.87**	**0.92**	0.79	0.86	0.85	0.90	**0.87**	**0.91**	0.81	0.89

4.2 Classification Performance Comparison Between Models Trained in Translated PROMISE and Original PROMISE (RQ3)

To answer RQ3, we compared the results of our first experiment with other empirical studies that used PROMISE NFR in its original language (English). Seven studies from the list shown in Table 1 were chosen because they used the same algorithms and a similar dataset, and most of them reported results of well-known performance metrics. As our comparison was carried out in terms of the macro f1-score metric, we contacted the authors of two studies [12,32] to clarify how the f1-score metric presented in their works had been calculated. Other studies reported the f1-score metric, and we calculated and obtained the corresponding values from other performance measures for those that did not report it. Table 5 presents the macro f1-score values and the difference between the results of the studies compared with our results. The column "Own" means the results of our experiment.

We found slight differences between our results and other studies concerning Shallow ML algorithms. Regarding the DL algorithms, the DL ensemble model proposed by [33] obtained an f1-score value of 0.94, a higher value for our CNN models (CNN and CNN+fastText). We compared the architecture of the CNN model with ours, finding a main difference in the use of two dropout layers, one before the convolutional layer and the other before the last dense layer; while in our model only one dropout layer after the convolutional layer. Finally, fastText [39] and NoRBERT [15] (developed with some custom code) get better results for FR/NFR classification than our models.

Table 4. Best hyperparameters for DL algorithms

	Dropout rate	Feature maps	Optimizer	Learning rate	Epochs
CNN+fastText	0.75	100	Adam	0.01	10
CNN+BETO	0.75	100	Adam	0.001	300
CNN	0.75	100	Adam	0.001	20
BETO	0.2	–	Adam	1e−05	15
fastText	–	–	–	0.101	100

Table 5. Metrics comparison between previous studies

Reference	Dataset	Algorithm	Macro F_1	Own	Difference	Own**	Dif**
[12]	B	NB	0.91	0.92	−0.01		
[12]	B	LR	0.91	0.91	0		
[12]	B	SVM	0.91	0.91	0		
[22]	A	SVM	0.93	0.91	0.02		
[32]	B	SVM	0.90	0.91	−0.01		
[30]	A	CNN	0.77	0.89	−0.12	0.90	−0.13
[33]	A	CNN	0.93	0.89	0.06	0.90	0.03
[33]*	A	CNN	0.94	0.89	0.07	0.90	0.04
[39]	A	fastText	0.92	0.89	0.03		
[15]	A	BERT	0.92	0.85	0.07		

Legend: A: PROMISE (English), B: PROMISE_exp (English), *ensemble model, **CNN+fastText.

4.3   Testing Trained Models on the Second Dataset (RQ4)

Models Performance. We tested the trained models derived from the first experiment on the second dataset. The results are shown in Table 6 for Shallow ML algorithms and Table 7 for DL algorithms. Contrary to the results obtained in the previous experiment, the BETO model gave the best results, with an f1-score of 0.84 for functional and 0.65 for non-functional classification, followed by the CNN with fastText model, with an f1-score of 0.77 and 0.51. In general, we found that the values for accuracy and the f1-score were lower than those obtained in experiment 1. It can be seen from the data in both Tables that the precision values for the functional class are above 0.85 in most of the models, but with a value below 0.51 for the non-functional class. This result indicates a problem in the prediction of NFR since less than 50% of the requirements that the models predict as non-functional are really non-functional. An analysis of several cases containing a false value of the non-functional class was performed to understand and identify the possible causes of these misclassifications.

The use of some specific technical words in FR could lead to confusion for the models. The word "servidor" ("server" in English) is a word normally used in NFR, such as server availability, server failure, etc. In the PROMISE NFR dataset, 15 requirements mention this word and they are all non-functional. But in the second dataset, one of the projects includes some requirements (considered functional by the author) that contain also the word "server" with other technical words (API, PC), which are incorrectly predicted as non-functional. Another example is the word "autenticar" ("authenticate" in English). Ten requirements in the second dataset contain that word (or similar) and are labeled as functional, but the models classify them as non-functional. Other examples are "base de datos" and "archivo" ("database" and "file" in English, respectively). Another cause could be the lack of detail in the second dataset for NFR, in comparison with the PROMISE NFR dataset. In a previous analysis of this dataset [27], some ambiguity in the requirements specification was identified as another cause for misclassifications of the NFRs.

Table 6. Results of testing the Shallow ML algorithms of experiment 1 on the second dataset (RQ4)

	NB + BoW Trigram Model 1		GNB+BoW Bigram Model 2		GNB+ TF-IDF Bigram Model 3		LR+BoW Bigram Model 4		LR+ TF-IDF Bigram Model 5		SVM+BoW Bigram Model 6		SVM+ TF-IDF Bigram Model 7	
	F	NF	F	NF	F	NF	F	NF	F	NF	F	NF	F	NF
A	0.64		**0.65**		0.58		**0.66**		0.46		0.64		0.64	
P	0.93	0.37	0.85	0.35	0.83	0.30	0.92	0.39	0.98	0.30	0.89	0.36	0.93	0.38
R	0.58	0.85	0.66	0.61	0.57	0.62	0.61	0.83	0.31	**0.98**	0.60	0.75	0.58	0.85
F_1	0.71	0.52	**0.74**	0.44	0.68	0.40	0.73	0.53	0.47	0.45	0.72	0.49	0.71	0.52

Table 7. Results of testing the DL algorithms of experiment 1 on the second dataset (RQ4)

	CNN Model 8		BETO Model 9		fastText Model 10		CNN+fastText Model 11		CNN+BETO Model 12	
	F	NF	F	NF	F	NF	F	NF	F	NF
A	0.61		**0.78**		0.61		**0.68**		0.60	
P	0.76	0.20	0.97	0.51	0.96	0.36	0.89	0.40	0.93	0.35
R	0.72	0.24	0.74	0.91	0.52	0.92	0.67	0.73	0.53	0.87
F_1	0.74	0.22	**0.84**	**0.65**	0.68	0.52	**0.77**	**0.51**	0.67	0.50

Models Comparison. We performed Cochran's Q and McNemar's test statistics to ensure the performance metrics comparison has statistical significance, testing the null hypothesis (all the model performance are similar), and the alternative hypothesis (all the model performance has significance difference). We combined some models from Table 6 and Table 7 to generate interesting groups for testing. The Cochran's Q test statistics for model groups are shown in Table 8.

These results show that there is a statistical significance in the difference between all the models' performance, between the performance of the models based on Shallow ML algorithms, and between the performance of the models based on DL algorithms.

We performed some pair-wise McNemar's tests on the models based on DL algorithms, shown in Table 9. We calculated a significance level of 0.001 using the Bonferroni alpha adjustment: considering $k = 10$, then the new value is $0.05/[k(k-1)/2] = 0.05/[10(9)/2] = 0.001$. The comparison of BETO with each of the other DL models demonstrated there are significant differences in the performance between BETO and these models. However, others pair-wise of CNN, fastText, CNN with fastText, and CNN with BETO obtained a p-value greater than 0.001, showing that these models did not show significant differences among them.

Table 8. Cochran's q test statistics

Models groups	Q value	P value
All models shown in Table 6 (Models 1–12)	122.677	<0.05
Models based on Shallow ML algorithms (Models 1–7)	78.935	<0.05
Models based on DL algorithms (Models 8–12)	42.559	<0.05

Table 9. McNemar's test statistics

Models pairs	χ^2 value	P value
BETO and fastText	34.018	**<0.001**
BETO and CNN	23.141	**<0.001**
BETO and CNN+fastText	10.208	**<0.001**
BETO and CNN+BETO	33.767	**<0.001**
CNN and fastText	0.004	0.945
CNN and CNN+fastText	4.148	0.041
fastText and CNN+fastText	6.318	0.019
CNN and CNN+BETO	0.005	0.943

5   Discussion

Several studies have shown that Shallow ML algorithms, like SVM, present good performance when the training is carried out in small and homogeneous datasets. We found Shallow ML algorithms give better results than DL algorithms (including fastText and BETO) using a translated PROMISE NFR (RQ1). This finding supports the work of other studies [25,28] that have demonstrated a better performance of Shallow ML algorithms in small datasets. Although the datasets used in this research are considered Large-Scale Requirements Engineering (i.e., more than 100 and until 1000 requirements [35]), they are still considered small in comparison to other datasets used in text classification with ML techniques, like the IMDb reviews and Reuters dataset presented in [25]. Structured data is another factor that also influences these results since it is easier to train a Shallow ML model. In this research, the requirements specifications from the PROMISE NFR dataset are (semi-)structured.

The fastText classifier was not only the easier algorithm to implement, but also one of the most effective DL algorithms. Besides, it was the fastest algorithm among the DL algorithms. In comparison, BETO was the one with the longest time, due to its architecture and computation complexity, expressed as a limitation in [13, 31]. In general, in experiment 1 the f1-score values are higher in the nonfunctional over the functional classification, which confirms other similar studies that use PROMISE NFR in its original language [1, 15]. A possible explanation for these results is that the PROMISE NFR dataset has a detailed requirement description, especially the use of some cardinal numbers for the NFRs. Kurtanovic and Maalej [22] found that cardinal numbers are the best single informative feature.

We also validate the impact of the use of fastText and BETO embeddings (RQ2). The fastText embedding increments the performance when used as weights in the embedding layer in the CNN architecture and decreases the number of epochs in training. It is therefore likely that the use of fastText embedding is an influential factor in classification performance when using with CNN, according to other results in text classifications [37, 42].

Regarding the multi-language aspect in requirements classification, we found some differences between the English and Spanish languages in the classification process itself. The dataset search was the more challenging. We had to expand the search for research papers on both languages, consider other not well-known repositories, and a translation process had to be carried out, due a dataset of labeled requirements in the Spanish language was not found. Even though the techniques were the same, some resources for the Spanish language (e.g. the Snowball stemmer for Spanish from NLTK library) had to be investigated. Furthermore, a comparative review of our metrics' values was performed with previous studies that used the same algorithms and the same dataset in its original language (RQ3). The metrics values of the shallow ML algorithms are observably close, but with higher values for the DL algorithms in the English language. We suspect that different architectures of the DL models could be one of the factors explaining these gaps. Considering these differences, it can be assumed that the process used in this study can be used for other languages similar to Spanish (Romance languages).

A second dataset in Spanish with different origins was used to validate the generalization of the models obtained in experiment 1 (RQ4). All the models decrease the f1-score values for a functional classification except for the BETO classifier. According to these results, we can claim that the BETO classifier, even trained with a semi-structured and balanced dataset, can offer a better capability to classify unseen data (generalization). These results reflect those of

Hao et al. (2019) [14] and Hey et al. (2020) [15], who also came up with the best-generalized models using BERT. Cochran's Q test and McNemar's test results confirm the higher performance of BETO over others models is statistically significant. Without considering BETO, the DL algorithms performed equally, but significantly less than BETO. It would be interesting to test our obtained BETO classifiers with other Spanish software repositories to confirm our results.

Finally, the results from experiment 2 also showed that the performance values are much lower in the non-functional classification, similar to the results obtained in [27]. This was mainly due to the low precision values. There may be several possible explanations for this result. One is that some technical words are used in the definitions of FR, misclassifying FR as NFR. Another factor could be the lack of detail in the definition of the requirements, compared with PROMISE NFR. The use of n-grams could also make the shallow ML models more restrictive [14], even the DL models also present the same problem.

6   Threats to Validity

In this section, we present the threats to the validity of our research as well as the actions that were taken for their reduction or (partial) mitigation.

1. **Internal validity:** A common problem in automatic requirements classification is the lack of high-quality labeled datasets. Given that the PROMISE NFR dataset has been already used in previous works, we considered a translation to Spanish using Google Translate. However, this translation process can introduce some bias (e.g. gender bias), which can not be mitigated. For instance, in the English Language, the word "user" is applied for both gender, but in our translation, "usuario" represents only a neutral/masculine gender. Furthermore, it's being considered in the future to use other translators such as DeepL[12], and perform comparisons to ensure greater consistency in the translation.

2. **Construct validity:** The classification performance was evaluated by means of very well-known metrics (i.e., precision, recall and f1-score). However, working with ML models and small datasets like the PROMISE NFR dataset, there is a risk of overfitting the models. To avoid this risk, we used the 10-fold cross-validation technique and tested with different parameters. The code and dataset were shared in Zenodo, so other developers and/or researchers can emulate this work.

3. **External validity:** To increase the external validity of our results obtained from the first experiment, a second dataset created with different software projects to PROMISE was used to validate our obtained models. The requirements specifications from this dataset represent three project types (web, android, and broker) and several domains (e.g. Education, Media, Medical, Finance, etc.) to validate the generalizability of the trained models on PROMISE NFR. The experiment has shown that BETO (and to a lesser

---

[12] https://www.deepl.com/translator.

extent CNN+fastTest) can achieve a good generalization performance on classifying requirements for different project types and domains. However, considering the diversity of the Spanish language in Spanish-speaking countries, more tests in these models need to be undertaken to ensure the generalization to other Spanish variants and other languages.

7   Conclusions

The purpose of the current study is to investigate the performance of the pre-trained models fastText and BETO in comparison with other traditional algorithms for automatic functional and non-functional requirements classification in Spanish datasets, through two experiments. Our findings from our first experiment, using the translated PROMISE NFR dataset, reveal that most of the Shallow ML outperform the DL algorithms. However, BETO performed significantly better on the second dataset (second experiment). This suggests that pre-trained language models can obtain better generalizability, i.e., can have a better capability for classifying new requirements, even when their specifications could not be very well detailed. Finally, even though the BETO classifier obtained the best performance in generalizability, the fastText classifier overperformed the others DL algorithms in time processing. Further research might continue to explore requirements classification in the Spanish language. As our results found that BETO offers the best generalized model, we plan to replicate the experiment but including other Spanish BERT-based algorithms. Another natural progression of this work is expanding our Spanish dataset with requirements from the industry. However, in order to identify potential companies interested in sharing their data (if exist), we plan to conduct some interviews in order to understand the actual practices regarding requirements documentation in Spanish-speaking organizations.

Acknowledgement. This research was partially funded by Xunta de Galicia/ FEDER-UE ED413C 2021/53 (Database Lab, UDC) and Galician Ministry of Culture, Education, Professional Training, and University (grants ED431G2019/04, ED431C2022/19).

References

1. Abad, Z.S.H., Karras, O., Ghazi, P., Glinz, M., Ruhe, G., Schneider, K.: What works better? A study of classifying requirements. In: Proceedings - 2017 IEEE 25th International Requirements Engineering Conference, RE 2017 (2017). https:// doi.org/10.1109/RE.2017.36
2. AlDhafer, O., Ahmad, I., Mahmood, S.: An end-to-end deep learning system for requirements classification using recurrent neural networks. Inf. Softw. Technol. **147**, 106877 (2022)
3. Alrumaih, H., Mirza, A., Alsalamah, H.: Toward automated software requirements classification. In: 2018 21st Saudi Computer Society National Computer Conference (NCC), pp. 1–6. IEEE (2018)

4. Apaza, R.D.G., Barrios, J.E.M., Becerra, D.A.I., Quispe, J.A.H.: ERS-TOOL: hybrid model for software requirements elicitation in Spanish language. In: Proceedings of the International Conference on Geoinformatics and Data Analysis, pp. 27–30 (2018)
5. Plaza-del Arco, F.M., Molina-González, M.D., Urena-López, L.A., Martín-Valdivia, M.T.: Comparing pre-trained language models for Spanish hate speech detection. Expert Syst. Appl. **166**, 114120 (2021)
6. de Arriba, A., Oriol, M., Franch, X.: Applying transfer learning to sentiment analysis in social media. In: 2021 IEEE 29th International Requirements Engineering Conference Workshops (REW), pp. 342–348. IEEE (2021)
7. Cañete, J., Chaperon, G., Fuentes, R., Ho, J.H., Kang, H., Pérez, J.: Spanish pre-trained BERT model and evaluation data. In: PML4DC at ICLR 2020 (2020)
8. Cleland-Huang, J., Mazrouee, S., Liguo, H., Port, D.: NFR [data set], March 2007. https://doi.org/10.5281/zenodo.268542
9. Dalal, M.K., Zaveri, M.A.: Automatic text classification: a technical review. Int. J. Comput. Appl. **28**(2), 37–40 (2011)
10. De Arriba, A., Oriol, M., Franch, X.: Merging datasets for emotion analysis. In: 2021 36th IEEE/ACM International Conference on Automated Software Engineering Workshops (ASEW), pp. 227–231. IEEE (2021)
11. Devlin, J., Chang, M.W., Lee, K., Toutanova, K.: BERT: pre-training of deep bidirectional transformers for language understanding. arXiv preprint arXiv:1810.04805 (2018)
12. Dias Canedo, E., Cordeiro Mendes, B.: Software requirements classification using machine learning algorithms. Entropy **22**(9), 1057 (2020)
13. Han, X., et al.: Pre-trained models: past, present and future. AI Open **2**, 225–250 (2021)
14. Hao, Y., Dong, L., Wei, F., Xu, K.: Visualizing and understanding the effectiveness of BERT. arXiv preprint arXiv:1908.05620 (2019)
15. Hey, T., Keim, J., Koziolek, A., Tichy, W.F.: NoRBERT: transfer learning for requirements classification. In: 2020 IEEE 28th International Requirements Engineering Conference (RE), pp. 169–179. IEEE (2020)
16. Hussain, A., Mkpojiogu, E.O., Kamal, F.M.: The role of requirements in the success or failure of software projects. Int. Rev. Manag. Mark. **6**(7S), 306–311 (2016)
17. Instituto Cervantes: El español una lengua viva (2021). https://cvc.cervantes.es/lengua/espanol_lengua_viva/. Accessed 30 Nov 2021
18. Janiesch, C., Zschech, P., Heinrich, K.: Machine learning and deep learning. Electron. Mark. **31**(3), 685–695 (2021). https://doi.org/10.1007/s12525-021-00475-2
19. Jindal, R., Malhotra, R., Jain, A.: Techniques for text classification: literature review and current trends. Webology **12**(2) (2015)
20. Joulin, A., Grave, E., Bojanowski, P., Mikolov, T.: Bag of tricks for efficient text classification. arXiv preprint arXiv:1607.01759 (2016)
21. Kim, Y.: Convolutional neural networks for sentence classification. In: Proceedings of the 2014 Conference on Empirical Methods in Natural Language Processing (EMNLP), pp. 1746–1751. ACL, Doha, October 2014. https://doi.org/10.3115/v1/D14-1181
22. Kurtanovic, Z., Maalej, W.: Automatically classifying functional and non-functional requirements using supervised machine learning. In: Proceedings - 2017 IEEE 25th International Requirements Engineering Conference, RE 2017 (2017). https://doi.org/10.1109/RE.2017.82
23. Li, G., Zheng, C., Li, M., Wang, H.: Automatic requirements classification based on graph attention network. IEEE Access **10**, 30080–30090 (2022)

24. Li, L.F., Jin-An, N.C., Kasirun, Z.M., Chua, Y.P.: An empirical comparison of machine learning algorithms for classification of software requirements. Int. J. Adv. Comput. Sci. Appl. **10**(11) (2019)

25. Li, Q., et al.: A survey on text classification: from shallow to deep learning. arXiv preprint arXiv:2008.00364 (2020)

26. Lima, M., Valle, V., Costa, E., Lira, F., Gadelha, B.: Software engineering repositories: expanding the promise database. In: Proceedings of the XXXIII Brazilian Symposium on Software Engineering, pp. 427–436 (2019)

27. Limaylla-Lunarejo, M.I., Condori-Fernandez, N., Luaces, M.R.: Towards an automatic requirements classification in a new Spanish dataset. In: 2022 IEEE 30th International Requirements Engineering Conference (RE), pp. 270–271. IEEE (2022)

28. Liu, S.: Sentiment analysis of yelp reviews: a comparison of techniques and models. arXiv preprint arXiv:2004.13851 (2020)

29. López-Úbeda, P., Plaza-del Arco, F.M., Díaz-Galiano, M.C., Martín-Valdivia, M.T.: How successful is transfer learning for detecting anorexia on social media? Appl. Sci. **11**(4), 1838 (2021)

30. Navarro-Almanza, R., Juarez-Ramirez, R., Licea, G.: Towards supporting software engineering using deep learning: a case of software requirements classification. In: 2017 5th International Conference in Software Engineering Research and Innovation (CONISOFT), pp. 116–120. IEEE (2017)

31. Qiu, X.P., Sun, T.X., Xu, Y.G., Shao, Y.F., Dai, N., Huang, X.J.: Pre-trained models for natural language processing: a survey. Sci. China Technol. Sci. **63**(10), 1872–1897 (2020). https://doi.org/10.1007/s11431-020-1647-3

32. Quba, G.Y., Al Qaisi, H., Althunibat, A., AlZu'bi, S.: Software requirements classification using machine learning algorithm's. In: 2021 International Conference on Information Technology (ICIT), pp. 685–690 (2021). https://doi.org/10.1109/ICIT52682.2021.9491688

33. Rahimi, N., Eassa, F., Elrefaei, L.: One-and two-phase software requirement classification using ensemble deep learning. Entropy **23**(10), 1264 (2021)

34. Raschka, S.: Model evaluation, model selection, and algorithm selection in machine learning. arXiv preprint arXiv:1811.12808 (2018)

35. Regnell, B., Svensson, R.B., Wnuk, K.: Can we beat the complexity of very large-scale requirements engineering? In: Paech, B., Rolland, C. (eds.) REFSQ 2008. LNCS, vol. 5025, pp. 123–128. Springer, Heidelberg (2008). https://doi.org/10.1007/978-3-540-69062-7_11

36. Sainani, A., Anish, P.R., Joshi, V., Ghaisas, S.: Extracting and classifying requirements from software engineering contracts. In: 2020 IEEE 28th International Requirements Engineering Conference (RE), pp. 147–157. IEEE (2020)

37. Santos, I., Nedjah, N., de Macedo Mourelle, L.: Sentiment analysis using convolutional neural network with fastText embeddings. In: 2017 IEEE Latin American Conference on Computational Intelligence (LA-CCI), pp. 1–5. IEEE (2017)

38. Sayyad Shirabad, J., Menzies, T.: The PROMISE repository of software engineering databases. School of Information Technology and Engineering, University of Ottawa, Canada (2005). https://promise.site.uottawa.ca/SERepository

39. Tiun, S., Mokhtar, U., Bakar, S., Saad, S.: Classification of functional and non-functional requirement in software requirement using Word2vec and fast text. J. Phys. Conf. Ser. **1529**, 042077 (2020)

40. Tripathy, A., Agrawal, A., Rath, S.K.: Classification of sentiment reviews using N-Gram machine learning approach. Expert Syst. Appl. **57**, 117–126 (2016)

41. Úbeda, P.L., Díaz-Galiano, M.C., López, L.A.U., Martín-Valdivia, M.T., Martín-Noguerol, T., Luna, A.: Transfer learning applied to text classification in Spanish radiological reports. In: Proceedings of the LREC 2020 Workshop on Multilingual BIO 2020, pp. 29–32 (2020)
42. Umer, M., et al.: Impact of convolutional neural network and fastText embedding on text classification. Multimedia Tools Appl. **82**, 1–17 (2022)
43. Vanjani, M., Aiken, M.: A comparison of free online machine language translators. J. Manag. Sci. Bus. Intell **5**, 26–31 (2020)
44. Virtanen, A., et al.: Multilingual is not enough: BERT for Finnish. arXiv preprint arXiv:1912.07076 (2019)
45. Xu, R., Yang, Y.: Cross-lingual distillation for text classification. arXiv preprint arXiv:1705.02073 (2017)
46. Zhao, L., et al.: Natural language processing (NLP) for requirements engineering: a systematic mapping study. arXiv preprint arXiv:2004.01099 (2020)

RE for Artificial Intelligence

Exploring Requirements for Software that Learns: A Research Preview

Marie Farrell[1]($\boxtimes$), Anastasia Mavridou[2]($\boxtimes$), and Johann Schumann[2]

[1] Department of Computer Science, The University of Manchester, Manchester, UK
marie.farrell@manchester.ac.uk
[2] KBR Inc at NASA Ames Research Center, Mountain View, USA
anastasia.mavridou@nasa.gov

Abstract. Context & Motivation: The development of software that learns has revolutionized how many systems perform. For the most part, these systems are neither safety- nor mission-critical. However, as technology and aspirations advance, there is an increased desire and need for Machine Learning (ML) software in safety- and mission-critical systems, e.g., driverless cars or autonomous space robotics. **Problem:** In these domains, reliability is crucial and systems have to undergo much scrutiny in terms of both the developed artefacts and the adopted development process. Central to the development of such systems is the elicitation and definition of software requirements that are used to guide the design and verification process. The addition of software components that learn, and the associated capability for unforeseen behavior, makes defining detailed software requirements especially difficult. **Principal ideas/results:** In this paper, we identify unique characteristics of software requirements that are specific to ML components. To this end, we collect and examine requirements from both academic and industrial sources. **Contribution:** To the best of our knowledge, this is the first work that presents real-life, industrial patterns of requirements for ML components. Furthermore, this paper identifies key characteristics and provides a foundation for developing a taxonomy of requirements for software that learns.

1 Introduction

The design of critical systems begins with the definition of natural-language objectives and high-level requirements. Once defined, high-level objectives and requirements are subsequently decomposed into detailed system- and module-level requirements. Any subsequent verification and validation effort must support traceability of requirements and provide evidence that they are upheld. In fact, requirements traceability is prescribed by a number of international standards including DO-178C for the aerospace domain [21].

This process is well-understood for traditional systems since these systems typically operate in constrained, well-defined environments and thus usually exhibit predictable behavior. As a result, requirement analysis for such systems

This is a U.S. government work and not under copyright protection in the U.S.; foreign copyright protection may apply 2023
A. Ferrari and B. Penzenstadler (Eds.): REFSQ 2023, LNCS 13975, pp. 179–188, 2023.
https://doi.org/10.1007/978-3-031-29786-1_12

returns absolute answers, i.e., absolute success (requirements are satisfied by the system) or absolute failure (requirements are not satisfied by the system).

Non-traditional systems, on the other hand, such as ones that rely on Machine Learning (ML) components, bring uncertainty that impacts requirements elicitation and analysis. Requirements for ML components often use probabilities to quantify uncertainties about system behavior and the environment. Furthermore, their analysis may return either absolute or probabilistic answers.

Similar challenges are encountered and must be addressed for autonomous systems that often rely on AI components, such as ML for key behaviors. In general, systems consist of both hardware and software components and requirements must be defined for both of these aspects. In this paper, we are primarily concerned with software requirements, although we recognize that sometimes software and hardware requirements cannot be completely separated.

Since accurate requirements provide the underlying properties against which a system should be verified, this research preview paper provides a starting point for those interested in verifying ML systems. Understanding the nature of requirements for systems that incorporate ML components is important, particularly if these systems are to operate in critical domains e.g., aerospace.

Our ultimate goal is to extend NASA's Formal Requirements Elicitation Tool (FRET)[1] [12,19] for capturing, formalizing, and analyzing requirements for software that learns. To achieve this, we must first develop an understanding of the nature of such requirements and what distinguishes them from requirements for classical systems. Thus, our research plan involves: (1) examine existing requirements, (2) determine the commonalities and differences between requirements for classical systems and those that learn, (3) identify the specific unique characteristics of requirements for systems that learn, (4) develop a taxonomy of requirements, (5) classify and examine those identified in (1), and (6) make appropriate extensions to FRET to support requirements for software that learns. This research preview paper describes our initial findings from exploring a corpus of such requirements, focusing on steps (1)–(3) above.

We observed an increasing number of research papers discussing aspects and challenges of the requirements engineering discipline in the development of autonomous, AI-based systems [5,15,16,23]. The aspects identified in these papers are usually discussed at a high level without showcasing real-life requirements. We differ from the above related work by providing specific requirements and requirement patterns from industrial case studies and missions. The requirements that we have gathered are primarily from the aerospace domain with some autonomous driving examples. We distill the common features amongst these requirements and provide a basis for developing a taxonomy of ML requirements.

2   Requirements for Autonomous Systems

According to a recent survey [17], requirements engineering is one of the most challenging activities for ML-related system development. This is primarily due

---

[1] https://github.com/NASA-SW-VnV/fret.

Table 1. Requirement examples from literature review and the R-RAV project.

Req ID	Requirement	Source
Literature Studies		
[LR-001]	The aircraft location does not exceed a specified lateral offset from the runway centerline during taxiing	[2, 4]
[LR-002]	The aircraft does not veer off the sides of the runway during taxiing	[2, 4]
[LR-003]	All bounding boxes produced shall be no more than 10% larger in any dimension than the minimum sized box capable of including the entirety of the pedestrian	[14]
[LR-004]	The ML component shall determine the position of the specified feature in each input frame within 5 pixels of actual position	[14]
[LR-005]	The ML component shall identify the presence of any person present in the defined area with an accuracy of at least 0.93	[14]
[LR-006]	The ML component shall perform as required in the defined range of lighting conditions experienced during operation of the system	[14]
[LR-007]	The ML component shall identify a person irrespective of their pose with respect to the camera	[14]
[LR-008]	When Ego is 50 m from the crossing, the object detection component shall identify pedestrians that are on or close to the crossing in their correct position	[11]
[LR-009]	In a sequence of images from a video feed any object to be detected should not be missed more than 1 in 5 frames	[11]
[LR-010]	Position of pedestrians shall be determined within 50cm of actual position	[11]
[LR-011]	The object detection component shall perform as required in all situations Ego may encounter within the defined ODD	[11]
[LR-012]	The object detection component shall perform as required in the face of defined component failures arising within the system	[11]
R-RAV Project		
[RRAV-001]	The neural network shall output the cross track distance error (perpendicular distance from the rover to the centerline.) Error to truth must not exceed X	[R-RAV]
[RRAV-002]	Neural network shall output cross track heading error (the angle between the rover heading and the centerline.) Error to truth must not exceed X	[R-RAV]
[RRAV-003]	Upon receiving an image, the Neural Network shall output the distance and the angle within X seconds (latency)	[R-RAV]
[RRAV-004]	Neural network shall output a sensible distance: the value must be between 0 and half the width of the taxiway plus X (buffer X so that it can still report if it is off the taxiway)	[R-RAV]
[RRAV-005]	Neural network shall output a sensible angle: the value must be between -90 and 90 degrees	[R-RAV]
[RRAV-006]	The neural network shall achieve a minimum of X% accuracy on training and Y% accuracy on testing	[R-RAV]
[RRAV-007]	(Local robustness) The neural network shall be robust to small perturbations in the image (pixels)	[R-RAV]
[RRAV-008]	(Semantic variations) The neural network shall be robust to irrelevant variations in the scene	[R-RAV]
[RRAV-009]	The neural network shall safely navigate intersections	[R-RAV]
[RRAV-010]	The magnitude of the cross track distance error shall drop below X m within T seconds and remain there	[R-RAV]
[RRAV-011]	The magnitude of the cross track heading error shall drop below X degrees within T seconds and remain there	[R-RAV]

to uncertainties around the exact input/output behavior of ML components. To better understand such uncertainties, we present requirements that we have gathered from different sources, as shown in Tables 1 and 2.

Table 2. Requirement patterns extracted from missions and industrial case studies.

Req ID	Requirement Pattern (source: NASA)
[IC-001]	The sw shall achieve an average PARAMETER value of X
[IC-002]	The sw shall estimate PARAMETER to within $+-X$ with a $Y\%$ confidence
[IC-003]	The sw shall estimate the confidence of the PARAMETER estimate
[IC-004]	The requirement shall be verified by measuring the average of the parameter over N repetitions
[IC-005]	The sw shall estimate PARAMETER with an $X\%$ confidence interval of no more than $+-Y$
[IC-006]	The sw shall calculate the PARAMETER confidence interval at an $X\%$ confidence level
[IC-007]	The sw shall calculate the PARAMETER as a probability distribution
[IC-008]	The sw shall determine PARAMETER with a high level of confidence
[IC-009]	The sw shall detect $X\%$ of occurrences of EVENT
[IC-010]	The risk-ratio requirements shall be verified using a statistically significant set of SCENARIOS
[IC-011]	The sw shall cause EVENT at a rate less than X times per Y DURATION
[IC-012]	The sw shall detect CONDITION that implies EVENT is probable
[IC-013]	The sw shall take action so that the risk ratio thresholds are satisfied

Studying the Literature: We examined literature from the assurance case domain [2–4,9,11,14]. Requirements **[LR-001]** and **[LR-002]** in Table 1 refer to the TaxiNet system [10], which uses a vision-based neural network to predict an aircraft's position on the runway relative to the center-line to enable autonomous runway taxiing. We recognise that **[LR-001]** and **[LR-002]** are quite high-level and might also apply to non-autonomous systems with similar goals. These are system-level requirements and likely have implications for both hardware and software. We consider them as specific to the ML component because TaxiNet is driven by a neural network whose output is directly related to the preservation of these requirements. The autonomous driving requirement **[LR-005]** is a functional requirement about the ML component.

R-RAV Project: Table 1, Part 2 shows requirements that we elicited in conjunction with developers as part of the R-RAV project at NASA Ames, which takes a similar approach to TaxiNet and is driven by a neural network. These requirements are likely more detailed than those in [2] because the focus was on the elicitation, formalization, and verification of detailed requirements rather than on assurance case development. **[RRAV-005]**, **[RRAV-006]** and **[RRAV-007]** specifically refer to the neural network. Conversely, **[RRAV-011]** describes the behavior of the neural network based control system. Such requirements about control system behavior can also be found in traditional systems.

Missions/Industrial Case Studies: Table 2, contains 13 sanitized requirement patterns, which were obtained after manually analyzing 770 requirements from missions and industrial case studies that use AI. We call these *patterns* as they do not contain specific system details. Upper case variables must be instantiated by actual names and values to yield a multitude of similar requirements. Many of the requirements shown in Table 2 specify constraints on the computed parameters (e.g., **[IC-001]**) and have a notion of confidence (e.g., **[IC-002]**) which we discuss in Sect. 3. Other requirements explicitly mention probabilities, e.g., that a particular event is probable once a specific condition is met (**[IC-012]**).

3 ML Requirement Attributes and Characteristics

By looking at the requirements in Tables 1 and 2, we observe that notions of *confidence, accuracy,* and *average value* are often used to describe probabilistic requirements. However, the semantics of these terms is not always clear; we elaborate below. We also consider other characteristics including robustness, data-driven learning and quality aspects.

3.1 Confidence, Criticality, and Risk Levels

Requirements **[LR-001]** and **[LR-002]** are associated with different *confidence levels.* Although these confidence levels are not part of the requirement text, they have been added in [2, 4] through separate requirement attributes. For example, **[LR-001]** must hold 95% of the time (lower confidence level), while **[LR-002]** must hold 100% of the time (higher confidence level), i.e., the TaxiNet system must avoid any runway excursion.

Confidence levels are tightly related to *risk levels.* For each hazard (e.g., lateral offset violation) a risk is calculated based on the likelihood and severity of the event [22]. Risk factors are usually defined in relevant standards, e.g., see risk ratios in [1]. The level of risk of an event determines the level of risk associated with the corresponding requirement and implies the required confidence level that must be achieved. E.g., **[LR-002]** is associated with a high risk level since it should not be violated under any circumstance. On the contrary, **[LR-001]** is lower on the risk scale, since it might be violated without unwanted consequences.

Confidence levels may consist of *quantitative* and *qualitative* components. A confidence level may indicate the amount of testing necessary for the result to be accepted, the accepted success rate, the type of analysis that must be performed to verify the property, etc. All of these characterize *the level of confidence that must be achieved in order to assure that the corresponding requirement is met.* Consider, e.g., **[LR-001]**, where the quantitative component of the confidence level requires: (1) a 95% success rate for TaxiNet to be within a specified lateral offset and (2) a way of ensuring (during V&V) that this measure is achieved. Qualitative components do not contain explicit numbers. For a requirement with a high criticality level the qualitative component might recommend specific (combinations of) verification techniques to ensure greater coverage and rigor.

Although confidence levels for [LR-001] and [LR-002] are not part of the requirement text, confidence levels do appear in others. For example, in [IC-002] the parameter estimation must hold with "Y%" confidence. Confidence levels are frequently used to inform the choice and/or combination of verification methods. For example, requirements with low confidence levels may be only tested (up to a desired level of coverage) whereas requirements with high confidence levels may require testing alongside formal verification methods and runtime monitoring [18]. For critical systems, assurance cases typically contain an argument that the confidence level is met by sufficiently rigorous verification and implemented mitigations.

3.2   Accuracy as a Measure of Functional Correctness

The notion of accuracy is frequently encountered in ML system requirements [6]. Accuracy is quantifiable and evaluates the functional correctness of an ML model (accuracy during training) or the correctness of the output of an ML component (accuracy during testing/execution). Recent work defines and provides a way of calculating the accuracy of an ML system based on the results obtained while in training and during testing/deployment [20]. Essentially, accuracy defines the rate at which the ML must answer correctly. In this respect, accuracy might be viewed as a quantitative aspect of confidence, e.g., the required success rate of an ML system.

Specifically, [RRAV-006] defines the required levels of accuracy during both training and testing of a neural network. [LR-005] requires an accuracy level of at least '0.93' during execution — this is the required rate that the ML component correctly identifies the presence of a person. [IC-002] incorporates both accuracy and confidence. It requires an accuracy interval, a bound within which a specific parameter should be estimated, but it also includes a percentage for the confidence level that needs to be achieved, i.e., an acceptable success rate.

We recognise that accuracy is also important in classical control systems that often rely on (potentially noisy) sensors (e.g. aircraft engine controllers). For ML components, accuracy also refers to the decision-making process. For example, an inaccurate control system produces a value outside of a correct range, it is not capable of misidentifying the presence of a human in a roadway. Further, it is possible to enumerate the outputs (even erroneous ones) of a control system but the output behaviours of a system driven by ML can often not be completely predetermined (as new behaviours may be learned). The control system inaccuracy is also more straightforwardly reproduced than ML inaccuracies since the system continues to learn.

3.3   Achievement of Average Value

We also frequently encountered the word *average* in our requirement examples (e.g., [IC-004]). The *achievement of an average value* measure evaluates the *consistency* of the output data of an ML system. The requirement provides the target average value (X) of the ML output data and testing/analysis techniques need to assure that, on average, the ML component meets the target value (X). This notion of an average value is frequently encountered in requirements for ML systems but usually less often in requirements for traditional systems, which typically deal with mode logic or embody implicit state machines rather than numerical algorithms.

3.4 Robustness

ML components, such as deep neural networks, are vulnerable to adversarial perturbations in the input. This means that small changes to the input may cause the network to produce erroneous output and has implications for both safety and security [13]. A related property is *local robustness* which specifies that all points within a particular distance from one another be given the same label. E.g., [**RRAV-007**] requires that each input similar to one encountered during training produces a similar output. This kind of requirement is unique to ML systems since it refers to the training and the test set as well as previously unseen data.

3.5 Data-Driven Learning

Most ML systems are data-driven. In particular, neural networks are *trained* using *training data* (e.g. set of images) which can often contain synthetic data. After training, ML systems are tested on *testing data* that should be separate from the training data and is expected to be representative of the physical deployment environment. Once deployed, the system may continue to learn using previously unseen data that were not present in either the training or testing data. Although not discussed in this paper, requirements are necessary to specify data coverage, specific conditions (e.g., weather conditions for TaxiNet), data accuracy and validation, as well as data acquisition and processing. Care should be taken to ensure that all of the data in the training set have been accurately labelled during a pre-processing phase [14]. Further, the data used should be relevant, complete, accurate and balanced [11]. Details on requirements for ML data can be found in the proposed EASA guidelines [8].

3.6 Quality Aspects

The presented requirements are formulated as detailed and often low level. They tend to refer to the specific way that the system or component should achieve its goal and the allowable bounds for particular variables/computations. In addition to a new flavour of requirements, developers and users of ML systems are interested in quality aspects such as *explainability, transparency, fairness,* and *ethical* requirements. We have not encountered such examples but we understand that, as this field progresses, they will become necessary, in particular, for ML systems that are deployed in close proximity to humans [7].

4   Uncertainty in ML Requirements

Clearly, many of the characteristics identified in Sect. 3 relate to uncertainty and probability plays a distinguishing role in ML requirements. As such, below we examine the probabilistic nature of requirements for ML systems.

Not Always Probabilistic: ML systems contain uncertainty. However, some of the requirements and the way that the system functions follow deterministic behavior as is usual for traditional systems. E.g., requirements [**LR-002**] (confidence level 100%) and [**RRAV-005**] describe specific behaviors that do not, at least at a high-level, create or rely on uncertain behavior. To this end, even though the addition of ML increases uncertainty, there are still more traditional requirements that the system should uphold. These requirements can be captured with existing requirement engineering (RE) tools that do not necessarily support probabilities and be verified using existing methods.

Probabilities Within Requirements: The majority of the requirements that we collected *contain* probabilities. This feature is unique to ML systems since more traditional systems usually exhibit deterministic behavior and so probabilities are not often present in their requirements. Examples include: [**RRAV-006**] and [**IC-012**], both are formulated as logical statements with explicit probabilities. To capture such requirements, RE tools must support probabilities.

Confidence Levels or Probabilities about Requirements: In several cases, there was explicit mention of *confidence levels* or of probabilities *about* how often a specific requirement should hold. Such requirements were prevalent in the mission/industrial case studies requirement patterns. Confidence levels may appear inside (e.g., [**IC-002**]) or outside of the requirement text (e.g., [**LR-001**]) through an associated attribute/tag. To support the specification of confidence levels, tags can be used in RE tools to account for both quantitative and qualitative confidence components. Such tags allow us to separate, in some cases, formalization from analysis concerns and may inform the choice of verification method, as outlined in the previous section.

5   Conclusion, Limitations, and Future Work

Threats to Validity: In this paper, we provide a collection of requirements for software that learns. The selection of specific requirements poses a threat to *conclusion validity*, since our efforts were primarily focused on the aerospace domain and it is possible that other domains would reveal additional characteristics. We intend to explore this as future work. It is also possible that our derived patterns are incomplete and examining other use cases will help to identify gaps. We examined a significant number of requirements, both from the literature and other projects, but given that applications of ML software are fast-evolving, it is possible that this sample is not large enough to be representative. This poses a threat to *external validity* of the generalizability of our results. Finally, we do not explicitly distinguish requirements of the ML component from requirements of systems that may be put in place (e.g. monitors) that maintain oversight and invoke mitigations if the ML system produces erroneous or dangerous results.

Conclusion: As ML becomes more prevalent, it is necessary to create new/modify existing methods to verify systems incorporating ML. This paper focuses on providing and analysing multiple, detailed, real-life requirements for ML components. We observed that such requirements fall into recurring patterns. Table 2 shows 13 reusable patterns, which were derived from 770 mission/industrial requirements. We studied their common characteristics, and provided a classification in terms of their probabilistic nature. We focused primarily on requirement specification, rather than verification, but in practice these requirements would be verified. Exploring the available methods that are capable of expressing probabilistic properties for verification will be necessary as this work progresses.

Our ultimate goal is to extend FRET, but we believe that the presented work is relevant in general for RE tools and approaches. The collection of identified requirement patterns can be integrated as requirement templates in RE tools to guide developers in writing ML requirements, which can be challenging. These patterns, coupled with the key characteristics that we identify, provide a foundation for developing a taxonomy of requirements for software that learns.

Acknowledgements. The authors thank Thomas Pressburger and Irfan Sljivo for requirement examples, insightful feedback and discussions. Thanks also to the anonymous reviewers who provided detailed improvement suggestions. Marie Farrell was supported by a Royal Academy of Engineering Research Fellowship. Anastasia Mavridou and Johann Schumann were supported by NASA contract 80ARC020D0010.

References

1. Standard Specification for Detect and Avoid System Performance Requirements. ASTM International (2020)
2. Asaadi, E., et al.: Assured integration of machine learning-based autonomy on aviation platforms. In: Digital Avionics Systems, pp. 1–10. IEEE (2020)
3. Asaadi, E., Denney, E., Menzies, J., Pai, G.J., Petroff, D.: Dynamic assurance cases: a pathway to trusted autonomy. Computer **53**(12), 35–46 (2020)
4. Asaadi, E., Denney, E., Pai, G.: Quantifying assurance in learning-enabled systems. In: Casimiro, A., Ortmeier, F., Bitsch, F., Ferreira, P. (eds.) SAFECOMP 2020. LNCS, vol. 12234, pp. 270–286. Springer, Cham (2020). https://doi.org/10.1007/978-3-030-54549-9_18
5. Belani, H., Vuković, M., Car, Ž.: Requirements engineering challenges in building AI-based complex systems (2019)
6. Berry, D.M.: Requirements engineering for artificial intelligence: what is a requirements specification for an artificial intelligence? In: Gervasi, V., Vogelsang, A. (eds.) REFSQ 2022. LNCS, vol. 13216, pp. 19–25. Springer, Cham (2022). https://doi.org/10.1007/978-3-030-98464-9_2

7. Chuprina, T., Mendez, D., Wnuk, K.: Towards artefact-based requirements engineering for data-centric systems. In: Workshop on Requirements Engineering for Artificial Intelligence, vol. 2857. CEUR-WS (2021)
8. European Union Aviation Safety Agency (EASA): EASA Concept Paper: First usable guidance for Level 1 machine learning applications (2021)
9. Fremont, D.J., Chiu, J., Margineantu, D.D., Osipychev, D., Seshia, S.A.: Formal analysis and redesign of a neural network-based aircraft taxiing system with VERIFAI. In: Lahiri, S.K., Wang, C. (eds.) CAV 2020. LNCS, vol. 12224, pp. 122–134. Springer, Cham (2020). https://doi.org/10.1007/978-3-030-53288-8_6
10. Frew, E., et al.: Vision-based road-following using a small autonomous aircraft. In: IEEE Aerospace Conference, vol. 5, pp. 3006–3015 (2004)
11. Gauerhof, L., Hawkins, R., Picardi, C., Paterson, C., Hagiwara, Y., Habli, I.: Assuring the safety of machine learning for pedestrian detection at crossings. In: Casimiro, A., Ortmeier, F., Bitsch, F., Ferreira, P. (eds.) SAFECOMP 2020. LNCS, vol. 12234, pp. 197–212. Springer, Cham (2020). https://doi.org/10.1007/978-3-030-54549-9_13
12. Giannakopoulou, D., Mavridou, A., Rhein, J., Pressburger, T., Schumann, J., Shi, N.: Formal requirements elicitation with FRET (2020)
13. Gopinath, D., Katz, G., Păsăreanu, C.S., Barrett, C.: DeepSafe: a data-driven approach for assessing robustness of neural networks. In: Lahiri, S.K., Wang, C. (eds.) ATVA 2018. LNCS, vol. 11138, pp. 3–19. Springer, Cham (2018). https://doi.org/10.1007/978-3-030-01090-4_1
14. Hawkins, R., Paterson, C., Picardi, C., Jia, Y., Calinescu, R., Habli, I.: Guidance on the assurance of machine learning in autonomous systems (AMLAS). arXiv preprint arXiv:2102.01564 (2021)
15. Heyn, H.-M.: Requirement engineering challenges for AI-intense systems development. In: 2021 IEEE/ACM 1st Workshop on AI Engineering-Software Engineering for AI (WAIN), pp. 89–96. IEEE (2021)
16. Horkoff, J.: Non-functional requirements for machine learning: challenges and new directions. In: Requirements Engineering, pp. 386–391. IEEE (2019)
17. Ishikawa, F., Yoshioka, N.: How do engineers perceive difficulties in engineering of machine-learning systems? - Questionnaire survey. In: International Workshop on Conducting Empirical Studies in Industry and International Workshop on Software Engineering Research and Industrial Practice, pp. 2–9 (2019)
18. Luckcuck, M., Farrell, M., Dennis, L.A., Dixon, C., Fisher, M.: Formal specification and verification of autonomous robotic systems: a survey. ACM Comput. Surv. 52(5), 1–41 (2019)
19. Mavridou, A., et al.: The ten lockheed martin cyber-physical challenges: formalized, analyzed, and explained. In: Requirements Engineering, pp. 300–310 (2020)
20. Nakamichi, K., et al.: Requirements-driven method to determine quality characteristics and measurements for machine learning software and its evaluation. In: Requirements Engineering, pp. 260–270. IEEE (2020)
21. Rierson, L.: Developing Safety-Critical Software: A Practical Guide for Aviation Software and DO-178C Compliance. CRC Press (2017)
22. Ross, R.S.: Guide for conducting risk assessments. Technical report, National Institute of Standards and Technology, September 2012. SP 800–30 Rev. 1
23. Vogelsang, A., Borg, M.: Requirements engineering for machine learning: perspectives from data scientists (2019)

Requirements Engineering for Automotive Perception Systems: An Interview Study

Khan Mohammad Habibullah[1(✉)], Hans-Martin Heyn[1], Gregory Gay[1],
Jennifer Horkoff[1], Eric Knauss[1], Markus Borg[2], Alessia Knauss[3],
Håkan Sivencrona[3], and Jing Li[4]

[1] Chalmers — University of Gothenburg, Gothenburg, Sweden
{khan.mohammad.habibullah,hans-martin.heyn,jennifer.horkoff,
eric.knauss}@gu.se, greg@greggay.com
[2] Lund University, Lund, Sweden
markus.borg@cs.lth.se
[3] Zenseact AB, Gothenburg, Sweden
{alessia.knauss,hakan.sivencrona}@zenseact.com
[4] Kognic AB, Gothenburg, Sweden
polly.jing.li@kognic.com

Abstract. Background: Driving automation systems (DAS), including autonomous driving and advanced driver assistance, are an important safety-critical domain. DAS often incorporate perceptions systems that use machine learning (ML) to analyze the vehicle environment. **Aims:** We explore new or differing requirements engineering (RE) topics and challenges that practitioners experience in this domain. **Method:** We have conducted an interview study with 19 participants across five companies and performed thematic analysis. **Results:** Practitioners have difficulty specifying upfront requirements, and often rely on scenarios and operational design domains (ODDs) as RE artifacts. Challenges relate to ODD detection and ODD exit detection, realistic scenarios, edge case specification, breaking down requirements, traceability, creating specifications for data and annotations, and quantifying quality requirements. **Conclusions:** Our findings contribute to understanding how RE is practiced for DAS perception systems and the collected challenges can drive future research for DAS and other ML-enabled systems.

Keywords: Machine learning · Requirements engineering · Perception systems · Driving automation systems · Autonomous driving

1 Introduction

Driving automation systems (DAS), including both autonomous driving (AD) and advanced driver assistance systems (ADAS), are software systems designed

© The Author(s), under exclusive license to Springer Nature Switzerland AG 2023
A. Ferrari and B. Penzenstadler (Eds.): REFSQ 2023, LNCS 13975, pp. 189–205, 2023.
https://doi.org/10.1007/978-3-031-29786-1_13

to augment or automate aspects of vehicle control [15]. DAS have long been a domain of interest. However, the increased capabilities and usability of machine learning (ML) have subsequently improved the capabilities of—and interest in—such systems. Research advances have produced improved comfort and safety, and reduced fuel and energy consumption, emissions, and travel time [15].

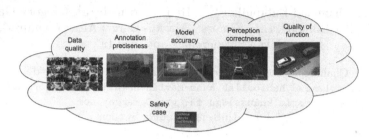

Fig. 1. Conceptual model of quality transitions from data collection to the quality of the automotive function.

DAS functionality depends on the correctness and the integrity of perception systems that blend ML-based models and traditional signal processing[1]. The usage of ML for perception relies on a large quantity of data. Data quality, context, and attributes—as well as annotation quality—have a significant impact on the resulting system quality. However, it is difficult to make direct connections between data, annotation, ML model quality and the resulting functional quality of a perception system (e.g., between the boxes in Fig. 1). The inherent uncertainty of ML—coupled with the desired levels of data quality and coverage—creates substantial process and requirements engineering (RE) challenges in perception system development [7].

RE is an important foundational element of quality assurance and safety engineering. RE plays a critical role in perception system development by enabling explicit capture of safety and quality requirements, supporting communication, recording functional expectations, and ensuring that standards are followed. Recent research has explored RE challenges for ML systems, e.g., [6,22]. However, such challenges have not been thoroughly explored in the context of perception systems for DAS. Addressing this gap is necessary to advance practices in both this domain and in the broader context of RE for AI.

To explore important topics and challenges for perception systems, we have conducted an interview study with 19 expert interviewees from five companies working in various DAS roles. We analyzed interview data using thematic coding to produce eight major themes: perception, requirements engineering, systems and software engineering, AI and ML models, annotation, data, ecosystem and

---

[1] In this paper, we focus specifically on ML-based perception systems for DAS, but often use the term *perception systems* as shorthand.

business, and quality. Here, we analyze data collected as part of the RE theme, and explore critical RE topics and challenges for perception systems[2].

Our findings indicate that practitioners have difficulty breaking down specifications for the ML components. In practice, individuals report that they use scenarios, operational design domains (ODDs), and simulations as part of RE. Practitioners experience RE challenges related to uncertainty, ODD detection, realistic scenarios, edge case specification, traceability, creating specifications for data and annotations, and quantifying quality requirements.

By summarizing the views and challenges of different experts on RE for ML-enabled perception systems, our results are valuable for practitioners working to advance this area. Additionally, our findings contribute to improving RE knowledge more broadly in other domains reliant on ML.

2   Related Work

RE for ML: Recent research has focused on how RE could or must change in the face of rising use of ML. Systematic mapping studies on RE for ML identified new contributions in this area, including approaches, checklists, guidelines, quality models, classifications and evaluations of quality models, taxonomies, and quality requirements [4,8,21]. Ahmad et al. investigated current approaches for writing requirements for AI/ML systems, identified tools and techniques to model requirements for AI/ML, and pointed out existing challenges and limitations in this area [3]. Belani et al. identified and discussed RE challenges for ML and AI-based systems, and reported that identifying NFRs throughout the software lifecycle is one of the main challenges [6]. Heyn et al. used three use cases of distributed deep learning to describe AI system engineering challenges related to RE [10], including context, defining data quality attributes, human factors, testing, monitoring and reporting.

RE for Vehicles and DAS: Significant research has been performed on RE for vehicles. Liebel et al. identified challenges in automotive RE with respect to communication and organization structure [13]. Pernstal et al. stated that RE is one of the areas most in need of improvement at automotive original equipment manufacturers (OEMs), and also identified the ability to communicate via requirements as important [16]. Allmann et al. also noted requirements communication as a major challenge for OEMs and their suppliers [5]. Mahally et al. identified that requirements are the main enablers and barriers of moving towards Agile for automotive OEMs [14].

Research has also looked specifically at RE for AD, e.g., providing an overview of AD RE techniques [19], Riberio et al. identified AD RE challenges addressed by the literature, and identified the languages and description styles used to describe AD requirements, with special attention given to NFRs [17].

---

[2] A recent submission has used the same study data, but focuses on the annotation, data, and ecosystems and business themes [9].

Heyn et al. investigated challenges with context and ODD definition in ML-enabled perception systems [11], including a lack of standardisation for context definitions, ambiguities in deriving ODDs, missing documentation, and lack of involvement of function developers while defining the context. Ågren et al. identified six aspects of RE that impact automotive development speed, moving toward AD [2].

3   Methodology

Our study is guided by the following research questions:

RQ1: What are the RE-related topics of interest for perception systems for DAS?

RQ2: What challenges are experienced in RE for perception systems for DAS?

To address these questions, we conducted seven group interviews with 19 expert participants from five companies that are currently working in ML-based perception systems for DAS. Figure 2 gives an overview of the interview study.

Data Collection: We used semi-structured group interviews with a set of predetermined open-ended questions[3] to keep enough freedom to add follow-up with additional questions. The interviews were conducted between December 2021 and April 2022 via Microsoft Teams, and lasted between 1 h and 30 min to 2 h. We recorded all interview sessions with the permission of all participants; then transcribed, and anonymized the recordings for analysis. At least three researchers were present in each interview, with the same two researchers in all interviews to maintain consistency.

Table 1. Overview of conducted interviews (same as [9])

Interview	Field of work	Participants
A	Object detection	Product owner
B	Autonomous Driving	Product owner, test engineer, ML engineer, software developer
C	Vision systems	System architect, product owner, requirement engineer, deep learning engineer
D	AD and ADAS	System engineer, manager AD
E	Testing and validation AD	System architect, two product owners, compliance officer, data scientist
F	Data annotations	AI engineer, data scientist
G	Autonomous Driving	System safety engineer

---

[3] The interview guide can be found at: https://doi.org/10.7910/DVN/HCMVL1.

A summary of the participants is shown in Table 1. We chose participants who posses experience with ML, perception systems for DAS, software and systems engineering, RE, or data science, or who were working in the DAS industry. The sampling method was a mix of purposive, convenience, and snowball sampling. We sent open calls to the Swedish automotive industry, and our known contacts, then we asked the interviewees for further contacts. Our participants work with different aspects of DAS.

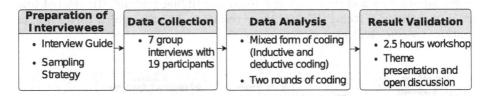

Fig. 2. Overview of interview study.

We started by asking for demographic information about the participants. We then showed them Fig. 1, asking for their feedback and using the figure to ground further discussions about how functional requirements relate to requirements on data and data annotation. We asked further questions about their requirements documentation, safety issues, and quality. Although we carefully chose interview participants, the opinions of the individual interviewees do not necessarily reflect the overall opinion of their companies. Due to the sensitive nature of information provided by interview participants and their respective companies, we are unable to disclose the raw interview data or specific details about ways of working. Finally, in a 2.5-hour workshop with roughly 20 participants, many of whom were interviewees, we presented and discussed our findings with illustrative quotes.

Data Analysis: We applied thematic analysis, as per Saladana [18]. We used a mixed form of coding, where we started with a number of high-level deductive codes based on the interview questions, then we started inductive coding, adding new codes while going through the transcripts. At least three of the researchers worked together to code each of the transcribed interviews. We observed saturation after five interviews, as not many new inductive codes emerged. In a second round of coding, a new group of at least two researchers reviewed the interview transcripts and verified the codes. Finally, we used pattern coding to identify emerging themes and sub-categories. To illustrate our points, we use a number of interview quotes. For increased anonymity, participants are assigned a random identifier, such that P1 does not necessarily match to interview A. In this paper, we focus on findings specifically in the RE theme. Heyn et al. have analyzed the ecosystem and business, data, and annotation themes [9]. Further themes will be analyzed in future work.

4   Results

Based on the thematic analysis, we divide the RE theme into sub-themes—
"Operational Design Domain (ODD), "Scenarios and Edge Cases", "Requirements Breakdown", "Traceability" and "Requirements Specification"—and
important topics within each sub-theme. The sub-themes and topics are summarized in Fig. 3. Our themes reflect both RE topics and challenges, addressing both
RQ1 and RQ2. We also note how many interviewees discussed the sub-theme.
These sub-themes and topics answer RQ1, identifying relevant RE-related topics in perception systems. We use these results to identify which topics are, or
contain, specific challenges (RQ2) in Sec. 5.

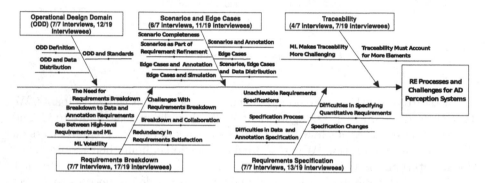

Fig. 3. Mind map illustrating relevant RE topics and challenges for DAS perception
systems.

4.1   Operational Design Domain (ODD)

An ODD is a description of a domain that a DAS will operate in—e.g., the road
or weather conditions. As part of RE, one needs to define not only requirements,
but assumptions about the domain, context, and scope of operation. Operational
context and scope for perception systems is particularly important as the inten-
sity of hazards depends upon the current ODD. ODD-related topics came up in
all interviews and were discussed by 63% of the participants.

ODD Definition: ODDs should be captured as part of the requirements specification. Several interviewees mentioned ODD detection—where the system
detects that a certain ODD is currently applicable for a DAS function—and
ODD exit detection—when the ODD is no longer applicable. ODD detection
requires information on what to detect and detection accuracy. For example, on
highways, DAS needs to detect different dynamic objects than in urban areas.

ODD and Standards: Interviewees state that ODDs are critical, and therefore,
it is desirable to follow a standard or process for specifying and defining ODDs.
This need has been recognized and new initiatives for the definition of ODD

exist, e.g., the interviewees mention the PAS-1883 standard, and we are aware of other standards (e.g., ISO 21448/SOTIF) that include ODDs.

ODD and Data Distribution: One interviewee stated that data distribution requirements are highly influenced by ODDs. For example, camera data can be classified according to descriptions in the ODD, and this mapping can reveal missing data, driving further data collection. As it is not feasible to collect data in all possible contexts, it is necessary to have an efficient sampling process covering the most common ODDs.

"If the performance of the model is not good enough in some part of the ODD, for instance during the night or snow weather and so on, then we can select more samples from those areas." - P16

Another interviewee pointed out that although ODDs drive data collection, collecting certain types of data required by the ODD can still be very difficult.

"... mining for specific use cases. For instance, it is not easy to collect data that contains animals in it. You need some way to mine and find those specific frames which will be sent for annotations and then be used during training." - P16

4.2   Scenarios and Edge Cases

Several interviewees described how scenarios are crucial as part of the requirements specification process. In this context, scenarios describe specific operational paths and conditions for a vehicle, and one ODD may include a number of scenarios. As such, although there are links to scenario-based requirements methods[20], there are also clear differences. Scenarios and edge cases came up in 86% of interviews and were discussed by 58% of participants.

Scenario Completeness: It is important that perception systems perform correctly and that the vehicle handles failures in as many scenarios as possible. As such, scenarios can help in requirements derivation.

"If we refer to the classic system engineering process, I think nowadays it's quite hard ... we are trying to use the scenario to derive the requirements. If we ... see the features or the distribution of the scenarios based on the data from the real world. Then we can derive the high-level requirements based on that data, the scenario database." - P4

One interviewee stressed the difficulty of defining and assessing coverage.

"How do you define coverage? ... What is the scenario space for pedestrian children? Is it based on how the area you have annotated looks inside of your bounding box? Do you parameterize it on the size of the bounding box, parameterized on conditions around you? How would you divide that space and define it in a way that allows even measures? Have I covered not just enough children, but also enough variety of children? " - P18

Scenarios and Annotation: Even if all important scenarios are reflected in training data, annotation errors may result in unsafe behavior—e.g., a perception system may recognize a human as a tree during a snowy or rainy day.

> *"We'll pick out some scenarios that we feel (are) likely not correct, for instance, if it's a rainy night, then maybe the annotator is not annotating (people) as accurately as in the day." - P8*

Scenarios as Part of Requirement Refinement: Our results show that testing through scenarios enables iterative requirements refinement. Engineers iteratively refine their expectations of correct behavior by examining scenarios and capturing observations from simulation or in the field.

> *"... we have to learn through testing, so probably it will start with some rough set of requirements, some obvious setting requirements. Then we will, through real-world testing, discover and learn exactly how we want to behave. " - P2*

> *"It seems like a test-driven development process ... we have the scenarios to drive the development and give more input and also we get the benefit of testing." - P4*

Edge Cases: Interviewees stated that, in addition to normal scenarios, it is crucial and challenging to deal with edge cases. The interviewees used subtly different terms, such as edge cases, rare cases, and cases that occurred very infrequently. We use the term "edge cases" for simplicity. These cases may be missed by studying data distributions, but are very critical to ensure safety.

> *"The cars ... will end up in situations that no one could predict, that we've never seen before, and somehow we need, even in this situation, one individual car needs to perform better than a human driver, and human drivers are real good at handling edge cases. The neural networks will not do that." - P13*

Edge Cases and Annotation: Edge cases cause issues by creating confusion among annotators. Data from edge cases is often annotated inconsistently. The topic of annotation is explored in more detail by Heyn et al. [9].

> *"We label whether a vehicle is in our lane or not. But how should you? You can think of so many corner cases when you are out driving. When you are doing a lane change. Which lane are you in then, and how would you then place all the other vehicles or lane lines? Maybe there are double lane lines and which is valid and which is not? This leads to a lot of confusion among annotators." - P17*

Scenarios, Edge Cases and Data Distribution: One interviewee pointed out that scenarios, and especially rarer edge cases, are important for driving data collection efforts as part of having an effective data distribution. How well edge cases are covered can be an important development metric.

Edge Cases and Simulation: Interviewees stated that collecting data points for particular scenarios from the real world is necessary, but is particularly difficult for edge cases. This makes simulation challenging, as for safety-critical edge cases, practitioners have difficulty safely gathering enough data to run realistic simulations. This makes the process of iterative requirements refinement, as described previously, difficult for requirements associated with edge cases.

4.3 Requirements Breakdown

Requirements breakdown can involve both refining or decomposing requirements. Requirements breakdown was brought up as a topic in all interviews and was discussed by 90% of participants.

The Need for Requirements Breakdown: We see evidence that a traditional requirements breakdown is followed for perception systems. At least one participant spoke of splitting the problem to reduce complexity.

> "*We need to split the problem. We can't do all work at the same time on the complete problem.*" - P12

Another participant described an architectural-oriented breakdown.

> "*Let us say you don't want to collide with an object more than once in a billion hours. This is your top requirement and then you need some kind of architecture or idea of what your system looks like. That should realize this safety goal. This is where we typically come up with a functional architecture, and we start to break down the requirements of the parts of that functional architecture. Then we work. We refine it. The functional architecture becomes a system or logical architecture and we break it down into smaller and smaller pieces.*" - P7

Others describe the importance of separation of high-level requirements from technical requirements to have an upper layer that is resilient to change.

> "*To me, at least the function level will be the same in 100 years because there's no need that you change it. If your function doesn't change, because today you satisfy that function by combustion engine, in the next 50 years by electric, and in the next, I don't know, 100 years by something more intelligent ... By changing your technical system level specifications, you still can satisfy your function.* " - P19

Challenges with Requirements Breakdown: Participants commented on the challenges of connecting high-level requirements to low-level requirements and general challenges with requirements breakdown in this context.

> "*I would say we're working with that challenge and, not that it's an easy one, but we do believe that it's necessary to connect the top-level requirements or the quality of the function, and to map that to quantitative or performance requirements on, for example, perception, precision, and control.*" - P13

> "*What you can do is interact the most closely with ... some component, maybe in perception, and these are the ones who would place direct requirements on the previous component, so it is to me a bit of a hierarchical model to approach the difficulties in breaking down the final safety goal to the early stages in our processing chain. I think one tricky thing is, that it's a hierarchical way in some ways, but you also have to go in both directions in that hierarchical model.* " - P6

Several interviewees report that traditional requirements breakdowns cannot be easily applied.

> *"For sure, we will not start with the classical software approach, where you start with some requirements and then keep breaking those down and through the V-Model because it will be impossible to capture the behavior of autonomous vehicle with requirements."* - P2

Breakdown to Data and Annotation Requirements: Interviewees explained that, although linking functional requirements to system accuracy is often possible, breaking functional requirements into data and annotation requirements is more difficult.

> *" Working with system level requirements, I can look at function requirements and figure out roughly what kind of accuracy we need ... That does not necessarily mean that I can tell how precisely annotation has to be, because I need to know how the software works to figure that out. Another translation needs to happen where I gave my requirements to the developers and they have to figure out what kind of accuracy they need from the data to meet the system requirements and with so many translations on the way, it is easy for things to get lost somewhere."* - P6

> *"...it is difficult to write good requirements on data quality and annotation preciseness and have those links all the way up to feature requirements (Fig. 1). Which I think is because of the dimensionality of the problem. The input space is so enormous that it's really tricky to get a single set of requirements there."* - P15

Breakdown and Collaboration: Challenges arise when teams collaborate to specify quality requirements.

> *" Creating one function would involve multi-team collaboration usually. I guess it's not as easy as evaluating your own system when other people are kind of involved, so you have to come up with scenarios and things to test your algorithms with and could try to come up with a plan. "* - P4

Frequent and direct interaction with the stakeholders can reduce this difficulty and help engineers to identify the requirements. In this case, stakeholders have internal roles in the perception system development.

> *"I think it is a lot of interaction with direct stakeholders in the end ... because the direct consumers of whatever you are producing know exactly what they need to fulfill their own requirements from their own stakeholders. So the negotiation across these interfaces is where the most interaction happens. "* - P9

Gap Between High-level Requirements and ML: When breaking down high-level requirements to very specific requirements on the ML-based perception system, results show that traditional RE practices are able to be applied up to a certain point - even though challenging. However, the breakdown for the ML based components is particularly challenging. As such, there are boundaries within the system where requirements methods change.

> *"If we talked about some other requirements or specifications not for the AD stack. ... those things still can follow the traditional way for critical system. ... if we distinguish those two parts, ... for the black box or part or AD business part, it's hard to follow, but for the rest we still can leverage the classic knowledge."* - P4

We see that it is difficult to specify requirements for the whole perception system. However, there are often still requirements—in terms of various performance metrics—at a high-level.

> " If we say the requirements were specified for the entire AD stack, I think it's quite hard to have very precise or detailed specifications for all functions, but actually, we have some high-level metrics like safety, performance, functionality, or traffic comfort metrics ... We have something, but they are very different from the traditional understanding of the specification." - P4

Redundancy in Requirements Satisfaction: One interviewee described how requirements are allocated to ensure redundancy in the solution.

> "We typically try to break down the problem to come up with redundant solutions. You would have one algorithm using one sensor, which has some capacity to detect the pedestrian, and then use another algorithm and another algorithm in parallel. And you use another sensor and ... decompose the problem such that ... it's very unlikely that all of them would miss this pedestrian. That's a way to try and get reasonable requirements on every perception component." - P6

ML Volatility: One interview pointed out, due to dependencies between components and the volatile nature of ML, changes in the ML model can cause drastic changes in other parts of the system.

> " People sometimes start setting requirements on sensors, and then start setting requirements on data, and calibration accuracy, and then also on annotation, preciseness, and that somehow should influence the model accuracy. Maybe one problem we have with ML is that, if there are things slightly off, it cannot just lead to a slight degradation, but to complete degradation of the entire system." - P17

4.4 Traceability

37% of interviewees, in 57% of the interviews, brought up points related to traceability in perception systems.

ML Makes Traceability More Challenging: Known traceability challenges are exacerbated by the use of ML and associated data. Interviewees described that when systems or modules fail to meet particular key performance indicators (KPIs), tracing the source of the issue is difficult due to the combination of ML models and traditional code. Traceability was discussed in four out of our seven interviews and by seven out of 19 participants.

> "I think what is important at the end is the KPIs on the rightmost features of the figure (Fig. 1). Then if you want to track down why it is not working, it's not very easy to find which module is not working as supposed to, or maybe it works, but in a combination of something else, it creates some kind of strange behavior. " - P14

Traceability Must Account for More Elements: It is important that traceability be maintained not just between code and requirements, but also with ML elements—e.g., models and datasets—that determine the overall functionality.

> " *I think it is important to keep track of exactly which data was used to train the model, and be able to also show that to the general public if needed, right? ... having traceability all the way through development is something we aim for.*" - P8

Typically, trace links would link to typical elements like requirements and safety goals, but now they should also link to scenarios.

> "*I don't want to say something that is wrong, you need this traceability, and then when you trace back you see that, OK, I had a safety goal that was talking about this specific scenario.* " - P19

4.5   Requirements Specification

Aspects of documentation and requirements specification were discussed in all interviews, and by 68% of participants.

Unachievable Requirements Specifications: Two interviewees mentioned that sometimes clients provide unachievable requirements, even though requirements specifications are clear and precise.

> "*Sometimes clients come to us with a very well written set of requirements, like we want this annotator and want this precision or accuracy ... Then they send us data. But when we start looking at the data, it turns out that, given this data, these requirements are basically impossible to meet.*" - P18

Difficulties in Specifying Quantitative Requirements: Due to confidentiality, interviewees were not able to elaborate on specific target levels for quantitative requirements. However, they did reflect generally about the difficulty in determining quantitative quality targets.

> "*... for model accuracy, what does success look like in functional safety? If you can recognize 99% rebounding boxes of possessions, is it good enough? If you have a recall of 100%, but your precision is only 50%, would that be good enough?*" - P17

Specification Process: One interviewee emphasized that documentation of the rationale and goals of the project can serve as a form of requirement specification.

> "*I think it's valuable to actually document after what principles you're working, document the problem you're trying to solve and that is basically a set of requirements, even if they're not necessarily traceable upwards all the way.*" - P15

Specification Changes: The uncertain and highly iterative nature of perception systems and their development environment means that specifications are particularly prone to change.

> "*Requirements at any level are not something that is static. They should reflect your current best interpretation. These things can change because your understanding or your development process changes or the environment changes because there are suddenly new demands on how something is supposed to perform or you learn something new about the system or its environment.* " - P15

Difficulties in Data and Annotation Specification: One interviewee said that specifying data requirements is difficult and different from functional specification, as it is hard to identify features and ensure data quality upfront.

> *"It's very different how you write a data specification ... it's hard to know what the future expects and what type of classes we want and how we want to combine certain objects ... we future proof our datasets quite well by specifying. We do specify a lot of classes."* - P5

Another interviewee reported that it is difficult to specify quality (nonfunctional) requirements on data and annotation, and to understand how qualities affect model performance.

> *" I work a lot with image quality before any ML is involved. Even that is very difficult to quantify. We can have very much right objectively measurable requirements on image quality, sharpness. Then how those translate to the actual performance of a ML algorithm is not at all linear."* - P16

Another participant described challenges in specifying requirements for data annotation when dealing with external partners. It is difficult to have an upfront, detailed specification of data classes and accuracy levels. Instead, data specification needs to be developed iteratively and experimentally with suppliers.

5   Summary and Discussion

RE Topics (RQ1): We have identified a number of RE topics in Sec. 4, as summarized by Fig. 3. These topics can be seen as a sort of check-list when working with ML-based perception systems—a list of issues that should be considered.

Our interviewees emphasize that the definition and limits of ODDs are an integral part of perception systems, and these ODDs have important impacts on data requirements and collection, confirming findings in Heyn et al. [11]. Similarly, perception systems development relies heavily on the use of scenarios and associated edge cases. Such scenarios play a key role in dictating annotation, data collection and simulation. As part of the RE process for perception, it is particularly important to capture edge case scenarios, and these edge cases also play an important role in annotation, simulation, and data collection.

RE for Perception System Challenges (RQ2): Our results indicate that **ODD detection and ODD exit detection** are challenging, as this requires information not only about what to detect in the environment, but also how to detect it and the accuracy of the detection. In addition, **data requirements** are highly influenced by the content of an ODD, therefore ODDs can be used to evaluate whether a data distribution is sufficient for good ML model performance. However, it is not always easy to **collect the data** specified by ODDs. Heyn et al. also emphasized the importance of ODDs in DAS, and noted the lack of a common definition for ODDs [10]. Our participants go further and mention the need for ODD standardization (and efforts in that regard).

One major challenge is that simulations should reflect **realistic scenarios**, echoed by Acuna et al. [1]. For ensuring safe perception, the collected data and scenarios must be thorough, and the perception system must avoid failure in all scenarios. In addition to covering normal scenarios, it is important to **specify edge cases** among scenarios, which are then used to determine data distributions. However, edge cases introduce challenges as they create **confusion among annotators** and are challenging to **test in reality** due to safety concerns.

Breaking down requirements for data and annotations can be very difficult, and additional challenges are introduced due to requirements dependencies and the need for multiple teams to collaborate. In general, we believe that the **gap between standard RE methods and ML components** is both a technical gap and a gap in training and backgrounds, as the ML components are often engineered by data scientists without a software background.

Difficulties in breakdown, ML opaqueness, as well as the the introduction of more elements to trace (e.g., ODDs, scenarios, training data), make it difficult to establish **traceability**. These challenges are in addition to the known challenges with motivating and using traceability in practice [23].

Creating **specifications for data and annotations** is challenging, as it is difficult to have an upfront specification for data classes, e.g., pedestrians and crosswalks. Furthermore, sometimes ML components are assigned unrealistic and **unachievable requirements**. Although requirements change is a frequently acknowledged RE problem [12], with perception systems, the **level of uncertainty and change** is particularly high due to uncertainty about the system, including ML, and the environmental targets. **Quantifying quality requirements** (e.g., accuracy) is also particularly challenging in perception systems, echoing the results of Vogelsang and Borg [22].

Some of these challenges are relatively new from an RE perspective (e.g., ODD detection, missing edge case), while others have been long recognized (e.g., traceability [23], specification changes [12]). As mentioned, three additional themes from the same study are reported and analyzed by Heyn et al. [9]. Although the article focuses on different themes, the qualitative topics covered in that work and our work here have some overlap, particularly in topics related to data and annotation. However, here, the topics of data and annotation are approached from an RE perspective, while the other article takes an ecosystems and process view on topics and challenges related to perception systems in DAS.

Although the focus of this work has been on perception systems, we believe that many of the RE practices and challenges found would apply more generally to other domains reliant on ML. For example, challenges breaking down specification would hold due to the volatility and opaqueness of ML. Further work should contrast RE challenges and practices in other ML-enabled domains.

5.1   Threats to Validity

Internal Validity: We internally peer-reviewed the interview guide and conducted a pilot interview to improve the guide and process. We sent a preparation email to all the interview participants with the details and purpose of the interview study. To maintain consistency in the interview process, at least three authors conducted each interview, with two authors present in all interviews.

All interviews were conducted in English, and the auto-generated transcripts were 'fixed' by authors by listening to audio recordings and correcting any transcription errors. Note that the working language of each company was English, so the language should not have created barriers.

Although qualitative coding always comes with some bias, we mitigated this threat by following established literature [18], coding in multiple rounds, using inductive and deductive codes, and having multiple authors participate in each round of coding, with in-depth discussion on code meanings and assignments.

External Validity: We used a mixture of purposive and snowball sampling. As our study needed a certain set of expertise to answer our questions, we could not conduct random sampling, using our networks and their contacts. Still, due to the size of the study, with participants covering a wide variety of roles with varying experience levels, covering differing company roles and sizes in the perception system ecosystem, we believe we have a relatively representative sample. Furthermore, we argue that we reached a sufficient point of saturation with our interview data, as we noticed a sharp decline in emerging codes after analyzing the fifth group interview.

Note that one cannot link participants to interviews and companies, this is done deliberately to protect the anonymity of our participants. Although this may affect transferability of our results, we feel this level of anonymity does not greatly hurt our results. Though our study results are limited to perception systems in DAS, we argue that some findings can apply to other safety-critical or perceptions systems. This applicability should be explored in future studies.

6   Conclusion

Our study investigated RE practices and challenges during the development of PS. We interviewed 19 participants from five companies and identified a number of RE practices and challenges that impact heavily the functional safety assurance of PS for DAS. The results of this study suggest future research directions in RE and ML to mitigate the challenges practitioners are facing.

Acknowledgements. Support for this project was provided by Vinnova pre-study 2021-02572. We thank all participants.

References

1. Acuna, D., Philion, J., Fidler, S.: Towards optimal strategies for training self-driving perception models in simulation. Adv. Neural. Inf. Process. Syst. **34**, 1686–1699 (2021)
2. Ågren, S.M., Knauss, E., Heldal, R., Pelliccione, P., Malmqvist, G., Bodén, J.: The impact of requirements on systems development speed: a multiple-case study in automotive. Requirements Eng. **24**(3), 315–340 (2019)
3. Ahmad, K., Bano, M., Abdelrazek, M., Arora, C., Grundy, J.: What's up with requirements engineering for artificial intelligence systems? In: 2021 IEEE 29th International Requirements Engineering Conference(RE), pp. 1–12. IEEE (2021)

4. Ali, M.A., Yap, N.K., Ghani, A.A.A., Zulzalil, H., Admodisastro, N.I., Najafabadi, A.A.: A systematic mapping of quality models for AI systems, software and components. Appl. Sci. **12**(17), 8700 (2022)
5. Allmann, C., Winkler, L., Kölzow, T., et al.: The requirements engineering gap in the oem-supplier relationship. J. Univer. Knowl. Manage. **1**(2), 103–111 (2006)
6. Belani, H., Vukovic, M., Car, Ž.: Requirements engineering challenges in building AI-based complex systems. In: 2019 IEEE 27th International Requirements Engineering Conference Workshops (REW), pp. 252–255. IEEE (2019)
7. Borg, M., et al.: Safely entering the deep: A review of verification and validation for machine learning and a challenge elicitation in the automotive industry. arXiv preprint arXiv:1812.05389 (2018)
8. Habibullah, K.M., Gay, G., Horkoff, J.: Non-functional requirements for machine learning: An exploration of system scope and interest. In: 2022 IEEE/ACM 1st International Workshop on Software Engineering for Responsible Artificial Intelligence (SE4RAI), pp. 29–36. IEEE (2022)
9. Heyn, H.M., et al.: Automotive perception software development: Data, annotation, and ecosystem challenges, (Submitted)
10. Heyn, H.M., et al.: Requirement engineering challenges for AI-intense systems development. In: 2021 IEEE/ACM 1st Workshop on AI Engineering-SE for AI (WAIN), pp. 89–96. IEEE (2021)
11. Heyn, H.-M., Subbiah, P., Linder, J., Knauss, E., Eriksson, O.: Setting AI in context: a case study on defining the context and operational design domain for automated driving. In: Gervasi, V., Vogelsang, A. (eds.) REFSQ 2022. LNCS, vol. 13216, pp. 199–215. Springer, Cham (2022). https://doi.org/10.1007/978-3-030-98464-9_16
12. Jayatilleke, S., Lai, R.: A systematic review of requirements change management. Inf. Softw. Technol. **93**, 163–185 (2018)
13. Liebel, G., Tichy, M., Knauss, E., Ljungkrantz, O., Stieglbauer, G.: Organisation and communication problems in automotive requirements engineering. Requirements Eng. **23**(1), 145–167 (2018)
14. M. Mahally, M., Staron, M., Bosch, J.: Barriers and enablers for shortening software development lead-time in mechatronics organizations: A case study. In: Proceedings of the 2015 10th Joint Meeting on Foundations of SE, pp. 1006–1009 (2015)
15. Mallozzi, P., Pelliccione, P., Knauss, A., Berger, C., Mohammadiha, N.: Autonomous vehicles: state of the art, future trends, and challenges. In: Automotive Systems and SE, pp. 347–367 (2019)
16. Pernståll, J., Gorschek, T., Feldt, R., Florén, D.: Software process improvement in inter-departmental development of software-intensive automotive systems – a case study. In: Heidrich, J., Oivo, M., Jedlitschka, A., Baldassarre, M.T. (eds.) PROFES 2013. LNCS, vol. 7983, pp. 93–107. Springer, Heidelberg (2013). https://doi.org/10.1007/978-3-642-39259-7_10
17. Ribeiro, Q.A., Ribeiro, M., Castro, J.: Requirements engineering for autonomous vehicles: a systematic literature review. In: Proceedings of the 37th ACM/SIGAPP Symposium on Applied Computing, pp. 1299–1308 (2022)
18. Saldaña, J.: The coding manual for qualitative researchers. The coding manual for qualitative researchers, pp. 1–440 (2021)
19. Staron, M.: Requirements engineering for automotive embedded systems. In: Automotive Systems and Software Engineering, pp. 11–28. Springer, Cham (2019). https://doi.org/10.1007/978-3-030-12157-0_2

20. Sutcliffe, A.: Scenario-based requirements engineering. In: Proceedings of 11th IEEE International Requirements Engineering Conference 2003, pp. 320–320. IEEE Computer Society (2003)
21. Villamizar, H., Escovedo, T., Kalinowski, M.: Requirements engineering for machine learning: A systematic mapping study. In: 2021 47th Euromicro Conference on SE and Advanced Applications (SEAA), pp. 29–36. IEEE (2021)
22. Vogelsang, A., Borg, M.: Requirements engineering for machine learning: Perspectives from data scientists. In: 2019 IEEE 27th International Requirements Engineering Conference on Workshops (REW), pp. 245–251. IEEE (2019)
23. Wohlrab, R., Steghöfer, J.P., Knauss, E., Maro, S., Anjorin, A.: Collaborative traceability management: Challenges and opportunities. In: 2016 IEEE 24th International Requirements Engineering Conference (RE), pp. 216–225. IEEE (2016)

An Investigation of Challenges Encountered When Specifying Training Data and Runtime Monitors for Safety Critical ML Applications

Hans-Martin Heyn[1,2]([⊠]) [iD], Eric Knauss[1,2] [iD], Iswarya Malleswaran[1],
and Shruthi Dinakaran[1]

[1] Chalmers University of Technology, SE-412 96 Gothenburg, Sweden
hans-martin.heyn@gu.se
[2] University of Gothenburg, SE-405 30 Gothenburg, Sweden

Abstract. [**Context and motivation**] The development and operation of critical software that contains machine learning (ML) models requires diligence and established processes. Especially the training data used during the development of ML models have major influences on the later behaviour of the system. Runtime monitors are used to provide guarantees for that behaviour. [**Question/problem**] We see major uncertainty in how to specify training data and runtime monitoring for critical ML models and by this specifying the final functionality of the system. In this interview-based study we investigate the underlying challenges for these difficulties. [**Principal ideas/results**] Based on ten interviews with practitioners who develop ML models for critical applications in the automotive and telecommunication sector, we identified 17 underlying challenges in 6 challenge groups that relate to the challenge of specifying training data and runtime monitoring. [**Contribution**] The article provides a list of the identified underlying challenges related to the difficulties practitioners experience when specifying training data and runtime monitoring for ML models. Furthermore, interconnection between the challenges were found and based on these connections recommendation proposed to overcome the root causes for the challenges.

Keywords: Artificial intelligence · Context · Data requirements · Machine learning · Requirements engineering · Runtime monitoring

1 Introduction

With constant regularity, unexpected and undesirable behaviour of machine learning (ML) models are reported in academia [9,24,26,53,54], the press, and

This project has received funding from the European Union's Horizon 2020 research and innovation program under grant agreement No 957197.

© The Author(s), under exclusive license to Springer Nature Switzerland AG 2023
A. Ferrari and B. Penzenstadler (Eds.): REFSQ 2023, LNCS 13975, pp. 206–222, 2023.
https://doi.org/10.1007/978-3-031-29786-1_14

by NGOs[1]. These problems become especially apparent, and reported upon, when ML models violate ethical principles. Racial, religious, or gender biases are introduced through a lack of insight into the (sometimes immensely large set of) training data and missing runtime checks for example in large language models such as GPT-3 [1], or facial recognition software based on deep learning [37]. Unfortunately, improving the performance of deep learning models often requires an exponential growth in training data [3]. Data requirements can help in preventing unnecessarily large and biased datasets [50]. Due to changes in the environment, ML models can become "stale", i.e., the context changes so significantly that the performance of the model decreases below acceptable levels [5]. Runtime monitors collect performance data and indicate the need for re-training of the model with updated training data. However, these monitors need to be specified at design time. Data requirements can support the specification of runtime monitors [7]. The lack of specifications becomes specifically apparent with ML models that are part of *critical* software[2] because it is not possible to establish traceability from system requirements (e.g., functional safety requirements) to requirements set on the training data and the runtime monitoring [36].

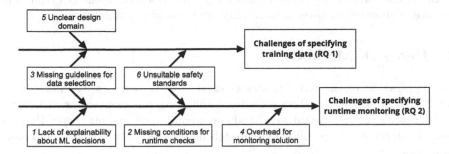

Fig. 1. Overview of identified challenge categories

Scope and Research Questions

The purpose of this study is to highlight current challenges experienced by practitioners in specifying training data and runtime monitoring for ML in safety critical software.

The paper contributes a practitioner's point of view on the challenges reported in academic literature. The aim is to identify starting-points for a future engineering research on the use of runtime monitors for critical ML systems. The following research questions guided this study:

---

[1] Non-governmental organisations, e.g., https://algorithmwatch.org/en/stories/.

[2] We define critical software as software that is safety, privacy, ethically, and/or mission critical, i.e., a failure in the software can cause significant injury or the loss of life, invasion of personal privacy, violation of human rights, and/or significant economic or environmental consequences [31].

RQ1: What are challenges encountered by practitioners when specifying training data for ML models in safety critical software?
RQ2: What are challenges encountered by practitioners when specifying runtime monitors especially in relation to fulfilling safety requirements?

Figure 1 shows the main themes we found in answering the research questions. Concerning RQ1, the interviewees reported on several problems: the data selection process is nontransparent and guidelines especially towards defining suitable measures for data variety are missing. There are no clear context definitions that help in defining data needs, and current safety standards provide little guidance. Concerning RQ2, we found that the problem of defining suitable metrics and the lack of guidance from safety standards also inhibits the ability to specify runtime monitors. Furthermore, practitioners reported on challenges regarding explainability of ML decisions, and the processing and memory overhead caused by runtime monitors in safety critical embedded systems.

The remaining sections of this paper are structured as follows: Sect. 2 outlines and argues for the research methods of this study; Sect. 3 presents the results amd answers to the research questions; Sect. 4 discusses the findings, provides recommendations to practitioners and for further research, identifies related literature, elaborates on threats to validity, and provides a conclusion.

2   Research Method

We applied a qualitative interview-based survey with open-ended semi-structured interviews for data collection. Following the suggestions of Creswell and Creswell [13] the qualitative study was conducted in four steps: Preparation of interviews, data collection through interviews, data analysis, and result validation.

Preparations of Interviews. Based on the a-priori formulated research questions, two of the researchers of this study created an interview guide[3] which was validated and improved by the remaining two researchers. The interview guide contains four sections of questions: the first section includes questions about the interviewees' current role, background and previous experiences. The second section focuses on questions that try to understand challenges when specifying and selecting training data for ML models and how training data affect the performance of these models. The third section investigates challenges when ML models are incorporated in critical systems and how they affect the ability to specify training data. The fourth section concentrates on the run time monitoring aspect of the ML model and contains questions on challenges when specifying runtime monitors.

Sampling Strategy: We chose the participants for this study purposefully using a maximum variation strategy [14]. We were able to recruit interviewees from

---
[3] The interview guide is available at https://doi.org/10.7910/DVN/WJ8TKY.

five different companies, ranging from a local start-up to a multinational world leading communication company. An overview is given in Table 1.

A selection criteria for the company was that they must work with safety-critical systems and ML. Within the companies we tried to find interview candidates with different roles and work experiences to obtain a view beyond the developers' perspective. Besides function developers and ML model developers, we were interested in interviewing requirement engineers and product / function owners because they represent key roles in deriving system or function specifications. We provided the companies with a list of roles that we identified beforehand as interesting for interviewing[4]. Additionally, we interviewed two researchers from academia who participate in a joint industry EU Horizon 2020 project called VEDLIoT[5]. Both researchers worked also with ML models in industry before. Therefore, they could provide insights into both the academic and the industry perspective. A list of the ten interviewees for this study is provided in Table 2.

Table 1. Companies participating in the study

Company	Area of operations	Employees	Countries
1	Telecommunication networks	> 10.000	World
2	Automotive OEM	> 10.000	World
3	Automatic Driving	> 1.000	Europe
4	Industrial camera systems	> 1000	USA
5	Deep Learning optimisation for IoT	> 100	Sweden

Table 2. Participants of the study

Interviewee	Role	Experience
A	Researcher (Academic)	Functional Safety for ADAS (5 years)
B	Function developer	Sensor and perception systems (20 years)
C	Principal engineer	ML model integration (10 years)
D	ML model developer	Distributed and edge systems (3 years)
E	Function owner	ADAS perception functions (8 years)
F	Function developer and test engineer	Automatic driving systems (25 years)
G	Data Scientist	Distributed systems (12 years)
H	Requirement Engineer	Perception systems (8 years)
I	Researcher (Academic)	Neural Network development (8 years)
J	Functional Safety Manager	Sensor systems (20 years)

ADAS: Advanced Driver Assistance Systems

---

[4] The list included functional safety experts, requirement engineers, product owners or function owners, function or model developers, and data engineers.
[5] Very efficient deep learning in the Internet of Things.

Data Collection Through Interviews. All interviews were conducted remotely using either the conference software Zoom or Microsoft Teams and took between 60 - 90 min. The a-priori defined interview guide was only available to the interviewers and was not distributed to the participants beforehand. Each participant was interviewed by two interviewers who alternated in asking questions and observing. At the start of each interview, the interviewers provided some background information about the study's purpose. Then, the interview guide was followed. However, as we encouraged discussions with the interviewees, we allowed deviations from the interview guide by asking additional questions, or changing the order of the questions when it was appropriate [30]. All interviews were recorded and semi-automatically transcribed. The interviewers manually checked and anonymised the results.

Data Analysis. The data analysis followed suggestions by Saldana [42] and consisted of two cycles of coding and validation of the themes through a workshop and member checking.

First Coding Cycle: Attribute coding was used to extract information about the participants' role and previous experiences. Afterwards, the two interviewers independently applied structural coding to collect phrases in the interviews that represent topics relevant to answering the research questions. The researchers compared the individually assigned codes and applied descriptive coding with the aim of identifying phrases that describe common themes across the interviews.

Theme Validation: In a focus group, the identified themes were presented and discussed. Thirteen researchers from both industry and academia in the VEDLIoT project participated. Three of the participants also were interviewed for this study. The aim of the focus group was to reduce bias in the selection of themes and to identify any additional themes that the researchers might have missed.

Second Coding Cycle: After the themes were identified and validated, the second coding cycle was used to map the statements of the interviewees to the themes, and consequently identify the answers to the research questions. The second cycle was conducted by the two researchers who did not conduct the first cycle coding in order to reduce confirmation bias. The mapping was then confirmed and agreed upon by all involved researchers.

Result Validation. Member checking, as described in [14, Ch. 9] was used to validate the identified themes that answer RQ 1 and RQ 2. Additionally, we presented the results in a 60 min focus group to an industry partner and allowed for feedback and comments on the conclusions we drew from the data.

3   Results

During the first coding cycle, structural coding resulted in 117 statements for RQ1 and 77 statements for RQ2. Through descriptive coding preliminary themes were found. The statements and preliminary themes were discussed during a workshop. Based on the feedback from the workshop, 117 statements for RQ1 were categorised into eight final challenge themes and three challenge categories relating to the challenge of specifying training data. Similar, the 77 original statements for RQ2 were categorised into 13 final challenge themes in five challenge categories relating to the challenge of specifying runtime monitoring. A total of six challenge categories emerged for both RQs, out of which two categories contain challenges relating to both training data and runtime monitoring specification, and three challenge themes base on statements from both RQs. The categories and final challenge themes are listed in Table 3. Additionally, for each challenge theme, we indicate the implication of the findings for requirements engineering.

3.1 Answer to RQ1: Challenges Practitioners Experience When Specifying Training Data

The interviewees were asked to share their experiences in selecting training data, the influence of the selection of training data on the system's performance and safety, and any experiences and thoughts on defining specifications for training

Table 3. Challenge groups (bold) and themes found in the interview data. Data.: Challenges related to specifying training data (RQ1). Monitor.: Challenges related to specifying runtime monitoring (RQ2).

ID	Challenge Theme	Relates to Data.	Relates to Monitor.	Related Literature
1	**Lack of explainability about ML decisions**		✓	
1.1	No access to inner states of ML models		✓	[18]
1.2	No failure models for ML models		✓	[51]
1.3	IP protection		✓	
2	**Missing conditions for runtime checks**		✓	
2.1	Unclear metrics and/or boundary conditions		✓	[11, 21, 43]
2.2	Unclear measure of confidence		✓	[17, 34]
3	**Missing guidelines for data selection**	✓	✓	
3.1	Disconnection from requirements	✓		[16, 42]
3.2	Grown data selection habits	✓		[20, 33]
3.3	Unclear completeness criteria	✓		[49]
3.4	Unclear measure of variety	✓	✓	[45, 50]
4	**Overhead for monitoring solution**		✓	
4.1	Limited resources in embedded systems		✓	[38]
4.2	Meeting timing requirements		✓	
4.3	Reduction of true positive rate		✓	
5	**Unclear design domain**	✓		
5.1	Design domain depends on available data	✓		[6]
5.2	Uncertainty in context	✓		[22]
6	**Unsuitable safety standards**	✓	✓	
6.1	Focus on processes instead of technical solution	✓	✓	[10]
6.2	No guidelines for probabilistic effects in software	✓		[28, 43]
6.3	Safety case only through monitoring solution		✓	[31, 46]

data for ML. Based on the interview data, we identified three challenge groups related to specifying training data: missing guidelines for data selection, unclear design domain, and unsuitable safety standards

Missing Guidelines for Data Selection. Four interviewees reported on a lack of guidelines and processes related to the selection of training data. A reason can be that data selection bases on "grown habits" that are not properly documented. Unlike conventional software development, the training of ML is an iterative process of discovering the necessary training data based on experience and experimentation. Requirements set on the data are described as disconnected and unclear for the data selection process. For example, one interviewee stated that if a requirements is set that images shall contain a road, it remains unclear what specific properties this road should have. Six interviewees described missing requirements on the data variety and missing completeness criteria as a reason for the disconnection of requirements from data selection.

> "For example, we said that we shall collect data under varying weather conditions. What does that mean?" - Interview B

> "How much of it (the data) should be in darkness? How much in rainy conditions, and how much should be in snowy situations?" - Interview F

Another interviewee stated that it is not clear how to measure variety, which could be a reason why it is difficult to define requirements on data variety.

> "What [is] include[d] in variety of data? Is there a good measure of variety?" - Interview A

RE Implication 1: **RE research should uncover new ways to specify variety and completeness criteria for data collection.**

Unclear Design Domain. Three interviewees describe uncertainty in the design domain as a reason for why it is difficult to specify training data. If the design domain is unclear, it will be challenging to specify the necessary training data.

> "We need to understand for what context the training data can be used." - Interview J

> "ODD [(Operational Design Domain)]? Yes, of course it translates into data requirements." - Interview F

RE Implication 2: **RE research must provide better ways to specify the context, since data selection and completeness criteria depend on it.**

Unsuitable Safety Standards. Because we were specifically investigating ML in safety critical applications, we asked the participants if they find guidance in safety standards towards specifying training data. Five interviewees stated that current safety standards used in their companies do not provide suitable guidance for the development of ML models. While for example ISO 26262 provides guidance on how to handle probabilistic effects in hardware, no such guidance is provided for software related probabilistic faults.

> "The ISO 26262 gives guidance on the hardware design; [...] how many faults per hour [are acceptable] and how you achieve that. For the software side, it doesn't give any failure rates or anything like that. It takes a completely process oriented approach [...]." - Interview C

One interviewee mentioned that safety standards should emphasise more the data selection to prevent faults in the ML model due to insufficient training.

> "To understand that you have the right data and that the data is representative, ISO 26262 is not covering that right now which is a challenge." - Interview H

RE Implication 3: **RE methods and practices are needed to operationalise safety standards for the selection of training data.**

3.2   Answer to RQ2: Challenges Practitioners Experience When Specifying Runtime Monitors

We asked the interviewees on the role of runtime monitoring for the systems they develop, their experience with specifying runtime monitoring, and the relation of runtime monitoring to fulfilling safety requirements on the system. We identified five challenge groups related to runtime monitoring: lack of explainability about ML decisions, missing conditions for runtime checks, missing guidelines for data selection, overhead for monitoring solution, and unsuitable safety standards.

Lack of Explainability About ML. A reason why it is difficult to specify runtime monitors for ML models is the inability to produce failure models for ML. In normal software development, causal cascades describe how a fault in a software components propagates trough the systems and eventually leads to a failure. This requires the ability to break down the ML model into smaller components and analyse their potential failure behaviour. Four interviewees however reported that they can only see the ML model as a "black-box" with no access to the inner states of the ML model. As a consequence, there is no insight into the failure behaviour of the ML model.

> "[Our insight is] limited because it's a black box. We can only see what goes in and then what comes out to the other side. And so if there is some error in the behavior, then we don't know if it's because [of a] classification error, planning error, execution error?" - Interview F

The reason for opacity of ML models is not necessarily due to technology limitations, but also due to constraints from protection of intellectual property (IP).

> "Why is it a black box? That's not our choice. That's because we have a supplier and they don't want to tell us [all details]." - Interview F

RE Implication 4: **RE can play a crucial role in navigating the trade-off between protecting IP of suppliers and sharing enough information to allow for safety argumentation.**

Missing Conditions for Runtime Checks. A problem of specifying runtime monitors is the need for finding suitable monitoring conditions. This requires the definition of metrics, goals and boundary conditions. Five interviewees report that they face challenges when defining these metrics for ML models.

> "What is like a confidence score, accuracy score, something like that? Which score do you need to ensure [that you] classified [correctly]?" - Interview F

Especially the relation between correct behaviour of the ML model and measure of confidence is unclear, and therefore impede runtime monitoring specification.

> "We say confidence, that's really important. But what can actually go wrong here?" - Interview J

RE Implication 5: **RE is called to provide methods for identifying conditions for runtime checks.**

Missing Guidelines for Data Selection. The inability to specify the meaning of data variety also relates to missing conditions for runtime checks. For example, runtime monitors can be used to collect additional training data, but without a measure of data variety it is difficult to find the required data points.

Overhead for Monitoring Solution. An often overlooked problem seems to be the induced (processing) overhead from a monitoring solution. Especially in the automotive sector, many software components run on embedded computer devices with limited resources.

> "You don't have that much compute power in the car, so the monitoring needs to be very light in its memory and compute footprint on the device, maybe even a separate device that sits next to the device." - Interview F

And due to the limited resources in embedded systems, monitoring solutions can compromise timing requirements of the system. Additionally, one interviewee reported concerns regarding the reduction of the ML model's performance.

> "[...] the true positive rate is actually decreasing when you have to pass it through this second opinion goal. It's good from a coverage and safety point of view, but it reduces the overall system performance." - Interview F

RE Implication 6: **RE methods are needed to help finding suitable runtime checks that do not negatively impact the performance of the system.**

Unsuitable Safety Standards. Safety standards are mostly not suitable for being applied to ML model development. Therefore, safety is often ensured through non-ML monitoring solutions. Interviewees reported that it is not a good solution to rely only on the monitors for safety criticality:

> "[...] so the safety is now moved from the model to the monitor instead, and it shouldn't be. It should be the combination of the two that makes up safety." - Interview B

One reason is that freedom of inference between a non-safety critical component (the ML model), and a safety critical component (the monitor) must be ensured which can complicate the system design.

> "And then especially if you have mixed critical systems [it] means you have ASIL [(Automotive Safety Integrity Level)] and QM [(Quality Management)] components in your design [and] you want to achieve freedom from interference in your system. You have to think about safe communication and memory protection." - Interview J

RE Implication 7: **RE is called to provide traceability and requirements information models that allow a complete description of the system, its monitors, and their relationship to high-level requirements (such as safety).**

4   Discussion and Conclusion

The results reveal connections between the challenges. Not all theme groups relate exclusively to one of the two challenges. For example, themes in the groups *unsuitable safety standards* and *missing guidelines for data selection* relate to both challenges of specifying training data and runtime monitoring. Furthermore, we identified cause-effect relations between different themes and across different group of themes. For example *IP protection* is a cause for the *inability of accessing the inner states* and for *creating failure models for ML model*. We based this assessment on a semantic analyses of the words used in the statements relating to these themes. For example, Interviewee F stated that:

> "That neural network is something [of a] black box in itself. You don't know why it do[es] things. Well, you cannot say anything about its inner behavior" - Interview F

Later in the interview, the same participants states:

> "Why is it a black box? That's not our choice. That's because we have a supplier and they don't want to tell us [all details]." - Interview F

Fig. 2 illustrates the identified cause-effect relations, relations between the themes, and how the different themes relate to the challenges.

Recommendations to Practitioners and for Further Research. The identified root causes of the challenges described by the participants allowed us to formulate recommendations listed in Table 4 and implications towards RE practises stated after each challenge theme in the previous section. Because these recommendations try to solve root causes described by the participants of the interview study, we think they are a useful first step towards solving the challenges related to specifying training data and runtime monitoring.

4.1   Related Literature

The problem of finding the "right" data: For acquiring data, data scientists have to rely on data mining with little to no quality checking and potential biases [4]. Biased datasets are a common cause for erroneous or unexpected behaviour of ML models in critical environments, such as in medical diagnostic [8], in the juridical system [19,38], or in safety-critical applications [15,47].

There are attempts to create "unbiased" datasets. One approach is to curate manually the dataset, such as in the FairFace dataset [29], the CASIA-SURF CeFaA dataset [33], or Fairbatch [41]. An alternative road is to use data augmentation techniques to "rebalance" the dataset [27,46]. However, it was discovered that it is not sufficient for avoiding bias to use an assumed balanced datasets during training [20,51,52] because it is often unclear which features in the data need to be balanced. Approaches for curating or manipulating the dataset require information on the target domain, i.e., one needs to set requirements on the dataset depending on the desired operational context [6,16,22]. But deriving a data specification for ML is not common practise [25,34,43].

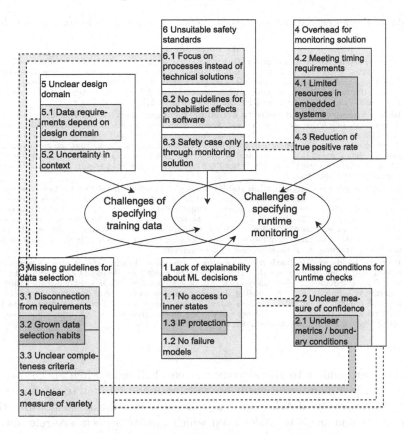

Fig. 2. Connection between the identified challenge themes. Enclosed themes have been identified as causes for the surrounding themes. Furthermore, dotted lines indicate relations between different themes.

The Problem of Finding the "Right" Runtime Monitor: Through clever test strategies, some uncertainty can be eliminated in regards to the behaviour of the model [11]. However, ML components are often part of systems of systems and their behaviour is hard to predict and analyse at design time [49]. DevOps principles from software engineering give promising ideas on how to tackle remaining uncertainty at runtime [35,48]. An overview of MLOps can be for example found in [32]. As part of the operation of the model, runtime models that "augment information available at design-time with information monitored at runtime" help in detecting deviations from the expected behaviour [17]. These runtime models for ML can take the form of model assertions, i.e., checking of explicitly defined attributes of the model at runtime [28]. However, the authors state that "bias in training sets are out of scope for model assertion". Another model based approach can be the creation of neuron activation patterns for runtime monitoring [12]. Other approaches treat the ML model as "black-box", and only check for anomalies and drifts in the input data [40] the output [44], or both

Table 4. Recommendations for practitioners and suggestions for further research

ID	Recommendation
I	**Avoid restrictive IP protection.** IP protection is a cause for the inability of accessing the inner states of the ML models (black-box model). This causes a nontransparent measure of confidence, and an inability to formulate failure models. To our knowledge, no studies have yet been performed on the consequences of IP protection of ML models on the ability to monitor and reason (e.g., in a safety case) for the correctness of ML model decisions.
II	**Relate measures of confidence to actual performance metrics.** For runtime monitoring, the measure of confidence is often used to evaluate the reliability of the ML model's results. But without understanding and relating that measure to clearly defined performance metrics of the ML model first, the measure of confidence provides little insight for runtime monitoring. In general, defining suitable metrics and boundary conditions should become an integral part of RE for machine learning as it affects both the ability to define data requirements and runtime monitoring requirements.
III	**Overcome grown data selection habits.** Grown data selection habits have been mentioned as a reason for a lack of clear completeness criteria and a disconnection from requirements. Based on our results, we argue that more systematic data selection processes need to be established in companies. This would allow for a better connection of the data selection process to requirement engineering and it creates a traceability between system requirements, completeness criteria and data requirements. Additionally, it might also reduce the amount of data needed for training, and therefore cost of development.
IV	**Balance hardware limitation in embedded systems.** Runtime monitoring causes a processing and memory overhead that can compromise timing requirements and reduce the ML model's performance. Today, safety criticality of systems with ML is mostly ensured through monitoring solutions. By decomposing the safety requirements instead onto both the monitoring and the ML model, the monitors might become more resource efficient, faster, and less constraining in regards to the decisions of the ML model. However, safety requirements on the ML models might trigger requirements on the training data.

[18]. However, similar to the aforementioned challenges when specifying data for ML, runtime monitoring needs an understanding on how to "define, refine, and measure quality of ML solutions" [23], i.e., in relation to non-functional requirements one needs to understand which quality aspects are relevant, and how to measure them [21]. Most commonly applied safety standards emphasise processes and traceability to mitigate systematic mistakes during the development of critical systems. Therefore, if the training data and runtime monitoring cannot be specified, a traceability between safety goals and the deployed system cannot be established [10].

For many researchers and practitioners, runtime verification and monitoring is a promising road to assuring safety and robustness for ML in critical software [2,11]. However, runtime monitoring also creates a processing and memory overhead that needs to be considered especially in resource-limited environments such as embedded devices [39].

The related work has been mapped to the challenges identified in the interview study in Table 3.

4.2   Threats to Validity

A lack of rigour (i.e., degree of control) in the study design can cause confounding which can manifest bias in the results [45]. The following mechanisms in this study tried to reduce confounding: The interview guide was peer-reviewed by an independent researcher, and a test session of the interview was conducted. To reduce personal bias, at least two authors were present during all interviews, and the authors took turn in leading the interviews. To confirm the initial findings from the interview study and reduce the risk of researchers' bias, a workshop was organised which was also visited by participants who were not part of the interview study. Another potential bias can arise from the sampling of participants. Although we applied purposeful sampling, we still had to rely on the contact persons of the companies to provide us with suitable interview candidates. We could not directly see a list of employees and choose the candidates ourselves. Regarding generalisability of the findings, the limited number of companies involved in the study can pose a threat to external validity. However, two of the companies are world-leading companies in their fields, which, in our opinion, gives them a deep understanding and experience of the discussed problems. Furthermore, we included companies from a variety of different fields to establish better generalisability. Furthermore, our data includes only results valid for the development of safety-critical ML models. We assume that the findings are applicable also to other forms of criticality, such as privacy-critical, but we cannot conclude on that generalisability based on the available data.

4.3   Conclusion

This paper reported on a interview-based study that identified challenges related to specifying training data needs and runtime monitoring for safety critical ML models. Through interviews conducted at five companies we identified 17 challenges in six groups. Furthermore, we performed a semantic analysis to identify the underlying root-causes. We saw that several underlying challenges affect both the ability to specify training data and runtime monitoring. For example, we concluded that restrictive IP protection can cause an inability to access and understand the inner states of a ML model. Without insight into the ML model's state, the measure of confidence cannot be related to actual performance metrics. Without clear performance metrics, it is difficult to define the necessary degree of variety in the training data. Furthermore, grown data selection impedes proper requirement engineering for training data. Finally, safety requirements should be distributed on both the ML model which can cause requirements on the training data, and on runtime monitors which can reduce the overhead by the monitoring solution. These recommendations will serve as starting point for further engineering research.

References

1. Abid, A., Farooqi, M., Zou, J.: Persistent anti-muslim bias in large language models. In: Proceedings of the 2021 AAAI/ACM Conference on AI, Ethics, and Society, pp. 298–306 (2021)
2. Ashmore, R., Calinescu, R., Paterson, C.: Assuring the machine learning lifecycle: Desiderata, methods, and challenges. ACM Comput. Surv. **54**(5), 1–39 (2021)
3. Banko, M., Brill, E.: Scaling to very very large corpora for natural language disambiguation. In: Proceedings of the 39th Annual Meeting of the Association for Computational Linguistics, pp. 26–33 (2001)

4. Barocas, S., Selbst, A.D.: Big data's disparate impact. Calif. L. Rev. **104**, 671 (2016)

5. Bayram, F., Ahmed, B.S., Kassler, A.: From concept drift to model degradation: An overview on performance-aware drift detectors. Knowl. Based Syst. 108632 (2022)

6. Bencomo, N., Guo, J.L., Harrison, R., Heyn, H.M., Menzies, T.: The secret to better ai and better software (is requirements engineering). IEEE Softw. **39**(1), 105–110 (2021)

7. Bencomo, N., Whittle, J., Sawyer, P., Finkelstein, A., Letier, E.: Requirements reflection: requirements as runtime entities. In: Proceedings of the 32nd ACM/IEEE International Conference on Software Engineering, vol. 2, pp. 199–202 (2010)

8. Bernhardt, M., Jones, C., Glocker, B.: Potential sources of dataset bias complicate investigation of underdiagnosis by machine learning algorithms. Nat. Med. 1–2 (2022)

9. Blodgett, S.L., Barocas, S., Daum'e, H., Wallach, H.M.: Language (technology) is power: A critical survey of "bias" in nlp. In: ACL (2020)

10. Borg, M., et al.: Safely entering the deep: A review of verification and validation for machine learning and a challenge elicitation in the automotive industry. J. Automotive Softw. Eng. **1**(1), 1–19 (2018)

11. Breck, E., Cai, S., Nielsen, E., Salib, M., Sculley, D.: The ml test score: A rubric for ml production readiness and technical debt reduction. In: 2017 IEEE International Conference on Big Data, pp. 1123–1132. IEEE (2017)

12. Cheng, C.H., Nührenberg, G., Yasuoka, H.: Runtime monitoring neuron activation patterns. In: 2019 Design, Automation & Test in Europe Conference & Exhibition, pp. 300–303. IEEE (2019)

13. Creswell, J.W., Creswell, J.D.: Research design: Qualitative, quantitative, and mixed methods approaches. Sage publications (2017)

14. Creswell, John W.; Poth, C.N.: Qualitative Inquiry and Research Design: Choosing Among Five Approaches, 4th edn. Sage Publishing (2017)

15. Fabbrizzi, S., Papadopoulos, S., Ntoutsi, E., Kompatsiaris, I.: A survey on bias in visual datasets. arXiv preprint arXiv:2107.07919 (2021)

16. Fauri, D., Dos Santos, D.R., Costante, E., den Hartog, J., Etalle, S., Tonetta, S.: From system specification to anomaly detection (and back). In: Proceedings of the 2017 Workshop on Cyber-Physical Systems Security and PrivaCy, pp. 13–24 (2017)

17. Giese, H., et al.: Living with uncertainty in the age of runtime models. In: Bencomo, N., France, R., Cheng, B.H.C., Aßmann, U. (eds.) Models@run.time. LNCS, vol. 8378, pp. 47–100. Springer, Cham (2014). https://doi.org/10.1007/978-3-319-08915-7_3

18. Ginart, T., Zhang, M.J., Zou, J.: Mldemon: Deployment monitoring for machine learning systems. In: International Conference on Artificial Intelligence and Statistics, pp. 3962–3997. PMLR (2022)

19. Goodman, B., Flaxman, S.: European union regulations on algorithmic decision-making and a "right to explanation". AI Mag. **38**(3), 50–57 (2017)

20. Gwilliam, M., Hegde, S., Tinubu, L., Hanson, A.: Rethinking common assumptions to mitigate racial bias in face recognition datasets. In: Proceedings of the IEEE CVF, pp. 4123–4132 (2021)

21. Habibullah, K.M., Horkoff, J.: Non-functional requirements for machine learning: understanding current use and challenges in industry. In: 2021 IEEE 29th RE Conference, pp. 13–23. IEEE (2021)

22. Heyn, H.-M., Subbiah, P., Linder, J., Knauss, E., Eriksson, O.: Setting AI in context: a case study on defining the context and operational design domain for automated driving. In: Gervasi, V., Vogelsang, A. (eds.) REFSQ 2022. LNCS, vol. 13216, pp. 199–215. Springer, Cham (2022). https://doi.org/10.1007/978-3-030-98464-9_16

23. Horkoff, J.: Non-functional requirements for machine learning: Challenges and new directions. In: 2019 IEEE 27th RE Conference, pp. 386–391. IEEE (2019)

24. Humbatova, N., Jahangirova, G., Bavota, G., Riccio, V., Stocco, A., Tonella, P.: Taxonomy of real faults in deep learning systems. In: 2020 IEEE/ACM 42nd International Conference on Software Engineering, pp. 1110–1121 (2020)

25. Ishikawa, F., Yoshioka, N.: How do engineers perceive difficulties in engineering of machine-learning systems?-questionnaire survey. In: 2019 IEEE/ACM Joint 7th International Workshop on Conducting Empirical Studies in Industry, pp. 2–9. IEEE (2019)

26. Islam, M.J., Nguyen, G., Pan, R., Rajan, H.: A comprehensive study on deep learning bug characteristics. In: 2019 ACM 27th European Software Engineering Conference, pp. 510–520 (2019)

27. Jaipuria, N., et al.: Deflating dataset bias using synthetic data augmentation. In: Proceedings of the IEEE CVF, pp. 772–773 (2020)

28. Kang, D., Raghavan, D., Bailis, P., Zaharia, M.: Model assertions for monitoring and improving ml models. Proc. Mach. Learn. Syst. **2**, 481–496 (2020)

29. Karkkainen, K., Joo, J.: Fairface: Face attribute dataset for balanced race, gender, and age for bias measurement and mitigation. In: Proceedings of the IEEE CVF, pp. 1548–1558 (2021)

30. King, N., Horrocks, C., Brooks, J.: Interviews in qualitative research. Sage (2018)

31. Knight, J.C.: Safety critical systems: challenges and directions. In: 24th International Conference on Software Engineering, pp. 547–550 (2002)

32. Kreuzberger, D., Kühl, N., Hirschl, S.: Machine learning operations (mlops): Overview, definition, and architecture. arXiv preprint arXiv:2205.02302 (2022)

33. Liu, A., Tan, Z., Wan, J., Escalera, S., Guo, G., Li, S.Z.: Casia-surf cefa: A benchmark for multi-modal cross-ethnicity face anti-spoofing. In: Proceedings of the IEEE CVF, pp. 1179–1187 (2021)

34. Liu, H., Eksmo, S., Risberg, J., Hebig, R.: Emerging and changing tasks in the development process for machine learning systems. In: Proceedings of the International Conference on Software and System Processes, pp. 125–134 (2020)

35. Lwakatare, L.E., Crnkovic, I., Bosch, J.: Devops for ai-challenges in development of ai-enabled applications. In: 2020 International Conference on Software, Telecommunications and Computer Networks, pp. 1–6. IEEE (2020)

36. Marques, J., Yelisetty, S.: An analysis of software requirements specification characteristics in regulated environments. J. Softw. Eng. Appli. (IJSEA) **10**(6), 1–15 (2019)

37. Mehrabi, N., Morstatter, F., Saxena, N., Lerman, K., Galstyan, A.: A survey on bias and fairness in machine learning. ACM Comput. Surv. **54**(6), 1–35 (2021)

38. Miron, M., Tolan, S., Gómez, E., Castillo, C.: Evaluating causes of algorithmic bias in juvenile criminal recidivism. Artifi. Intell. Law **29**(2), 111–147 (2021)

39. Rabiser, R., Schmid, K., Eichelberger, H., Vierhauser, M., Guinea, S., Grünbacher, P.: A domain analysis of resource and requirements monitoring: Towards a comprehensive model of the software monitoring domain. Inf. Softw. Technol. **111**, 86–109 (2019)

40. Rahman, Q.M., Sunderhauf, N., Dayoub, F.: Per-frame map prediction for continuous performance monitoring of object detection during deployment. In: Proceedings of the IEEE CVF, pp. 152–160 (2021)
41. Roh, Y., Lee, K., Whang, S., Suh, C.: Sample selection for fair and robust training. Adv. Neural. Inf. Process. Syst. **34**, 815–827 (2021)
42. Saldaña, J.: The coding manual for qualitative researchers. Sage Publishing, 2nd edn. (2013)
43. Sambasivan, N., Kapania, S., Highfill, H., Akrong, D., Paritosh, P., Aroyo, L.M.: "Everyone wants to do the model work, not the data work": Data cascades in high-stakes ai. In: 2021 Conference on Human Factors in Computing Systems, pp. 1–15 (2021)
44. Shao, Z., Yang, J., Ren, S.: Increasing trustworthiness of deep neural networks via accuracy monitoring. arXiv preprint arXiv:2007.01472 (2020)
45. Slack, M.K., Draugalis, J.R., Jr.: Establishing the internal and external validity of experimental studies. Am. J. Health Syst. Pharm. **58**(22), 2173–2181 (2001)
46. Uchôa, V., Aires, K., Veras, R., Paiva, A., Britto, L.: Data augmentation for face recognition with cnn transfer learning. In: 2020 International Conference on Systems, Signals and Image Processing, pp. 143–148. IEEE (2020)
47. Uricár, M., Hurych, D., Krizek, P., Yogamani, S.: Challenges in designing datasets and validation for autonomous driving. arXiv preprint arXiv:1901.09270 (2019)
48. Vierhauser, M., Rabiser, R., Grünbacher, P.: Requirements monitoring frameworks: A systematic review. Inf. Softw. Technol. **80**, 89–109 (2016)
49. Vierhauser, M., Rabiser, R., Grünbacher, P., Danner, C., Wallner, S., Zeisel, H.: A flexible framework for runtime monitoring of system-of-systems architectures. In: 2014 IEEE Conference on Software Architecture, pp. 57–66. IEEE (2014)
50. Vogelsang, A., Borg, M.: Requirements engineering for machine learning: Perspectives from data scientists. In: 2019 IEEE 27th International Requirements Engineering Conference Workshops, pp. 245–251. IEEE (2019)
51. Wang, A., et al.: Revise: A tool for measuring and mitigating bias in visual datasets. Int. J. Comput. Vis. 1–21 (2022)
52. Wang, T., Zhao, J., Yatskar, M., Chang, K.W., Ordonez, V.: Balanced datasets are not enough: Estimating and mitigating gender bias in deep image representations. In: Proceedings of the IEEE/CVF International Conference on Computer Vision (October 2019)
53. Wardat, M., Le, W., Rajan, H.: Deeplocalize: Fault localization for deep neural networks. In: 2021 IEEE/ACM 43rd International Conference on Software Engineering, pp. 251–262. IEEE (2021)
54. Zhang, X., et al.: Towards characterizing adversarial defects of deep learning software from the lens of uncertainty. 2020 IEEE/ACM 42nd International Conference on Software Engineering, pp. 739–751 (2020)

A Requirements Engineering Perspective to AI-Based Systems Development: A Vision Paper

Xavier Franch[1,2]([📧]) [iD], Andreas Jedlitschka[2] [iD], and Silverio Martínez-Fernández[1] [iD]

[1] Universitat Politècnica de Catalunya (UPC), Barcelona, Spain
{xavier.franch,silverio.martinez}@upc.edu
[2] Fraunhofer IESE, Kaiserslautern, Germany
andreas.jedlitschka@iese.fraunhofer.de

Abstract. *Context and motivation*: AI-based systems (i.e., systems integrating some AI model or component) are becoming pervasive in society. A number of characteristics of AI-based systems challenge classical requirements engineering (RE) and raise questions yet to be answered. *Question*: This vision paper inquires the role that RE should play in the development of AI-based systems with a focus on three areas: roles involved, requirements' scope and non-functional requirements. *Principal Ideas*: The paper builds upon the vision that RE shall become the cornerstone in AI-based system development and proposes some initial ideas and roadmap for these three areas. *Contribution*: Our vision is a step towards clarifying the role of RE in the context of AI-based systems development. The different research lines outlined in the paper call for further research in this area.

Keywords: Requirements Engineering · Artificial Intelligence · Machine Learning · AI-based System · Vision Paper · RE · AI · ML

1 Introduction

AI-based systems, defined as software systems that integrate artificial intelligence (AI) models and components [22], are becoming increasingly pervasive in society. Being yet-another-type of software system, the development of AI-based systems requires following usual software engineering practices [20] and, in particular, requirements engineering (RE) is expected to be applicable in this context.

Still, RE in the context of AI-based systems (which is sometimes referred to as RE4AI[1]) has been reported as challenging by several authors. Some authors have focused on particular RE issues (e.g., a precise definition of satisfaction of a specification in the presence of AI [3]). Others analyse RE4AI from a wider perspective. For instance, Ishikawa and Yoshioka conducted a questionnaire-based survey with 278 responses and report that "decision making with the customers" is the dominant concern when building

---

[1] Other authors are more specific and talk about RE for machine learning (ML) systems. In this paper, we have adopted the widest AI perspective, which includes ML.

© The Author(s), under exclusive license to Springer Nature Switzerland AG 2023
A. Ferrari and B. Penzenstadler (Eds.): REFSQ 2023, LNCS 13975, pp. 223–232, 2023.
https://doi.org/10.1007/978-3-031-29786-1_15

ML-based systems [16]. Several works [1, 12, 22] enumerate a number of challenges related to RE, e.g., importance of context, consideration of data-related requirements and need to define new types of non-functional requirements, this latter aspect also mentioned by Horkoff's seminal paper on the topic [13].

These works, cited as examples, uncover a tension between the current practices of AI-based development and RE. This is partly motivated by the novel and fast emergence of AI in the software arena. The unprecedented evolving pace of new AI solutions and technologies puts the emphasis on creating new models and algorithms to solve all kinds of complex problems, disregarding methodological aspects required by the complexity to integrate these models and algorithms into a large software system [18]. This complexity calls for adopting well-established software engineering practices that have been largely ignored [20], RE being one of them. What are the requirements that apply to these models, to the data needed to build them, and to the algorithms to process them? Who is in charge of formulating these requirements? The answer to this type of questions will shape the form RE4AI will take in the future.

2 Background

From a technological stance, a cause of this tension is the data-oriented nature of AI-based systems. Data management has resulted in new **roles** involved in the development of AI-based systems. Besides, data lies at the heart of a major activity in AI-based system development, namely training, which may have its own requirements, different from those for the system-to-be, therefore yielding diverse **requirement scopes**. These new scopes may bring their particular perspectives on requirements, represented by new types of **non-functional requirements**, or redefinition of existing ones. In this paper, to make our vision concrete, we are going to focus on the three aforementioned aspects.

Roles. Based on a literature review, Pei et al. present an overview of the different roles involved in RE for ML systems, their RE-related concerns and challenges, and collaboration patterns among them [26]. Starting from the classical RE roles of Business Expert, Requirements Engineer and Software Engineer, they propose adding Domain Expert and Data Scientist. They model the collaboration among these actors using $i*$, although the proposed model does not include the requirements engineer, which makes the responsibilities and dependencies of this role implicit or even hidden. Collaboration among Requirements Engineer and Data Scientist is also stressed as a key factor by Ahmad et al. [1].

Adopting a more specific stance and through an interview-based survey, Vogelsang and Borg take the data scientist perspective, given the importance of this role in ML system development [32]. The paper focuses on the activities done, processes followed and challenges found by data analysts in the context of RE4AI and does not explore connections with other roles. Still, the authors make a clear point that data scientist decisions should be subordinated to the classical job of the requirements engineer.

Non-Functional Requirements (NFRs). Several authors have explored which NFRs apply to AI-based systems; in fact, according to a mapping study by Martínez-Fernández et al., this is the hottest topic in the RE4AI-related literature [22].

A good number of papers explore a designated type of NFR in detail, e.g., safety, performance [4][29]. Other authors adopt a holistic perspective and investigate which NFRs apply to AI-based systems. For instance, Habibullah and Horkoff conducted an interview-based survey with ten practitioners [11] to elicit NFR types, their priorities, and most relevant NFR-related challenges. In summary, they state: (1) NFR types can be grouped into thematically-relevant clusters; (2) there are a number of new NFR types specifically related to AI-based systems, or whose relevance excels in this context, such as trust, ethics and explainability [2]; (3) other traditional NFR types, such as usability, are not considered so prioritary (although as usual, there are conflicting views on the importance of this and other NFR types in AI-based systems [11]).

Requirements Scope. Some authors have already considered the need to identify the concrete system part, which is the target of a particular NFR. For instance, performance, as discussed in [29], refers to model performance. More generally, Siebert et al. propose a layered view approach to ML system quality, from Environment to System/Infrastructure and then to ML Components, embracing model and data [28]. This approach is also adopted by Habibullah et al., who argue that requirements (concretely, NFRs) over ML systems may apply to different scopes [10]. They propose as scopes: Training Data, ML Algorithm, ML Model, Results and the whole ML System. Then, they explore which NFRs apply to each scope. In some cases, application requires an adaptation of the standard definition (e.g., from a software system perspective to a data perspective).

At their turn, adopting an intentional viewpoint, Nalchigar et al. identify three perspectives in modelling ML requirements [24]: (1) Business view, expressing stakeholder requirements; (2) Analytics Design view, representing the design of ML solutions for addressing the former requirements; (3) Data Preparation view, conceptualising the design of data preparation tasks. The latter two views are related to some of the scopes identified in [10], although with emphasis on design consequences.

3   RE4AI: Vision and Roadmap

In this paper, we envision that **RE shall become the cornerstone that coordinates all roles, activities and artefacts that are involved in the development of AI-based systems**. We support this vision upon the following arguments:

- Requirements engineers possess a number of skills that make them well-suited for this new challenge, especially communication skills [25]. For instance, they know how to talk to people of different profiles and how to bring them together. Therefore, they are in a good position to mediate the communication gap amongst roles.
- "Classical" RE distinguishes different scopes for requirements, e.g. stakeholder requirements, system requirements, etc. [14]. Therefore, considering additional scopes as those mentioned in the Sect. 2, seems to fit naturally in the discipline.
- Lately, new NFRs have been incorporated in the RE body of knowledge, in different types of systems (e.g., mobile games [30]), or due to societal needs (e.g., sustainability [5]). Thus, RE is well-prepared to replicate the process for AI-significant qualities, and help in the processes of which and where apply to every context.

Building upon this vision and the background outlined in Sect. 2, we elaborate a roadmap for each of the three areas, which we are focusing on. The roadmap consists of a baseline research position followed by an enumeration of some research lines.

Roles. Our baseline research position aligns with Vogelsang and Borg' statement on the need of the requirements engineer to act as a bridge among the customer and technical roles as data scientist [32]. For this reason, we place the requirements engineer role in the centre of the scene (see Fig. 1). Surrounding it, we identify several other roles (see definitions in Table 1 and most relevant relationships in Fig. 1):

- We split the concept of Business Expert from [26] into Customer, Domain Expert, Ethics Manager, and Regulation Expert, recognizing the importance of adhering to all kinds of regulations and social demands when developing AI-based systems.
- We introduce the Software Engineer as a multi-facet role embracing all software engineering roles different from RE: software architect, developer, etc.
- We have decided to split the role of data scientist into two: *(i)* the Data Engineer, who takes care of all data-related aspects in the typical AI/ML pipeline (mining, harvesting, selecting, cleaning, annotating, enriching, augmenting, ...); *(ii)* the AI Expert, who knows the algorithms and models existing in the AI discipline, when they can be applied and what results do they bring. It is worth remarking that, as usual, a person may play more than one role, therefore our identification of two different roles does not preclude that a single person, who could be labelled as a Data Scientist, ultimately plays both of them together.

Table 1. Roles involved in RE4AI.

Stakeholder	Main responsibility
Customer	Has the vision of the AI-based system and provides feedback when requested
Domain Expert	Has knowledge on the domain (including the data in that domain) in which the AI-based system will operate
Ethics Manager	Ensures that the AI-based systems work according to ethical principles
Regulations Expert	Ensures that regulations on trustworthiness, inclusiveness, etc., are fulfilled
Requirements Engineer	Formulates the needs of the customer, collaborating with all other roles
Data Engineer	Gathers, manipulates, and tests data to make it usable by other roles
AI Engineer	Knows the best algorithm to be applied in every situation
Software Engineer	Designs, develops, tests, and deploys software as required

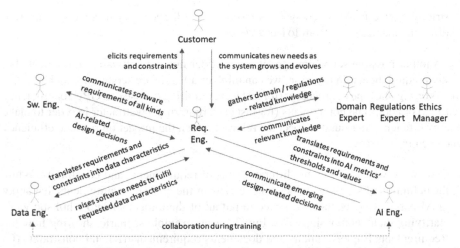

Fig. 1. RE4AI: roles and a representative sample of their relationships.

This baseline position opens a research roadmap along the following lines:

- To complete a catalogue of roles and their responsibilities. Concerning responsibilities, goal-oriented (intentional) models as proposed in [26] look as an appropriate approach, also because this type of models is well-suited to include NFRs as discussed below.
- Related to the previous item, it can be argued that the presented figure has a classical flavour, not completely agile. On the one hand, we are not including a role such as Product Owner. On the other hand, all interactions are proposed to go through the Requirements Engineer, who could eventually become a bottleneck. We can envisage more agile micro-interactions, where, e.g., the Data Engineer and the AI Engineer may directly collaborate during the training process to curate the data set to achieve the required values for accuracy (represented with dotted lines in Fig. 1).
- The central position of the Requirements Engineer requires additional knowledge compared to a more traditional setting. For instance, the Requirements Engineer needs to understand what are the data characteristics that matter to Data Engineers (e.g., size, balance, …) and how requirements relate to them.

Requirements Scope. We concur with Habibullah et al.'s vision on the existence of requirements scopes that distinguish software, data and AI algorithms. This baseline position opens a research roadmap along the following lines:

- Determine the full set of relevant scopes. For instance, some scope may be worth adding. Remarkably, we can think of adding a Data Engineering scope from the software perspective. For instance, when new data is needed, it may be necessary to develop some software component to gather this data from the source in appropriate quality, and this component should be developed according to its own requirements. Remarkably, such a Data Engineering scope could be useful in other contexts not strictly related to AI-based systems where it is still necessary to acquire data from different sources (e.g., from IoT devices).

Another possible scope emerges if we consider not just software requirements but system requirements. In this case, we can think of a Hardware scope for which requirements on e.g. the type of processor (for instance, requiring the use of a GPU for efficiency reasons) or additional components (for instance, requiring a wattmeter in order to make energy efficiency measurable) become relevant, given the impact on runtime efficiency and even in accuracy.

- Clarify the workflow among different types of requirements and constraints. While the definition of scopes provides a static view of the types of requirements that apply in AI-based systems, there is a need to put all of them together into a holistic view, clarifying their relationships. See Fig. 2 for an example scenario showing how the Requirements Engineer elicits and documents requirements (R) and constraints (C) from a Customer deploying an app for plant recognition.

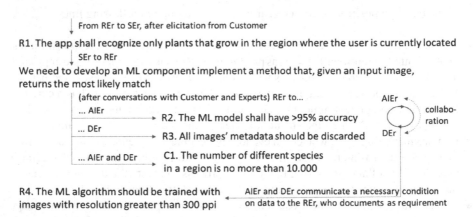

Fig. 2. RE4AI: example scenario showing the flow of requirements (roles identified by initials).

Non-Functional Requirements. Current approaches (*cf.* Background) consider all types of NFRs at the same level of abstraction, e.g., Habibullah and Horkoff's clusters [11]. We envision the convenience of hierarchizing NFR types. In particular, we propose as a baseline position to use the structure proposed in the ISO/IEC 25010 standard [15], which distinguishes quality in use and product quality models, with the former defined in terms of the latter. In addition, because a number of NFR types may not apply to all the requirement scopes, or their definition may vary from scope to scope [11], we propose to replicate this structure for every scope (see Fig. 3).

From this baseline position, we foresee the following research lines:

- The composition and relationships of the different quality models is a significant long-term milestone to achieve by the community. Of course, it may be argued that, because the use of standards is not widespread in the traditional RE context [7], it can be even harder to push for standards in this lively AI context, but still we believe that the structure that standards provide, entails a benefit *per se* to consolidate what is meant by RE4AI.

- In another vein, as hinted above in Fig. 3 and aligned with the terminology proposed, e.g., by the IREB association [14], we prefer to move from NFRs to quality requirements and constraints. The reason is, on the one hand, to adhere to current terminology promoted by certification bodies and other authors [27], and on the other hand the fact that constraints may play an important role when it comes to understand the limits of data in a particular context: a constraint may well limit the size of data, the period of availability, and other information that can be relevant to Data Engineers and AI Engineers to do their job.

- There are a number of concepts that have arisen in the AI community that relate to NFRs and quality, whose fit to this vision needs to be explored. Examples are: data smells [6], highly related to data requirements; Great Expectations (https://greatexpe ctations.io/), as an open standard for data quality; model cards [23], as an example of description of models which can serve to check whether requirements at the scope of ML Model are satisfied or not.

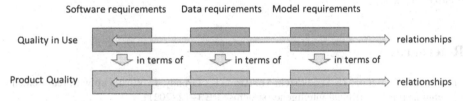

Fig. 3. RE4AI: different quality models.

4 Discussion

In this vision paper, we have reflected on the role of RE in the development of AI-based systems (RE4AI) and advocated that RE should articulate all activities and roles around. For space issues, we have focused the vision on three concrete major areas that directly relate to the data-oriented nature of AI-based systems, not considering other that can be equally important [3, 22]. For each area, we have envisaged a roadmap in the form of baseline position and research lines departing from this position.

These three areas have been presented as independent, but they are clearly interrelated. For instance, some NFR types will not apply to all scopes, or some scopes will not be of interest for all roles. In order to integrate these areas (and others that we are not addressing, e.g. verification and validation), we think of constructing conceptual models such as ontologies for knowledge representation [17] which can integrate all these concepts into a holistic model, as we have done in the field or architectures for AI-based systems [9]. Going further, we can think of linking requirements with design decisions (e.g., which algorithms work better for the elicited requirements) and apply situational method engineering with this purpose, as we have done in previous works related to data-driven methods for RE [8].

We think that the vision presented in this paper may impact future research and practice in RE4AI: concerning research, we have delineated a number of research lines, which may trigger investigation in the community; concerning practice, this vision may contribute to clarify practical aspects that arise in every AI project, by identifying responsibilities of different roles, defining scopes that are different than in traditional systems, and helping to understand quality requirements and constraints in the context of AI-based systems. We acknowledge that practical impact needs to be considered in the long-term, once research progresses more in the short- and mid-term through new results in the suggested research lines. To make this impact possible, we foresee different actions that the community can take. Some are low-hanging fruits, such as continuing the series of workshops related to the topic, notably AIRE and RE4AI, associated with conferences as REFSQ and IEEE RE, and to educational programs in software and systems engineering curricula. Others can be more ambitious, e.g. promoting a new RE certification program in the IREB association, which could have a high practical impact.

Acknowledgments. This paper is part of the project TED2021-130923B-I00, funded by MCIN/AEI/https://doi.org/10.13039/501100011033 and the European Union "NextGenerationEU"/PRTR.

References

1. Ahmad, K., Bano, M., Abdelrazek, M., Arora, C., Grundy, J.: What's up with requirements engineering for artificial intelligence systems? RE 1–12 (2021)
2. Balasubramaniam, N., Kauppinen, M., Hiekkanen, K., Kujala, S.: Transparency and explainability of AI systems: ethical guidelines in practice. In: Gervasi, V., Vogelsang, A. (eds.) REFSQ 2022. LNCS, vol. 13216, pp. 3–18. Springer, Cham (2022). https://doi.org/10.1007/978-3-030-98464-9_1
3. Berry, D.M.: Requirements engineering for artificial intelligence: what is a requirements specification for an artificial intelligence? In: Gervasi, V., Vogelsang, A. (eds.) REFSQ 2022. LNCS, vol. 13216, pp. 19–25. Springer, Cham (2022). https://doi.org/10.1007/978-3-030-98464-9_2
4. Burton, S., Gauerhof, L., Heinzemann, C.: Making making the case for safety of machine learning in highly automated driving. In: SAFECOMP, pp. 5–16 (2017)
5. Duboc, L., Penzenstadler, B., Porras, J., et al.: Requirements engineering for sustainability: an awareness framework for designing software systems for a better tomorrow. Requirements Eng. **25**, 469–492 (2020)
6. Foidl, H., Felderer, M., Ramler, R.: Data smells: categories, causes and consequences, and detection of suspicious data in AI-based systems. In: CAIN, pp. 229–239 (2022)
7. Franch, X., Glinz, M., Méndez, D., Seyff, N.: A Study about the knowledge and use of requirements engineering standards in industry. IEEE Trans. Software Eng. **48**(9), 3310–3325 (2022)
8. Franch, X., Henriksson, A., Ralyté, J., Zdravkovic, J.: Data-driven agile requirements elicitation through the lenses of situational method engineering. In: RE@Next, pp. 402–407 (2020)

9. Franch, X., Martínez-Fernández, S., Ayala, C., Gómez, C.: Architectural decisions in ai-based systems: an ontological view. In: QUATIC, pp. 18–27 (2022)
10. Mohammad Habibullah, K., Gay, G., Horkoff, J.: Non-Functional Requirements for Machine Learning: An Exploration of System Scope and Interest. CoRR abs/2203.11063 (2022)
11. Mohammad Habibullah, K., Horkoff, J.: Non-functional requirements for machine learning: understanding current use and challenges in industry. In: RE:, pp. 13–23 (2021)
12. Heyn, H.-M., Knauss, E., Pir Muhammad, A., et al.: Requirement engineering challenges for AI-intense systems development. In: WAIN, pp. 89–96 (2021)
13. Horkoff, J.: Non-functional requirements for machine learning: challenges and new directions. In: RE, pp. 386–391 (2019)
14. The International Requirements Engineering Board: IREB Certified Professional for Requirements Engineering – Foundation Level – Syllabus, v. 3.1.0 (2022)
15. ISO/IEC 25010:2011. Systems and software engineering — Systems and software Quality Requirements and Evaluation (SQuaRE) — System and software quality models
16. Ishikawa, F., Yoshioka, N.: How do Engineers perceive difficulties in engineering of machine-learning systems' questionnaire survey. In: CESSER-IP, pp. 2–9 (2019)
17. Jurisica, I., Mylopoulos, J., Yu, E.: Ontologies for knowledge management: an information systems perspective. Knowl. Inf. Syst. **6**(4), 380–401 (2004). https://doi.org/10.1007/s10115-003-0135-4
18. Khomh, F., Adams, B., Cheng, J., Fokaefs, M., Antoniol, G.: Software engineering for machine-learning applications: the road ahead. IEEE Softw. **35**(5), 81–84 (2018)
19. Kuwajima, H., Yasuoka, H., Nakae, T.: Engineering problems in machine learning systems. Mach. Learn. **109**(5), 1103–1126 (2020)
20. Lwakatare. L.E., Raj, A., Crnkovic, I., Bosch, J., Holmström Olsson, H.: Large-large-scale machine learning systems in real-world industrial settings: a review of challenges and solutions. Inf. Software Technol. **127**, 106368 (2020)
21. Martínez-Fernández, S., Franch, X., Jedlitschka, A., Oriol, M., Trendowicz, A.: Developing and operating artificial intelligence models in trustworthy autonomous systems. In: Cherfi, S., Perini, A., Nurcan, S. (eds.) RCIS 2021. LNBIP, vol. 415, pp. 221–229. Springer, Cham (2021). https://doi.org/10.1007/978-3-030-75018-3_14
22. Martínez-Fernández, S., Bogner, J., Franch, X., et al.: Software engineering for AI-based systems: a survey. ACM Trans. Software Eng. Methodol. **31**(2), 37e:1–37e:59 (2022)
23. Mitchell, M., Wu, S., et al.: Model cards for model reporting. In: FAT*, pp. 220–229 (2019)
24. Nalchigar, S., Yu, E., Keshavjee, K.: Modeling machine learning requirements from three perspectives: a case report from the healthcare domain. Requirements Eng. **26**(2), 237–254 (2021)
25. Paech, B.: What is a requirements engineer? IEEE Softw. **25**(4), 16–17 (2008)
26. Pei, Z., Liu, L., Wang, C., Wang, J.: Requirements engineering for machine learning: a review and reflection. In: REW, pp. 166–175 (2022)
27. Pohl, K.: Requirements Engineering - Fundamentals, Principles, and Techniques. Springer (2010). https://doi.org/10.1007/978-3-642-12578-2
28. Siebert, J., et al.: Towards towards guidelines for assessing qualities of machine learning systems. In: Shepperd, M., Brito e Abreu, F., Rodrigues da Silva, A., Pérez-Castillo, R. (eds.) QUATIC 2020. CCIS, vol. 1266, pp. 17–31. Springer, Cham (2020). https://doi.org/10.1007/978-3-030-58793-2_2
29. Tuncali, C.E., Fainekos, G., Prokhorov, D., Ito, H., Kapinski, J.: Requirements-driven test generation for autonomous vehicles with machine learning components. IEEE Transactions on Intelligent Vehicles **5**(2), 265–280 (2020)
30. Valente, L., Feijó, B., Leite, J.C.S.P.: Mapping quality requirements for pervasive mobile games. Requirements Eng. **22**(1), 137–165 (2017)

31. Villamizar, H., Escovedo, T., Kalinowski, M.: Requirements engineering for machine learning: a systematic mapping study. In: SEAA, pp. 29–36 (2021)
32. Vogelsang, A., Borg, M.: Requirements engineering for machine learning: perspectives from data scientists. In: REW, pp. 245–251 (2019)

Out-of-Distribution Detection as Support for Autonomous Driving Safety Lifecycle

Jens Henriksson[1], Stig Ursing[1], Murat Erdogan[2], Fredrik Warg[3], Anders Thorsén[3(✉)], Johan Jaxing[4], Ola Örsmark[5], and Mathias Örtenberg Toftås[1]

[1] Semcon, Department Software and Emerging Tech, Gothenburg, Sweden
{jens.henriksson,stig.ursing,mathias.ortenberg-toftas}@semcon.com
[2] Veoneer, Linköping, Sweden
murat.erdogan@veoneer.com
[3] RISE Research Institutes of Sweden, Borås, Sweden
{fredrik.warg,anders.thorsen}@ri.se
[4] Agreat, Gothenburg, Sweden
johan.jaxing@agreat.com
[5] Comentor, Gothenburg, Sweden
ola.orsmark@comentor.se

Abstract. [**Context and Motivation**] The automotive industry is moving towards increased automation, where features such as automated driving systems typically include machine learning (ML), e.g. in the perception system. [**Question/Problem**] Ensuring safety for systems partly relying on ML is challenging. Different approaches and frameworks have been proposed, typically where the developer must define quantitative and/or qualitative acceptance criteria, and ensure the criteria are fulfilled using different methods to improve e.g., design, robustness and error detection. However, there is still a knowledge gap between quality methods and metrics employed in the ML domain and how such methods can contribute to satisfying the vehicle level safety requirements. [**Principal Ideas/Results**] In this paper, we argue the need for connecting available ML quality methods and metrics to the safety lifecycle and explicitly show their contribution to safety. In particular, we analyse Out-of-Distribution (OoD) detection, e.g., the frequency of novelty detection, and show its potential for multiple safety-related purposes. I.e., as (a) an acceptance criterion contributing to the decision if the software fulfills the safety requirements and hence is ready-for-release, (b) in operational design domain selection and expansion by including novelty samples into the training/development loop, and (c) as a runtime measure, e.g., if there is a sequence of novel samples, the vehicle should consider reaching a minimal risk condition. [**Contribution**] This paper describes the possibility to use OoD detection as a safety measure, and the potential contributions in different stages of the safety lifecycle.

This research has been supported by the Strategic vehicle research and innovation (FFI) programme in Sweden, via the project SALIENCE4CAV (ref. 2020-02946) and by the Wallenberg AI, Autonomous Systems and Software Program (WASP) funded by Knut and Alice Wallenberg Foundation.

© The Author(s), under exclusive license to Springer Nature Switzerland AG 2023
A. Ferrari and B. Penzenstadler (Eds.): REFSQ 2023, LNCS 13975, pp. 233–242, 2023.
https://doi.org/10.1007/978-3-031-29786-1_16

Keywords: Automotive safety · Out-of-Distribution detection · Machine learning · Automated driving systems · Safety requirements

1   Introduction

Machine learning (ML) techniques are increasingly used in many domains to solve problems where rule-based algorithms are difficult or impractical to construct or do not scale. One common use is for object detection and classification in images/video, an area where large advances have been made in the last decade. This is often used for perception systems in robotics as a way to create a world model the machine can use to make decisions on future actions. In some cases, ML is used not only for environment perception but also for decision making. A challenge is determining how well ML will perform in a given application, especially for open-world problems with a virtually infinite variation of environmental conditions, edge cases and potentially adversarial attacks [18].

One such application is automated driving systems (ADS) [15], which are seen as a key technology for more efficient, available and safe mobility. However, for such safety-critical systems, the challenge to determine ML performance also becomes a safety issue. While there are several frameworks and standards, see Sect. 2, for ML in safety-critical applications, and many methods available for improving specification, testing and robustness, or performing run-time error detection, we argue that there is still a gap when it comes to analysing exactly how the available methods can contribute to fulfilling the safety requirements.

As a step towards a better understanding of how to combine the available methods and metrics for ML performance with the need for safety assurance, this paper analyses Out-of-Distribution (OoD) detection in light of a safety lifecycle. The general concept of OoD detection is to distinguish known objects/samples from unknown, e.g., detecting if an input to the ML model is similar enough such that the model may provide an accurate prediction, something that is less likely for a sample of an unknown distribution. Based on this analysis, we identified the following potential roles for OoD detection in a safety lifecycle:

- **Development phase:** During development it may be used to identify limitations in the training dataset through highlighting scenarios where the detection rate is below the required pass/fail threshold. Alternatively, it can suggest operational design domain (ODD) [15] reduction if certain scenarios are not fulfilling the allocated safety requirements.
- **Shadow mode:** From a continuous experimentation and deployment perspective, it could be used to test the expansion of ODD boundaries and highlight scenarios where more training data is needed.
- **Operational phase:** During run-time, OoD detection may help to identify uncertainties in the deep neural network (DNN) and trigger safe fallback routines if the uncertainty in the model goes above pre-defined threshold.

The remainder of this paper connects OoD detection to state-of-the-art safety-related research, followed by a description on how to incorporate OoD detection as a safety measure for different stages of the lifecycle of ML models in an ADS. Our vision is that this work will inspire more experimental work that demonstrates the effect on safety for various ML improvement techniques.

2   Related Work

The functional safety standard *ISO 26262* [10] is essential in automotive development. It deals with hazards caused by technical failures due to random and systematic faults in the system's hardware or software. However, the existing version does not fully cover safety requirements of ADS features relying on environment perception or ML [17]. A major issue with ML based systems is that they are not fully specifiable, while ISO 26262 implicitly assumes that all functionality is specified [8,16,17].

Increased automation raised the need to complement ISO 26262 with the safety of the intended functionality standard *ISO 21448 (SOTIF)* [12] dealing with hazards due to functional insufficiencies in the absence of system failures. Potential hazardous behaviour includes the inability to correctly perceive the environment, the lack of robustness with respect to sensor input variations, and unexpected behaviour by the decision making algorithm. All factors relevant for ML based solutions. The product development activities specified in ISO 21448 can be carried out in parallel with the activities in ISO 26262 as illustrated in Fig, 1. Other relevant standardization works in progress related to safety assessment of road vehicles relying on ML include *ISO/IEC TR 24029, ISO/IEC AWI TR 5469, and ISO/AWI PAS 8800*.

Burton et al. presented a work about safety assurance of ML for perception functions including an analysis of ISO 26262 and ISO 21448 [3]. The authors argue that due to the typical failure modes and performance limitations of ML, an absolute level of correctness is infeasible. Instead, in line with ISO 21448, quantitative assurance targets are required defining an acceptable limit to the probability that guarantees cannot be met. Mohseni et al. presented an extensive review and propose a taxonomy of ML safety that maps state-of-the-art ML techniques to key

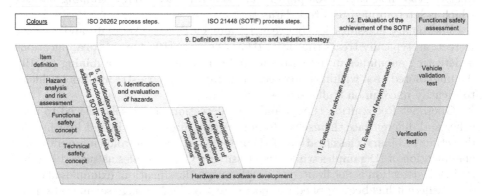

Fig. 1. Possible combined ISO 26262 and ISO 21448 development cycle (based on [12]).

safety engineering strategies [13]. Due to the lack of verification techniques for deep neural networks, ML model validation commonly relies on accuracy measurements on different large tests sets to cover the targeted ODD. This is an important metric of the success of the algorithm, but it fails to capture the model's confidence in its predictions. For example, if an image classifier trained to identify pedestrians and road signs was presented with a billboard showcasing a person, and the model had not encountered such examples during training, it could output class probabilities from anywhere between 100% human to 100% road sign. What these probabilities fail to show is the model's lacking confidence.

Borg et al. [1] studies an autonomous emergency breaking (AEB) system constructed in a simulator. The approach incorporates a *safety-cage* implemented through a variational autoencoder to detect samples that are OoD, and uses the simulator to do rigorous testing such that the system passes safety requirements. The requirements are assigned to the ML model that operates on the front looking camera. The same approach is used in this paper.

OoD detection is one way of dealing with the fact that data used to train an ML model rarely covers all possible scenarios the model will face when put into production. OoD deals with this open world assumption by dividing data into In-Distribution (ID) and OoD, where ID is typically data close to what was seen during training and OoD is unrecognized data. In the article [19] Yang et al. presents a unified framework for generalized OoD detection. This framework will be used as a base for discussing the potential of OoD detection in the ML-lifecycle.

Yang et al. [19] divides the larger subject of general OoD detection into sub-topics, namely: Anomaly detection, Novelty detection, Open set recognition/OoD detection, and outlier detection. The split is motivated with four dichotomies: whether covariate or semantic shift is detected; whether the ID set is treated as a single or multiple classes; whether ID classification is required; and whether the method uses inductive or transductive reasoning. Covariate and sematic shift refers to a difference between the data distributions of the training and testing sets, the two shifts differ in whether the distribution change occurs in the input data or the labels respectively. Inductive reasoning makes use of the entire dataset to make specific inferences on said data, this differs from trans-ductive reasoning that attempts to learn general rules from a training dataset which is later applied to a test dataset.

The binary classification between ID and OoD, i.e., Anomaly and Novelty Detection, typically treats the entire ID as one class [19]. One of its applications is detecting anomalies in data streams and can be used for data mining [14,19]. An unexplored area within this concept is monitoring data streams from vehicles to identify uncommon traffic scenes.

Open Set Recognition (OSR) and OoD detection are closely related to each other and differentiate themselves from the other approaches by not limiting the detection to binary classification of ID or OoD, but instead allowing multi-class classification of ID samples while still classifying OoD samples as such [19]. OSR and OoD detection is not limited to simple image classification, examples of other areas where it has been applied is object detection [5,7], image segmentation [4], and 3D object detection [9].

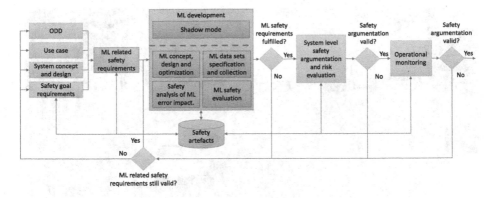

Fig. 2. View of lifecycle for ML based system (inspired by [3,11,12]).

3 Methodology

A possible methodology covering the safety lifecycle for an ML based system is shown in Fig. 2. The lifecycle is inspired by the work of Burton et al. [3] and is compatible with the ISO 26262 and ISO 21448 development cycle shown in Fig. 1. On the left side system level requirements are shown including safety goal requirements. In the next step these are broken down to ML related safety requirements that are input to the ML development phase.

A safety lifecycle typically starts by deriving safety goals, which are the top-level safety requirements, using hazard analysis and risk assessment. These safety goals are addressed through refined safety requirements introduced at different stages of the lifecycle. Figure 2 visualizes how the safety lifecycle can be extended for ML components using an iterative process that aims to extract safety artefacts and refine safety requirements throughout the development process to ensure correct behaviour during operation.

An issue with safety goals is the abstraction level. As they will typically express safe behaviour on vehicle level, refinement to safety requirements for different subsystems of the ADS will be necessary. Our suggestion is to construct a system abstraction to aid the refinement of relevant safety requirements to the respective systems. The abstraction is further elaborated in Sect. 3.1.

Incorporating ML into ADS has been summarized as challenging due to the inherent lack of understanding the ML behavior. Safety requirements can still be applied to systems that incorporate ML, as described in SOTIF. However, determining which methods, such as pass or fail criteria, to use in order to show compliance with the safety requirements are still lacking, both from academia and industry. We propose in this paper the OoD detection method as one potential criteria to determine the compliance level of ML safety requirements. We describe how this technique can be used in the development stage of the lifecycle in Sect. 3.2 and in the verification and operational stages of the lifecycle in Sect. 3.3.

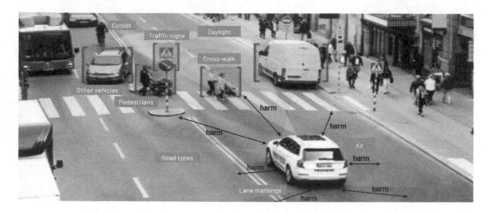

Fig. 3. Photo of an AEB equipped vehicle approaching a cross walk, illustrating a scene in the use case. Brackets indicate objects that must be included in the ODD. Arrows are used to illustrate possible sources of harm.

3.1 Autonomous Emergency Breaking Use-Case

As a relatively simple example to demonstrate the ML safety lifecycle we use an Autonomous Emergency Breaking (AEB) system. Among other goals, this system aims to avoid collisions with pedestrians by performing emergency breaks when said collisions are predicted. A scene from such a use case is shown in Fig. 3 with brackets indicating some of the objects that must be included in the ODD description for the AEB. Note that a complete ODD description must include much more information [6].

As mentioned, common practice in the automotive domain to define the high level safety goals and a safety concept is to follow the process shown in Fig. 1, starting with an item definition followed by a hazard analysis and risk assessment. For the AEB example, one safety goal resulting from this analysis could be to avoid harm to pedestrians. This is illustrated in Fig. 3 with arrows and the text 'harm' pointing to objects critical to avoid.

For a systematical breakdown of the top level safety goal to concrete safety requirements we propose to define the abstraction levels visualized in Fig. 4. The abstraction levels are defined as:

- ODD: The operational design domain defined as a set of operating conditions. Some relevant conditions are visualized in Fig. 3.
- AV: Autonomous Road Vehicle equipped with an AEB feature.
- AEB: The autonomous emergency braking system.
- Sensor/Perception System: Forward Looking Camera containing the DNN based ML algorithm.

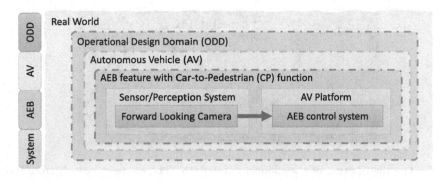

Fig. 4. A figure showing our proposed abstraction levels, in this case specified to our AEB use-case.

The high-level safety requirements are refined and allocated to the different systems in Fig. 4 after a fault tree analysis and system decomposition. We analyzed and designed the forward looking camera in the 'Sensor/Perception Platform' to satisfy the allocated safety requirements. For that purpose we developed a technical safety concept and derived technical safety requirements towards the perception model within the forward looking camera, which uses a DNN algorithm for object detection. Here we have used and modified safety requirements allocated to machine learning from [1] alongside the lifecycle in Fig. 2.

Below are some examples from the derived ML safety-related requirements that are allocated to the ML component [1]:

– SYS1-FLC-PER-REQ 1: The false negative rate of the perception algorithm within the forward looking camera shall not exceed 7% within 50ms.
– SYS1-FLC-PER-REQ 2: The false positive per image of the perception algorithm within the forward looking camera shall not exceed 0.1% within 80ms.

3.2   Machine Learning Development

With the initial safety requirements for the ML component in place, the ML development block in Fig. 2 can be pursued. Within the model optimization block, the process for training the model is defined, which encompasses the labeling format to fulfill the functional requirements, the model architecture type, and the optimization methods (such as the optimization strategy, training length, and loss evaluation). The block also marks the first use of OoD detection to identify underrepresented areas in the ODD and set the threshold for classifying data as ID or OoD.

If the training results are insufficient, two options are available: either improving the model performance or adjusting the functional scope. To enhance the model, acquiring additional training data or applying better architectures, techniques, or mitigation strategies may help. However, if the performance still falls short, adjusting the functional scope through refining the requirements or reducing the size of the ODD, particularly in weakly covered areas identified by the

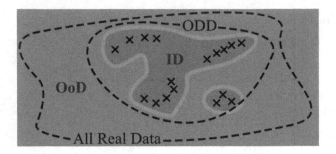

Fig. 5. The distribution of the models input data. The crosses mark discrete data points, the blue region corresponds ID regions, and the red region represents the OoD region. (Color figure online)

OoD detection, becomes necessary. Figure 5 offers a visual representation of the relationship between the ODD, the OoD, and the ID. In the figure, crosses denote discrete data points in the input data space, while the blue and red regions indicate areas that the model has comprehended (ID) or failed to (OoD). For the depicted case, the model has not covered all regions inside the ODD, therefore, the ODD must either be reduced or more data must be collected.

To determine how ML errors may impact the safety, we followed the safety analysis of Burton et al. [3]. Some of their conclusions can be seen in Fig. 6, these helped inform us how to formulate our safety requirements. One of the main causes of these safety errors lies in distributional shift, which can be mitigated using OoD as described in this section.

3.3 Remaining Lifecyle

Once the AEB System that incorporates ML has demonstrated acceptable performance during the ML development phase, the system level integration and verification and validation process can begin. During this step, the system level performance is evaluated to confirm that it complies with the previously generated safety requirements, thus ensuring that the system level safety argumentation is valid.

As part of verifying that the safety requirements are fulfilled, one can use shadow mode operation, i.e., the feature is deployed in the field but not allowed to control the vehicle. Instead the feature continuously collects and categorizes data into ID or OoD in the background. This allows for the identification of scenarios that are still challenging for the model, which in turn provides insights into how the current ML component can be improved through more focused data collection or ODD modification.

Subsequently, the AV may be deployed. It is at this point that operational monitoring commences, through the utilization of OoD detection to identify scenarios where the system is operating in uncertain conditions, such as those that were not well represented during the training phase and may not have been

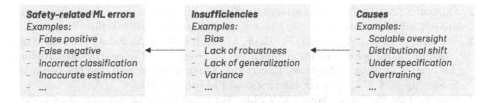

Fig. 6. Fault-model of safety-related ML errors (based on [2]).

generalized appropriately or at the edge of the ODD. If such scenarios arise, the system may revert to a safe state, such as achieving a minimal risk condition (where the vehicle is at a standstill) or transferring control to a human driver.

4   Conclusion and Outlook

This paper has introduced system abstraction levels for an ADS that allows for a breakdown of safety goals into safety requirements that can directly be allocated to the necessary systems, especially with a focus on how to derive safety requirements for subsystems making use of ML. This allows for separation of the existing concerns in the system, e.g., faults, errors or insufficiencies in the functionality.

OoD detection is one technique that is beneficial to validate the system to be within acceptable error margins. The technique allows for risk mitigation in ML components as it aims to reduce the false positive and negative samples. It is shown how this method can be used for varying purposes at different stages of the safety lifecycle.

Going forward, more experiments are needed on safety assurance techniques for ML to establish the actual performance of the system. For instance, while we have outlined how one can use OoD in the safety lifecycle, the extent of the contribution from this technique towards fulfilling the safety requirements is still unknown and will need more extensive experiments. An aim with this work is also to highlight the need to combine runtime evaluation techniques to combat the uncertainty that exist within the ML component.

Our vision and hope is to inspire more experimental evaluation and analysis into how OoD and other methods from the ML domain can be applied in a safety lifecycle, in particular how each method can help fulfill safety requirements allocated to ML components. Such knowledge will be crucial for determining how and when ML can be used in the design of safety-critical systems.

References

1. Borg, M., et al.: Ergo, smirk is safe: A safety case for a machine learning component in a pedestrian automatic emergency brake system. arXiv preprint arXiv:2204. 07874 (2022)

2. Burton, S.: A causal model of safety assurance for machine learning. arXiv preprint arXiv:2201.05451 (2022). https://doi.org/10.48550/arXiv.2201.05451

3. Burton, S., Hellert, C., Hüger, F., Mock, M., Rohatschek, A.: Safety Assurance of Machine Learning for Perception Functions. In: Deep Neural Networks and Data for Automated Driving: Robustness, Uncertainty Quantification, and Insights Towards Safety, pp. 335–358. Springer International Publishing (2022)

4. Cen, J., Yun, P., Cai, J., Wang, M.Y., Liu, M.: Deep metric learning for open world semantic segmentation. In: Proceedings of the IEEE/CVF International Conference on Computer Vision, pp. 15333–15342 (2021)

5. Du, X., Wang, X., Gozum, G., Li, Y.: Unknown-aware object detection: Learning what you don't know from videos in the wild. In: Proceedings of the IEEE/CVF Conference on Computer Vision and Pattern Recognition, pp. 13678–13688 (2022)

6. Gyllenhammar, M., et al.: Towards an operational design domain that supports the safety argumentation of an automated driving system. In: Proceedings of ERTS 2020. Toulouse, France (2020)

7. Hendrycks, D., Basart, S., Mazeika, M., Mostajabi, M., Steinhardt, J., Song, D.: Scaling out-of-distribution detection for real-world settings. arXiv preprint arXiv:1911.11132 (2019)

8. Hoss, M., Scholtes, M., Eckstein, L.: A Review of Testing Object-Based Environment Perception for Safe Automated Driving. Autom. Innov. 5(3), 223–250 (2022). https://doi.org/10.1007/s42154-021-00172-y

9. Huang, C., et al.: Out-of-distribution detection for lidar-based 3d object detection. arXiv preprint arXiv:2209.14435 (2022)

10. ISO: 26262:2018 Road Vehicles - Functional Safety. ISO (2018)

11. ISO: ISO/TR 4804:2020 Road Vehicles - Safety and Cybersecurity for Automated Driving Systems - Design, Verification and Validation. ISO (2020)

12. ISO: 21448:2022 Road Vehicles - Safety of the Intended Functionality. ISO (2022)

13. Mohseni, S., Wang, H., Yu, Z., Xiao, C., Wang, Z., Yadawa, J.: Taxonomy of Machine Learning Safety: A Survey and Primer. arXiv:2106.04823 [cs] (Mar 2022)

14. Ramachandra, B., Jones, M., Vatsavai, R.R.: A survey of single-scene video anomaly detection. IEEE Trans. Pattern Analysis Mach. Intell. 44, 2293–2312 (2020)

15. SAE: J3016 Taxonomy and Definitions for Terms Related to Driving Automation Systems for On-Road Motor Vehicles. Tech. Rep. J3016:2021, SAE Int. (Apr 2021)

16. Salay, R., Czarnecki, K.: Using Machine Learning Safely in Automotive Software: An Assessment and Adaption of Software Process Requirements in ISO 26262. arXiv:1808.01614 [cs, stat] (Aug 2018)

17. Salay, R., Queiroz, R., Czarnecki, K.: An Analysis of ISO 26262: Using Machine Learning Safely in Automotive Software. Arxiv preprint 1709.02435. (2017)

18. Tencent Keen Security Lab: Experimental Security Research of Tesla Autopilot. Tech. rep., (Mar 2019), https://keenlab.tencent.com/en/whitepapers/Experimental_Security_Research_of_Tesla_Autopilot.pdf

19. Yang, J., Zhou, K., Li, Y., Liu, Z.: Generalized out-of-distribution detection: A survey. arXiv preprint arXiv:2110.11334 (2021)

Crowd RE

Automatically Classifying Kano Model Factors in App Reviews

Michelle Binder[1], Annika Vogt[1], Adrian Bajraktari[2],
and Andreas Vogelsang[2]

[1] University of Cologne, Cologne, Germany
{mbinder1,avogt16}@smail.uni-koeln.de
[2] Computer Science, University of Cologne, Cologne, Germany
{bajraktari,vogelsang}@cs.uni-koeln.de

Abstract. **[Context and motivation]** Requirements assessment by
means of the Kano model is common practice. As suggested by the original authors, these assessments are done by interviewing stakeholders
and asking them about the level of satisfaction if a certain feature is
well implemented and the level of dissatisfaction if a feature is not or not
well implemented. **[Question/problem]** Assessments via interviews are
time-consuming, expensive, and can only capture the opinion of a limited
set of stakeholders. **[Principal ideas/results]** We investigate the possibility to extract Kano model factors (basic needs, performance factors,
delighters, irrelevant) from a large set of user feedback (i.e., app reviews).
We implemented, trained, and tested several classifiers on a set of 2,592
reviews. In a 10-fold cross-validation, a BERT-based classifier performed
best with an accuracy of 92.8%. To assess the classifiers' generalization,
we additionally tested them on another independent set of 1,622 app
reviews. The accuracy of the best classifier dropped to 72.5%. We also
show that misclassifications correlate with human disagreement on the
labels. **[Contribution]** Our approach is a lightweight and automated
alternative for identifying Kano model factors from a large set of user
feedback. The limited accuracy of the approach is an inherent problem of
missing information about the context in app reviews compared to comprehensive interviews, which also makes it hard for humans to extract
the factors correctly.

Keywords: Requirements Analysis · Kano Model · App Store
Analytics · Machine Learning · NLP

1 Introduction

Figuring out which features and related requirements are important to stakeholders and, thus, should be implemented first or with special care is one of the
core activities in Requirements Engineering (RE). There is a plethora of requirements prioritization techniques, in which usually costs and benefits are weighed
up either by expert assessment or by stakeholder involvement [4,13].

© The Author(s), under exclusive license to Springer Nature Switzerland AG 2023
A. Ferrari and B. Penzenstadler (Eds.): REFSQ 2023, LNCS 13975, pp. 245–261, 2023.
https://doi.org/10.1007/978-3-031-29786-1_17

One of the most well-known and applied techniques in requirements prioritization is the Kano model [17]. It is based on the two-factor theory by Herzberg et al. [14], which says that a factor that leads to satisfaction does not necessarily lead to dissatisfaction if absent and vice versa. The Kano model categorizes product features into a set of five factors, which have different satisfaction–dissatisfaction profiles.

Several studies unanimously report that scalability is one of the major limitations of requirements prioritization techniques [1,4,16]. This also holds for the assessment of product features according to the Kano model. The categorization of product features into the five Kano factors is done by interviewing or surveying stakeholders. This process is laborious and limited to the set of available stakeholders and a set of predefined requirements under investigation.

We investigate the possibility to identify Kano model factors automatically from a large set of user feedback to increase scalability and broaden the focus to a large set of users and their specific feedback. We implemented, trained, and tested several classifiers to explore their ability to identify Kano model factors in app reviews. The resulting categorization is, first of all, a categorization of user feedback, which may later be related to product features either manually or automatically [2,11].

We trained and evaluated our classifiers on two datasets, which were collected by independent research groups and which we labeled manually. To do so, we created a labeling guideline and labeled a large part of the data independently by two labelers. We used random undersampling to create a balanced set of 2,592 app reviews from the larger of the two datasets. We used this dataset for training and testing the classifiers in terms of a 10-fold cross validation. We used the second dataset to test the classifiers on 1,622 unseen and independently collected app reviews. Finally, we compared misclassifications of the classifiers with the initial labels of the two human labelers and whether they agreed or disagreed initially. In this paper, we make the following contributions:

- We find that Kano model factors can, to some degree, be automatically identified in app reviews.
- We find that our Kano model classifiers still lack sufficient generalization to other datasets.
- We find that misclassifications, to some extent, correlate with human disagreement on the labels.
- We publish two datasets with 8,126 app reviews overall, with labels representing Kano model factors.

Our approach is a lightweight and automated alternative for identifying Kano model factors from a large set of user feedback. The limited accuracy of the approach is an inherent problem of missing information about the context in app reviews compared to comprehensive interviews, which also makes it hard for humans to extract the factors correctly.

Availability of Data and Code: The datasets including the Kano model labels and the code for all classifiers and analysis procedures are publicly available.[1]

---

[1] https://doi.org/10.6084/m9.figshare.21618858.

2   Background and Related Work

In this section, we will introduce the Kano model and present related work in the field of app store analytics and natural language processing (NLP).

2.1   Kano Model

The Kano model [17] describes the relationship between customer satisfaction and the implementation status of quality characteristics of a product. Kano distinguishes between five categories:

- **Basic features** are perceived by users as intrinsic and "normal" for a product type. They only become aware of them when they are not available or not working (implicit expectations). Their presence does not lead to any satisfaction, while their absence leads to dissatisfaction.
- **Performance features** lead to dissatisfaction if not or poorly implemented, while leading to satisfaction if fully implemented.
- **Delighters** are features the customer is not expecting. Their presence leads to satisfaction, while they do not lead to dissatisfaction if not implemented.
- **Irrelevant features** lead neither to satisfaction nor to dissatisfaction.
- **Rejection features** lead to dissatisfaction if they are implemented.

In this paper, we ignore rejection features since their definition in the literature is ambiguous. The effect of their absence is described as causing dissatisfaction or as causing neither satisfaction nor dissatisfaction. Many sources only mention the first three factors of the Kano model.

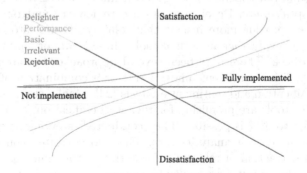

Fig. 1. Visualization of the Kano model.

Figure 1 shows a common representation of the model including rejection features according to the former interpretation. Features are rarely universal, i.e., what is a basic feature for one customer may be a performance feature for another, etc. In addition, there is a temporal aspect to these categories. As time goes on, delighters become performance features and performance features become basic features.

In his original work, Kano proposed to conduct interviews with stakeholders to categorize features into these classes. However, due to the massive amount of users of apps and the amount of reviews in app stores, neither interviews nor manual categorization are feasible for app developers.

2.2 Crowd-Based RE and App Store Analytics

User feedback is an important asset in the development of apps. A study [25] on the usage of analytics tools in app stores showed that tools only providing sales, download, and demographic data are not of high interest for developers. However, developers perceive tools that support app review analytics as helpful. Wang et al. [31] did a systematic mapping study on crowd-sourced requirements engineering using user feedback. They found that, in many works, user feedback has been used in requirements elicitation and requirements analysis mainly, but also in requirements management. Wouters et al. [34] created a method to integrate crowd-based requirements engineering into development. The crowd is responsible for generating, voting, and discussing ideas, while the remaining activities are done by the development team. Lim et al. [21] did a systematic literature review on data-driven requirements elicitation. They found that there are seven main sources of data used in the literature: online reviews, blogs, forums, software repositories, usage data, sensor readings and mailing list. Further, the main methods used were categorized as machine learning, rule-based classification, model-oriented, topic modeling and clustering.

Reviews in app stores are a rich source of user feedback for crowd-based RE [10]. Pagano and Maalej [27] conducted an empirical study on user feedback in app stores, showing that app stores can serve as communication channels between users and developers, allowing to continuously receive bug reports, feature requests, praise, etc. Developers can use reviews to understand new user needs since they provide more insight than plain statistics into how apps are actually used. They further find that tools should support automatic analyses of user feedback. There have been several approaches to automatic feature extraction, e.g., using sentiment analysis [11] or a combination of NLP, metadata, text classification and sentiment analysis [24]. Maalej et al. [23] found that review analytics tools are promising for review classification, as a classification accuracy of 85% to 92% is possible. They conducted interviews with nine practitioners to evaluate their analytics tool. They do not often consider reviews, as the manual extraction of relevant information is too time consuming. Also, they usually gather input from multiple sources, e.g., emails, test groups, etc. However, they perceive user reviews as a promising source of information when assisted by tools to filter and categorize them automatically.

We found two papers, where the authors suggested automatic approaches to identify Kano model factors in user feedback. AlAmoudi et al. [2] analyzed app store reviews and categorized them according to the Kano model by using NLP techniques and clustering. They achieved high precision but low recall for basic features, high recall but low precision for delighter features and mid to low results on performance features. Lee et al. [20] used sentiment analysis to categorize hotel service ratings into Kano factors. They achieved rather low results and concluded that the reviews tend to be more about personal experience with a service, rather than an overall evaluation.

2.3 NLP and Machine Learning

Natural language, due to its easy to write and comprehend nature, is the traditional way to document requirements. In the last decades, requirement engineers studied many aspects of NLP, ranging from modeling and abstracting key elements to automatic classification and clustering [6]. Two key challenges when using NLP for RE are availability of proper datasets and domain adaptation of models [6]. Applying NLP tools to RE task has developed from using traditional machine learning techniques on hand-crafted features like bag of words [24] to deep learning techniques where the input is encoded with word embeddings [32]. Recently, transfer learning approaches that work on large pretrained language models (e.g., BERT) showed the best results for many RE tasks [9,12,15,29].

3   Research Methodology

Our research is exploratory in the sense that we did not investigate any specific hypotheses. Instead, we propose and implement several automatic solutions and evaluate their performance using a quantitative evaluation study. Figure 2 shows an overview of our research design.

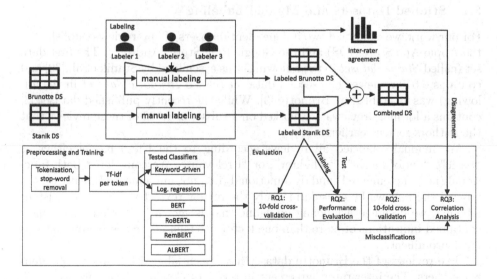

Fig. 2. Overview of research design

We used two public datasets that we labeled manually. More details on the labeling process will be provided in Sect. 3.1. We used these datasets to train and test several classifiers: two simple solutions that served as baselines (keyword-driven and logistic regression) and four variations of transfer learning classifiers (BERT, RoBERTa, RemBERT, and ALBERT). We used these classifiers from the BERT family since they showed good performance in similar RE tasks [9,12, 15,29] and they are conveniently offered by ML libraries[2]. More details on the classifiers will be given in Sect. 4. With this research design, we want to answer the following research questions:

RQ1: What performance do automatic classifiers achieve in identifying and distinguishing Kano model factors in app reviews? To answer this research question, we evaluate several classifiers (simpler and BERT-based) and compare their performance.

RQ2: How well do the classifiers generalize when applied to unseen app reviews? To answer this research question, we evaluate each model on a dataset that is completely different from the training data with respect to the contained reviews and to the time period in which they were collected. Further, we perform a 10-fold cross-validation on a combination of both datasets to investigate whether a more diverse dataset improves the classifiers' performance.

RQ3: Does misclassification of automatic classifiers correlate with disagreement of human judgement? To answer this research question, we assess the relationship between the cases that the classifier labeled (in)correctly and the cases where the human annotators (dis)agreed.

3.1   Studied Datasets and Manual Labeling

For our work, we examined two independent datasets of app reviews published in the Apple App Store (iOS) and the Google Play Store (Android). The first data set (called *Stanik dataset* in the following) was assembled by Stanik et al. [30] and contains 6,070 reviews. The second data set (called *Brunotte dataset* in the following) was assembled by Brunotte [3]. While the recently published dataset [3] contains a lot more reviews, we worked on a subset of 1,622 of these reviews that the authors sent us earlier.

We manually labeled all reviews according to the Kano model either as "basic", "performance", "delighter", or "irrelevant". We considered both functional (i.e., implemented) and dysfunctional (i.e., not implemented) features. We labeled the reviews by considering only the review text. To ensure a consistent labeling, we designed a labeling guideline that is shown in Table 1. If a review contained indications of more than one factor, we focused on the factor that was most prominent.

The reviews of the Brunotte dataset have been labeled by two independent researchers. The inter-rater agreement in terms of Cohen's Kappa was $\kappa = 0.7$, which may represent a "substantial agreement" [19]. If the two labelers disagreed, we involved a third labeler as a tie breaker. In the labeled dataset, we marked

---

[2] We used the Simple Transformers library: https://github.com/ThilinaRajapakse/simpletransformers.

Table 1. Kano labeling guideline

Basic	Performance
– user discontinues using the app or switches to alternative	– moderate amount of expressed joy / annoyance
– app is not usable (e.g., crashes, log in not possible)	– constructive criticism / suggestions
– lack of a basic feature results in a bad rating	
Delighter	**Irrelevant**
– app is favored and recommended over similar apps due to these features	– cannot be labeled as any other category
– user is a long term user due to these features	– no clear reference to a distinct feature
– praise or suggestion for addition	

whether the label of a review was unanimously assigned or if there was a need for a tie breaker. The resulting distributions of labels in the two datasets are depicted in Table 2.

Table 2. Distribution of labels in the datasets

Dataset	Total	Basic	Performance	Delighter	Irrelevant
Stanik dataset [30]	6,070	1,440	1,530	648	2,452
Brunotte dataset [3]	1,622	1,102	395	95	30

3.2 General Data Preprocessing

Data preprocessing for all classifiers consisted of the following steps: (1) We removed duplicates, non-English reviews, and reviews that consisted only of characters but no words. (2) We converted the labels into numerical values. Further preprocessing relevant only for specific classifiers is described in the sections of the classifiers.

3.3 Evaluation Strategy

To answer the research questions, we trained and tested several classifiers. For both RQ1 and RQ2, we used the Stanik dataset (or a subset of it) as training set. To mitigate the class imbalance, we performed random undersampling[3] to create a balanced dataset with 2,592 reviews (648 reviews per class). To mitigate the

---

[3] Random undersampling deletes examples from the majority class randomly until all classes have equally many samples.

random effect of undersampling, we did this five times and report the average performance metrics achieved with the five training samples. We report standard evaluation metrics (accuracy, precision, recall, F_1).

For RQ1 (general performance), we performed a 10-fold cross-validation on the undersampled Stanik dataset.

For RQ2 (performance on unseen data), we used the undersampled Stanik dataset for training and the Brunotte dataset for testing. Additionally, we combined both original datasets, undersampled the combined dataset to create a balanced dataset of 2,936 reviews (743 per class), and performed a 10-fold cross-validation on this combined dataset.

For RQ3 (correlation between misclassification and human disagreement), we split the Brunotte dataset into two subsets: one containing the reviews where the two labelers initially agreed on the label, and one containing the reviews where the two labelers initially disagreed on the label. We analyze the accuracy of the classifiers for these two subsets. In addition, we calculate a coefficient for the correlation between initial human disagreement and misclassifications of the classifiers.

4   Kano Factor Classifier

In this section, we describe the classifiers we implemented and tested.

4.1   Baseline Algorithms

Besides the preprocessing described in Sect. 3.2, we performed tokenization, stopword removal, and computed tf-idf values for both baseline classifiers. For each non-stopword term t in the training set D and each label $\ell \in \{\texttt{basic}, \texttt{delighter}, \texttt{irrelevant}, \texttt{performance}\}$, we calculated the term frequency-inverse document frequency $\texttt{tf-idf}(t, D_\ell, D)$, where D_ℓ is the set of all reviews in the training set that are labeled as ℓ. For both approaches we used functionalities of the scikit-learn library [28].

Keyword-Driven Classifier. We want a classifier that classifies a review based on the distribution of keywords among each class. For each review R and each label ℓ, we calculate the sum of the tf-idf values of all terms contained in the review: $M_\ell = \sum_{t \in R} \texttt{tfidf}(t, D_\ell, D)$ and then categorize the review by the label $\arg\max_\ell M_\ell$, which maximizes the sum of the tf-idf values.

Logistic Regression. As a second baseline algorithm, we implemented a simple tf-idf based logistic regression, where we provide the document-term matrix containing all tf-idf values per review and non-stopword as input.

4.2 Transfer Learning Classifiers

Transfer learning approaches use language models that have been pretrained on large sets of textual data (unsupervised learning). These language models are afterwards *finetuned* with labeled data from the task and domain they are supposed to be transferred to (supervised learning). We implemented four transfer learning classifiers that use different pretrained language models namely:

- BERT [8], a language representation model for pretraining deep bidirectional representations from unlabeled text.
- RoBERTa [22], an optimized version of BERT with more training data, slightly different training parameters, and different masking procedure.
- RemBERT [5], a pretrained language model with decoupled embeddings.
- ALBERT [18], a lite variant of BERT with less parameters resulting in faster training and less memory consumption.

By using the Simple Transformers library[4] to implement these classifiers, additional preprocessing was kept to a minimum, as this is already handled by the library. We used the default values for all hyperparameters.

5 Results

In this section, we present the results of our evaluation. For this, we divide the section into three parts, each covering one of our research questions. We performed each run on five different undersampled sets. The values in the tables are averages of each reported metric of the five runs.

5.1 RQ1: Performance of Classifiers

Table 3. Performance results from a 10-fold cross-validation on Stanik dataset in terms of accuracy (Acc.) and precision (Prec.), recall (Rec.), and F_1-score (F_1) for each label.

Classifier	Acc.	Basic			Performance			Delighter			Irrelevant		
		Prec.	Rec.	F_1	Prec.	Rec.	F_1	Prec.	Rec.	F_1	Prec.	Rec.	F_1
Keyword-Driven	.514	.589	.793	.675	.494	.521	.505	.411	.552	.467	.810	.194	.303
Logistic Regression	.663	.790	.824	.805	.568	.535	.549	.587	.559	.568	.698	.735	.714
RoBERTa	.918	.951	.967	.959	.880	.851	.864	.875	**.928**	.899	**.972**	.924	.947
BERT	**.928**	**.960**	**.972**	**.966**	**.896**	**.871**	**.883**	**.894**	.927	**.910**	.964	**.941**	**.952**
RemBERT	.633	.586	.673	.626	.518	.550	.534	.538	.630	.580	.596	.683	.637
ALBERT	.838	.893	.928	.909	.760	.721	.740	.796	.823	.809	.901	.878	.888

Table 3 shows the performance metric scores each classifier achieved in a 10-fold cross-validation on the Stanik dataset. The BERT classifier outperformed both baseline classifiers and RemBERT by a magnitude, with an accuracy of 92.8%.

BERT performs best across almost all labels and metrics, but RoBERTa's scores are very close to those of BERT. RoBERTa achieved slightly higher scores for recall in delighter features (+0.1%) and precision in irrelevant features (+0.8%). To answer RQ1, we can summarize that automated classifiers can identify and distinguish Kano model factors with an accuracy of up to 92.8%.

5.2   RQ2: Generalization to Unseen Data

Table 4 shows the performance results of all tested classifiers when trained on the entire Stanik dataset and tested on the Brunotte dataset. RoBERTa generally performed best with an accuracy of 72.5%, but for irrelevant features, Rem-BERT achieved a better recall than any other approach and ALBERT achieved the highest precision. Also, for delighters, BERT achieved a better recall than RoBERTa. Table 5 shows the results of a 10-fold cross validation on the combination of the Stanik dataset and the Brunotte dataset. Here, BERT is the clear winner, as it performed best across all labels and metrics, with the only exception being that RoBERTa, with +0.1%, had a negligible higher precision on irrelevant features. Again, the scores of BERT and RoBERTa are very close together. BERT achieves a very good accuracy of 95.7%, which is a huge improvement over the logistic regression classifier as best-performing baseline, only achieving an accuracy of 60%.

Table 4. Performance results from a validation on the Brunotte dataset (training on Stanik dataset) in terms of accuracy (Acc.) and precision (Prec.), recall (Rec.), and F_1-score (F_1) for each label.

Classifier	Acc.	Basic			Performance			Delighter			Irrelevant		
		Prec.	Rec.	F_1	Prec.	Rec.	F_1	Prec.	Rec.	F_1	Prec.	Rec.	F_1
Keyword-Driven	.593	.800	.741	.770	.513	.208	.296	.143	.663	.235	.000	.000	.000
Logistic Regression	.600	.858	.689	.764	.451	.360	.400	.181	.726	.289	.075	.100	.086
RoBERTa	**.725**	**.896**	**.783**	**.836**	**.515**	**.599**	**.553**	.391	.781	**.520**	.353	.093	.147
BERT	.682	.895	.734	.806	.461	.556	.504	.336	**.804**	.473	.381	.066	.112
RemBERT	.488	.538	.475	.504	.354	.558	.433	.225	.423	.293	.304	**.233**	**.264**
ALBERT	.660	.885	.715	.789	.432	.523	.473	.302	.770	.433	**.466**	.040	.072

Comparing the data from Tables 4 and 5, we can see significant differences. In the 10-fold cross-validation setting, performed on the combination of both datasets, we achieved very good results. In the setting where we trained on the Stanik dataset and evaluated on the Brunotte dataset, we see significantly lower metric scores. This indicates that, while the classifiers perform good when applied to unseen but "similar" data, this is not the case when evaluated on "unsimilar" data. By "similar", we mean the characteristics of the datasets, as each have been collected in different time periods, both cover different apps, etc. We specifically see problems when applied to "unsimilar" data in terms of delighter and irrelevant features, which is not the case in the other two evaluation

Table 5. Performance results from a 10-fold cross validation on the combined dataset (Stanik dataset and Brunotte dataset) in terms of accuracy (Acc.) and precision (Prec.), recall (Rec.), and F_1-score (F_1) for each label.

Classifier	Acc.	Basic			Performance			Delighter			Irrelevant		
		Prec.	Rec.	F_1	Prec.	Rec.	F_1	Prec.	Rec.	F_1	Prec.	Rec.	F_1
Keyword-Driven	.482	.475	.826	.602	.436	.466	.445	.452	.428	.438	.822	.207	.330
Logistic Regression	.655	.723	.811	.764	.544	.499	.519	.600	.574	.586	.725	.728	.725
RoBERTa	.946	.965	.976	.969	.926	.900	.911	.914	.952	.932	**.984**	.960	.971
BERT	**.957**	**.971**	**.983**	**.978**	**.944**	**.923**	**.932**	**.940**	**.960**	**.949**	.983	**.972**	**.977**
RemBERT	.528	.416	.509	.458	.399	.468	.431	.413	.560	.474	.462	.580	.514
ALBERT	.891	.923	.946	.934	.839	.819	.829	.866	.879	.872	.934	.923	.928

settings. This may be due to the time span that lies between the collection of the Stanik dataset (2013–2015) and the Brunotte dataset (2021), as language and culture in app reviews might have evolved, but also in accordance with the Kano model, features that once were delighters now fall in different categories and once irrelevant features may have become relevant. In contrast, features, which then were basic features may still be basic features nowadays. To answer RQ2, we can say that generalization to unseen and dissimilar app reviews is moderate. Possible solution or mitigation approaches are discussed in Sect. 6.

5.3 RQ3: Correlation Between Misclassification and Human Disagreement

Table 6 shows the accuracy of the classifiers when trained on the Stanik dataset and tested on the reviews of the Brunotte dataset with consistent initial labels and inconsistent initial labels. Further, for each classifier we computed the *phi* coefficient (or mean square contingency coefficient) to denote the correlation between misclassification and ambiguity of manual labeling. As binary variables, we used *Mis* $\in \{0,1\} = \{1$ if classification is incorrect, 0 otherwise$\}$ and *Diff* $\in \{0,1\} = \{1$ if initial labels have been different, 0 otherwise$\}$. The phi coefficient is given in the last column of Table 6.

We see a significant difference in the accuracy of labels where the raters agreed vs. labels where they disagreed, both for the baselines and for the BERT-based classifiers. For RemBERT and the logisitic regression, we observed a negligible positive correlation (< 0.2). For the keyword-based classifier, RoBERTa, BERT and ALBERT, we observed a weak positive correlation (0.2–0.3).

Table 6. Accuracy results for classification on labels with initial agreement vs. initial disagreement.

Classifier	Accuracy (agreed labels)	Accuracy (disagr. labels)	Phi Coeff.
Keyword-Driven	.600	.292	.219
Logistic Regression	.642	.293	.192
RoBERTa	**.775**	**.437**	**.269**
BERT	.726	.428	.229
RemBERT	.509	.362	.094
ALBERT	.703	.405	.224

To answer RQ3, we can say that all classifiers performed worse in terms of accuracy on reviews that caused initial disagreement among human annotators. The phi coefficients, though, are not impressively high and thus a strong correlation is not clearly visible.

6   Discussion

In this section, we discuss our findings and its impact for research and practice. We provide a critical analysis of its strengths, weaknesses, and the threats to validity.

6.1   Impact in Practice

We think that our results are promising and show the potential for an automatic solution with sufficient performance. In comparison to the largely manual original Kano model analysis, an automatic approach is cheaper and scales better to large sets of user feedback. The results may support requirements engineering and decision making.

Comparison with Existing Approaches. Existing automated approaches [2, 20] try to mimic the original Kano analysis by identifying the sentiment in user feedback, clustering it according to latent topics, and finally relating it to certain product features. The reported evaluations, however, indicate low predictive performance. AlAmoudi et al. [2] report F_1-scores between 0.30 and 0.63 for three Kano model factors. Lee et al. [20] did not perform a quantitative evaluation but report that "some meaningful results are found [...] This resulted in the limitation of quantification of 'Topic Modeling' method." We follow a different approach and use supervised learning to predict the Kano model factors directly from the text. Our results showed F_1 scores above 0.9 for all Kano model factors in the 10-fold cross validation on our combined dataset. Since the authors of the two papers did not share their datasets, we were not able to perform a direct comparison but we are confident that our approach would outperform the existing approaches also in a direct comparison.

Reviews are Missing Contextual Information. Our analysis of the labeling process and the performance of the classifiers show that neither humans nor any of our classifiers were able to always predict the Kano labels correctly (w.r.t. to what our truth set defines as correct). This may be an inherent problem of our approach to assess the Kano factors purely based on the text of a review. App reviews are limited in size, lack contextual information, and do not offer possibilities for further inquiries [25]. Therefore, the tone and sentiment that is conveyed with the review also plays a role for assessing the factor. Apparently, some of our tested classifiers were able to incorporate this at least to some degree. On the other hand, assessing Kano model factors by analyzing app reviews may always be less clear than the original assessment via specific interviews or surveys.

Application and Usefulness of the Approach. While our study focuses on the feasibility and the performance of the classifiers, an open question is how our approach is perceived in a real world setting. We envision our approach to be used as an assistant tool for requirements analysts. Therefore, the effectiveness needs to be assessed in its context of use. Results from our performance evaluation may not be transferable directly to its context of use since the use of tools also affects working habits and perceptions of analysts (cf. [33]).

6.2   Impact for Research

Testing Generalization on Unseen Data is Important. Many studies indicate the need to test classifiers on unseen and unconsidered data. Still, this practice is not very common in software engineering research. Just recently, Dell'Anna et al. [7] showed the importance of this step by applying two recently published classifiers to unseen data and observed a significant decrease in performance. In our study, we have seen a similar degradation in performance when training on the Stanik dataset and testing on the Brunotte dataset, although the two datasets are conceptually very similar and both are already diversified by incorporating reviews of several apps. This result shows that more data may be needed to train a classifier that generalizes well.

Hard for Humans, Hard for the Machine. Our results show a correlation between misclassifications and human disagreement. This suggests that reviews that were hard to classify for humans tend to also be hard to classify for the machine. This is consistent for all classifiers that we tested. To illustrate this, consider the following review:

"Most convenient calories counter app I've ever used and I've probably used them all, super easy to add foods on your own and also cheap, super recommended"

We labeled this review as performance factor although one of the two labelers identified it as delighter. Most classifiers also classified it as delighter. The reviewer is very happy about the app (high satisfaction) and favors it over other apps mainly due to easier-to-use features and cheaper price, which are classical performance factors. However, the high level of excitement may also indicate that these feature are real delighters, which the reviewer has never experienced in other apps. Here is another example:

"Can you please update this with the map of Bhutan & Nepal.. I am going to drive to bhutan this October but I can't find any bhutan map which could be useful to work in offline.. Please update us quickly.."

We labeled this review as delighter although one of the two labelers identified it as performance factor. Some classifiers classified it as performance factor while others even predicted it to be a basic feature. The reviewer is asking for an urgently demanded but rare feature, which may indicate that this is really a delighter for the reviewer. However, the review is also phrased in a way that suggests some disappointment ("I can't find", "Please update us quickly.."). Also, it is not clear from the review whether other maps are available offline and, thus, adding specific regions creates proportionately more satisfaction (i.e., performance factor).

6.3   Threats to Validity

Here, we discuss potential threats to validity of our models and evaluation.

Construct Validity. The Kano model, as defined by Kano [17] consists of five classes: delighter, performance, basic, irrelevant, and rejection features. In this paper, we did not consider the rejection feature class, since they are fairly rare and their definition is ambiguous. Still, there may be reviews that fall into this class. Examples include features that are annoying to users, e.g., excessive advertisement, but also features that users perceive as threat to their privacy, e.g., app tracking or the "blue read-checkmark" in messenger apps.

Internal Validity. In our manual labeling process, we assigned each review exactly one label. This, however, might be too coarse grained, as some reviews contain more than one Kano factor, e.g., a user reports a problem with a basic feature but also suggests ideas that are delighters. Reviews containing more than one factor are problematic, as they can be classified ambiguously by the classifiers and human annotators. There are two main approaches to solve this problem: (1) *Separation:* We separate each aspect of a complex review as a distinct review. This, however, may break contextual links between aspects. Also, the boundary between them is often blurred. (2) *Multilabel classification:* When a review covers multiple Kano factors, we can assign multiple labels accordingly. This introduces a huge overhead and thus is only feasible when a significant number of reviews include multiple different factors. Usually, this is not the case. In our labeling process, we decided to assign labels according to the most prominent aspect of a review, e.g., if a user reports a bug that makes the app unusable and expresses their dissatisfaction, but also reports a delighter, we assigned the basic label.

Despite our efforts to make the labeling process as transparent and systematic as possible, there may still be some variability in the resulting gold standard, e.g., misinterpretation of the users intention, blurred boundaries of the Kano factors, too broad or too narrow judgement or human mistakes.

External Validity. Our results have shown that generalization of our tested classifiers is fairly moderate when applied to unseen, dissimilar test data. This may indicate that more data is needed to train a classifier that generalizes better.

Lastly, app reviews are not the only relevant source of user feedback. Nayebi et al. [26] mined 70 apps for six weeks on app store reviews and on Twitter. They found that Twitter provided 22.4% more feature requests and 12.9% more bug reports.

7   Conclusions

In this paper, we presented an automated approach to app review classification according to the Kano model. We evaluated several BERT-based models and found that, overall, BERT performed best with an accuracy of 92.8% to 95.7%. We compared our findings to two baseline approaches based on traditional machine learning techniques and found that most BERT-based classifier outperform them by magnitudes, i.e., by around 30%. We evaluated the generalization of our classifier to unseen app reviews and found that the performance of all classifiers dropped significantly. We conclude that more data is needed to achieve a classifier that performs well on unseen data. We also evaluated that misclassification of the classifiers does, to some degree, correlate with ambiguity in the manual labeling process, as accuracy differs by 33.8% between consistent and inconsistent labeling. What is still missing in our work is an evaluation of the approach in its context, i.e., in terms of a user study.

Acknowledgements. We want to thank the authors of the two datasets for permission to use parts of their dataset and the permission to publish our labeled dataset. We also want to thank Murat Sancak for his initial work on the topic in his Bachelor's thesis.

References

1. Achimugu, P., Selamat, A., Ibrahim, R., Mahrin, M.N.: A systematic literature review of software requirements prioritization research. Inf. Softw. Technol. **56**(6), 568–585 (2014). https://doi.org/10.1016/j.infsof.2014.02.001
2. AlAmoudi, N., Baslyman, M., Ahmed, M.: Extracting attractive app aspects from app reviews using clustering techniques based on kano model. In: IEEE International Requirements Engineering Conference Workshops (REW). IEEE (2022). https://doi.org/10.1109/REW56159.2022.00030

3. Brunotte, W.: App Store Rev. (2022). https://doi.org/10.5281/zenodo.7319510
4. Bukhsh, F.A., Bukhsh, Z.A., Daneva, M.: A systematic literature review on requirement prioritization techniques and their empirical evaluation. Comput. Standards
   Interfaces **69**, 103389 (2020). https://doi.org/10.1016/j.csi.2019.103389
5. Chung, H.W., Févry, T., Tsai, H., Johnson, M., Ruder, S.: Rethinking embedding coupling in pre-trained language models. In: 9th International Conference on
   Learning Representations (ICLR). OpenReview.net (2021)
6. Dalpiaz, F., Ferrari, A., Franch, X., Palomares, C.: Natural language processing
   for requirements engineering: The best is yet to come. IEEE Softw. **35**(5), 115–119
   (2018). https://doi.org/10.1109/ms.2018.3571242
7. Dell'Anna, D., Aydemir, F.B., Dalpiaz, F.: Evaluating classifiers in SE research:
   the ECSER pipeline and two replication studies. Empirical Softw. Eng. **28**(1),
   (2022). https://doi.org/10.1007/s10664-022-10243-1
8. Devlin, J., Chang, M.W., Lee, K., Toutanova, K.: BERT: Pre-training of deep
   bidirectional transformers for language understanding. In: Proceedings of the 2019
   Conference of the North American Chapter of the Association for Computational
   Linguistics: Human Language Technologies, Volume 1 (Long and Short Papers),
   pp. 4171–4186. Association for Computational Linguistics (2019). https://doi.org/
   10.18653/v1/N19-1423
9. Fischbach, J., et al.: Automatic creation of acceptance tests by extracting conditionals from requirements: NLP approach and case study. J. Syst. Softw. **197**,
   11159 (2023). https://doi.org/10.1016/j.jss.2022.111549
10. Groen, E.C., et al.: The crowd in requirements engineering: the landscape and
    challenges. IEEE Softw. **34**(2), 44–52 (2017). https://doi.org/10.1109/ms.2017.33
11. Guzman, E., Maalej, W.: How do users like this feature? a fine grained sentiment
    analysis of app reviews. In: IEEE 22nd International Requirements Engineering
    Conference (RE) (2014). https://doi.org/10.1109/re.2014.6912257
12. Henao, P.R., Fischbach, J., Spies, D., Frattini, J., Vogelsang, A.: Transfer learning
    for mining feature requests and bug reports from tweets and app store reviews.
    In: IEEE International Requirements Engineering Conference Workshops (REW).
    IEEE (2021). https://doi.org/10.1109/rew53955.2021.00019
13. Herrmann, A., Daneva, M.: Requirements prioritization based on benefit and cost
    prediction: An agenda for future research. In: IEEE International Requirements
    Engineering Conference (RE) (2008). https://doi.org/10.1109/re.2008.48
14. Herzberg, F., Mausner, B., Snyderman, B.: The motivation to work. Transaction
    Pub (1993)
15. Hey, T., Keim, J., Koziolek, A., Tichy, W.F.: NoRBERT: Transfer learning for
    requirements classification. In: IEEE International Requirements Engineering Con-
    ference (RE) (2020). https://doi.org/10.1109/re48521.2020.00028
16. Hujainah, F., Bakar, R.B.A., Abdulgabber, M.A., Zamli, K.Z.: Software require-
    ments prioritisation: a systematic literature review on significance, stakeholders,
    techniques and challenges. IEEE Access **6**, 71497–71523 (2018). https://doi.org/
    10.1109/access.2018.2881755
17. Kano, N., Seraku, N., Takahashi, F., Tsuji, S.: Attractive quality and must-be
    quality. J. Japanese Society Qual. Contr. **14**(2), 147–156 (1984)
18. Lan, Z., et al.: A lite bert for self-supervised learning of language representations
    (2019)
19. Landis, J.R., Koch, G.G.: The measurement of observer agreement for categorical
    data. Biometrics **33**(1), 159–174 (1977). https://doi.org/10.2307/2529310

20. Lee, H., Cha, M.S., Kim, T.: Text mining-based mapping for kano quality factor. ICIC Express Letters. Part B, Applications: an International J. Res. Surv. **12**(2), 185–191 (2021)
21. Lim, S., Henriksson, A., Zdravkovic, J.: Data-driven requirements elicitation: a systematic literature review. SN Comput. Sci. **2**, 16 (2021)
22. Liu, Y., et al.: Roberta: A robustly optimized bert pretraining approach. ArXiv abs/1907.11692 (2019)
23. Maalej, W., Kurtanović, Z., Nabil, H., Stanik, C.: On the automatic classification of app reviews. Requirements Eng. **21**(3), 311–331 (2016). https://doi.org/10.1007/s00766-016-0251-9
24. Maalej, W., Nabil, H.: Bug report, feature request, or simply praise? on automatically classifying app reviews. In: IEEE 23rd International Requirements Engineering Conference (RE) (2015). https://doi.org/10.1109/re.2015.7320414
25. Maalej, W., Nayebi, M., Johann, T., Ruhe, G.: Toward data-driven requirements engineering. IEEE Softw. **33**(1), 48–54 (2016). https://doi.org/10.1109/ms.2015.153
26. Nayebi, M., Cho, H., Farrahi, H., Ruhe, G.: App store mining is not enough. In: IEEE/ACM International Conference on Software Engineering Companion (ICSE-C) (2017). https://doi.org/10.1109/icse-c.2017.77
27. Pagano, D., Maalej, W.: User feedback in the appstore: An empirical study. In: IEEE International Requirements Engineering Conference (RE) (2013). https://doi.org/10.1109/re.2013.6636712
28. Pedregosa, F., et al.: Scikit-learn: machine learning in Python. J. Mach. Learn. Res. **12**, 2825–2830 (2011)
29. Sainani, A., Anish, P.R., Joshi, V., Ghaisas, S.: Extracting and classifying requirements from software engineering contracts. In: IEEE International Requirements Engineering Conference (RE). IEEE (2020). https://doi.org/10.1109/re48521.2020.00026
30. Stanik, C., Haering, M., Maalej, W.: Classifying multilingual user feedback using traditional machine learning and deep learning. In: IEEE International Requirements Engineering Conference Workshops (REW), pp. 220–226 (2019). https://doi.org/10.1109/REW.2019.00046
31. Wang, C., Daneva, M., van Sinderen, M., Liang, P.: A systematic mapping study on crowdsourced requirements engineering using user feedback. J. Softw.: Evol. Process **31**(10), e2199 (2019). https://doi.org/10.1002/smr.2199
32. Winkler, J., Vogelsang, A.: Automatic classification of requirements based on convolutional neural networks. In: IEEE International Requirements Engineering Conference Workshops (REW) (2016). https://doi.org/10.1109/rew.2016.021
33. Winkler, J.P., Vogelsang, A.: Using tools to assist identification of non-requirements in requirements specifications – a controlled experiment. In: Requirements Engineering: Foundation for Software Quality (REFSQ), pp. 57–71. Springer International Publishing (2018). https://doi.org/10.1007/978-3-319-77243-1_4
34. Wouters, J., Menkveld, A., Brinkkemper, S., Dalpiaz, F.: Crowdbased requirements elicitation via pull feedback: method and case studies. In: Requirements Engineering. Requirements Engineering (2022). https://doi.org/10.1007/s00766-022-00384-6

Data-Driven Persona Creation, Validation, and Evolution

Nitish Patkar[✉][iD] and Norbert Seyff[iD]

University of Applied Sciences and Arts Northwestern Switzerland,
Windisch, Switzerland
{nitish.patkar,norbert.seyff}@fhnw.ch

Abstract. [**Context and motivation**] Personas are a well-known technique to represent a particular user type and stimulate software development. [**Question/problem**] Personas are often based on findings from ethnographic studies, and their creation can be time and effort intensive. Furthermore, validating the correctness of the Personas is an open issue. [**Principal ideas/results**] We advocate a data-driven approach that relies on analyzing various kinds of user data, in particular, user feedback and monitoring data, to create, validate, and evolve Personas. [**Contributions**] In this research preview paper, we discuss the problem we want to address with our research, formulate research questions, describe the initial technical solution, and present our planned contributions through a fictional usage scenario. Furthermore, we provide initial research results from an ongoing interview study that analyzes the use of Personas in practice.

Keywords: Personas · Data-driven requirements engineering
(DDRE) · User-centered design · Stakeholder analysis

1 Introduction

A Persona is a fictional representation of a user group's common behavior, goals, and motivations, compiled into a single individual [2,4]. Personas can greatly aid the software development process by providing insight into user behavior, goals, and motivations. For instance, user experience (UX) designers can use this information to design user interfaces that better support users in achieving their goals. When done correctly, this can lead to the development of software that is more easily accepted and provide greater value to users.

Traditionally, qualitative methods, for instance, ethnographic studies, have been used to collect relevant information about potential users [9]. Typically, this data collection occurs at the beginning of the development process. Since ethnographic methods are often the primary source of information gathering, the task of creating Personas can be time and resource-intensive. To address this issue, the automated creation of Personas using data science methods and tools to gain insights into large user data has emerged under the umbrella term "Data-Driven

© The Author(s), under exclusive license to Springer Nature Switzerland AG 2023
A. Ferrari and B. Penzenstadler (Eds.): REFSQ 2023, LNCS 13975, pp. 262–271, 2023.
https://doi.org/10.1007/978-3-031-29786-1_18

Persona Development (DDPD)". Salminen *et al.* provide an extensive overview of existing research in DDPD [7]. For instance, recent studies have used data originating from analytics platforms (*e.g.,* clickstreams), social media platforms, as well as data from app stores, such as age, gender, mobile device version, and operating system, to create Personas [3,6,13]. Methods, such as K-means clustering and non-negative matrix factorization, are typically used to analyze data and create Personas [7]. In our opinion, the suggested DDPD approaches are still maturing, which is evident from the fact that they are not widely used in practice yet. As correctly noted by Salminen *et al.* existing research in DDPD assumes that there is sufficient data available for algorithmic analysis to create Personas. This implies that software is already in place and in use. By placing a strong emphasis on the creation of Personas, DDPD does not embrace today's software life cycle in full. We argue that the continuous validation and evolution of Personas need to be considered as well. One aim of our ongoing research is to bridge this gap.

Another motivating factor for our research is our observation that the current adoption of Personas in practice suffers from various issues. This observation is also supported by the literature. For instance, in a 2014 industry survey, Billestrup *et al.* reported the obstacles of using Personas during software development [1]. Among other things, they found that knowledge about Personas varies between companies and different roles within companies. They also found that companies often allocate "zero" resources to Persona development. This can result in superficial Personas that are unfit to be used for development. Furthermore, even when meaningful Personas are created, there is limited empirical evidence on their use, particularly when software maintenance and evolution take place. Little is known about whether and how Personas change and evolve in this time.

> The main goal of our ongoing research is to harness the power of data-driven requirements engineering (DDRE) to automate Persona creation, validation, and evolution as a continuous process.

The overall expected contributions of the planned work are (1) empirical evidence on the use of Personas in practice, (2) a novel data-driven approach including the development of methods, algorithms, and research prototypes, for creating, validating, and evolving Personas, and (3) an evaluation of the proposed solution.

To accomplish our goal, we have identified several research questions and potential research contributions in Sect. 2. In Sect. 3, we sketch the design and development of our envisioned solution and discuss anticipated challenges. We present a fictitious usage example in Sect. 4 to show how Personas could be created, validated, and evolved automatically with our solution. In Sect. 5, we discuss early results from an interview study to better understand the use of Personas in practice. Finally, in Sect. 6, we summarize our main contributions.

2   Research Road Map

Based on our goal to automate continuous Persona creation, validation, and evolution, we defined a road map including key research questions (RQs) and potential research contributions (RCs).

RQ_1: How are Personas created, validated, and evolved in software development projects?

With our work on RQ_1, we foresee the following RC.

RC 1.1 Empirical evidence on Persona usage in software projects. To gain a deeper understanding of the use of Personas, as well as practitioners' perspectives and expert insights on data-driven Persona creation, validation, and evolution, we are using a grounded theory approach [10]. As a first step, we are conducting expert interviews. A large-scale practitioner survey will follow the interviews to validate and generalize the findings. In Sect. 5, we present initial findings related to *RC 1.1*.

RQ_2: How to provide automated support for continuous Persona creation, validation, and evolution?

Based on the results of RQ_1 and utilizing existing work in DDRE, we anticipate three potential RCs. In the first phase of our planned research, our focus will be on validation (RC 2.1), evolution (RC 2.2), and the creation of Personas for existing software (RC 2.3). It is important to note that our proposed approach would not eliminate the need for human involvement. However, we foresee that our automated solution supports its users, *e.g.*, requirements engineers, with the necessary information to make informed decisions about Personas.

RC 2.1 Validation of Personas. We define Persona validation as the process of verifying the accuracy of existing Personas based on data gathered during system operation. This involves verifying the accuracy of the values for specific Persona attributes (*e.g.*, age and personality traits) against the available data. Regularly conducting validation activities can ensure that the Persona(s) continue to accurately represent the target users and that software meets their needs.

Salminen *et al.* conclude in their SLR that research on Persona validation tends to be mostly manual and informal [7]. For example, qualitative methods, such as open discussion groups or interviews, are often used to validate automatically generated Personas.

As a first step in our research, we are planning to automate Persona validation by leveraging existing research in DDRE. Applying design science, we plan to build a tool-supported method for Persona validation [11]. More specifically, our implementation would allow the requirements engineer to explore values for various Persona attributes. This allows for the comparison with existing values. In Sect. 3, we provide more details about the planned technical solution.

We also plan to evaluate this first solution in different case studies. In particular, we aim at evaluating the applicability, effectiveness, and usability of the proposed solution.

RC 2.2 Evolution of Personas. We define Persona evolution as the process of adapting the existing Persona(s) based on data gathered during system operation. This, for example, means adding new values to the existing attributes of a Persona as well as changing attributes over time.

Based on our literature review and the analysis by Salminen *et al.*, there is no existing research on automated, data-driven Persona evolution [7].

Adding a Persona evolution functionality to the developed solution will be the second step in our planned research. Following design science, we will further advance our technical solution to allow for the updating and refining of existing Personas. This includes keeping track of existing Personas and their historical evolution using versioning. The evolution functionality will assist the requirements engineer in accurately representing the current target user groups. Similar to Persona validation, we plan to analyze the effects of automated Persona evolution through case studies.

RC 2.3 Creation of Personas. We define Persona creation as the process of automatically generating values for key Persona attributes based on data gathered from relevant sources. This involves analyzing the data to reveal patterns and characteristics of the emerging user groups to provide insights for requirements engineers.

As stated earlier in the introduction, the field of data-driven Persona creation has a vast body of research available. Currently, research either requires manual data collection methods, such as ethnographic studies or assumes that a system or part of a system is already in place to gather data for automatically generating Personas.

In the third step of our research, we plan to create Personas automatically for existing software as well. However, we will be relying on data from DDRE, including user feedback and monitoring data, to achieve this goal. If needed, a new Persona will be created when validation shows a mismatch between the current Persona(s) and actual users.

In the fourth step, however, we would like to go beyond the limitations of current approaches and create Personas for software that do not yet exist. In particular, we foresee creating Personas based on data from similar applications. We are confident that methods and tools within DDRE, in particular regarding the analysis of user feedback and reviews from marketplaces, provide the potential to support the development of such a solution.

3   The Persona Engine – A Solution Idea

In this section, we present our plan for the development of the "Persona Engine", a proposed technical research prototype. It will be built upon the "FAME" framework, which allows for simultaneous collection and processing of user feedback and monitoring data [5]. The FAME architecture already includes components for data acquisition, storage, and combination. Our goal is to extend FAME to support Persona validation (RC 2.1, Step 1), evolution (RC 2.2, Step 2), and the identification and creation of Personas for existing software (RC 2.3, Step 3). Currently, we are not planning to address the creation of Personas for software that does not yet exist (RC 2.3, Step 4), which calls for a new, even more, innovative technical solution. Our plan is to enhance and augment FAME by incorporating the following key capabilities.

3.1   Extending the Data Storage and Combination Component

FAME processes user feedback and monitoring data based on an underlying ontology. Our goal is to enhance this ontology to incorporate common Persona attributes as defined by Salminen et al. [8], without limiting ourselves to a specific template. Additionally, we plan to extend FAME's *Combiner component* for calculating Persona attribute values, which may require expanding FAME's data analysis capabilities. However, its current abilities already allow for generating values for common Persona attributes.

- Simple attributes, such as "age", "gender", and "location" can be directly based on data from user profiles.
- For attributes, such as "personality and character traits", a more sophisticated data analysis is necessary. Sentiment analysis of user feedback, which is already available in FAME, for instance, can assist in understanding users' level of criticism. Furthermore, the number of critical feedback can help to finally determine whether the users of a software system deserve to be called critical.
- Other attributes, such as "experience with the product" or "product-related behavior" can be generated based on monitoring data. Based on this data, we know which features are used by users, how often, and for how long they work with software.
- We also have identified Persona attributes where we currently do not see the possibility of generating values with the help of FAME. This includes attributes, such as "lifestyle".

3.2   Adding a "Persona Explorer and Visualizer" Component

We intend to add a *Data Explorer and Visualizer* component to FAME to enable the requirements engineer and other project members to explore information regarding Personas in an intuitive and easily understandable manner. This new component has the potential to include several features, but further discussions with practitioners are needed to determine their specific needs. Based on the generated data, this component could provide visual aids, such as charts, graphs, and tables, to give an overview of users, attribute values, and trends. It could also allow for data filtering based on criteria, such as geographical location or demographics, for more targeted analysis.

A more advanced feature could be recommendations based on rules, where the system not only allows for data exploration but also recommends values for Persona attributes. For example, a rule could dictate that if over 60% of users fall within a particular age range, the average of that range should be proposed as the value for the Persona's age attribute. Similar rules can be created for other attributes, such as "gender", and "location". The Data Visualizer component could also include a user-friendly dashboard, allowing the requirements engineer to view the most critical and actionable data at a glance, facilitating decision-making.

The Persona Explorer and Visualizer Component will keep the requirements engineer in charge of decision-making. However, it is important to note that the final decisions regarding Persona creation, validation, and evolution may not solely be based on the data available in our solution. We acknowledge that other factors, such as business, political, or social considerations, can influence these decisions.

3.3 Anticipated Challenges

The implementation of the technical solution presents several challenges, including the need for clear rules for processing data of varying types and sources, and for determining values for the various Persona attributes. This involves complex tasks like pattern recognition in data, assigning appropriate weights to different data sources, and evaluating the accuracy of existing Personas in representing the target user group through comparison with new data. Moreover, presenting the results in a manner that is both comprehensive and actionable, while being easily understandable by the requirements engineer, is another significant challenge.

Another concern that must be addressed is data privacy. Gathering data from users and later (RC 2.3, Step 4) from other platforms (*e.g.,* marketplaces) may raise privacy concerns, which must be addressed responsibly.

Finally, it is important to fully understand the needs of the different stakeholders, including requirements engineers and UX designers. To gain a deeper understanding of their needs and concerns, an interview study has been initiated, providing us with valuable insights into their requirements.

4   Our Solution at Work – A Fictional Use Case

In this section, we describe a couple of scenarios to show how Personas – once our contributions are in place – could be used in the future. Imagine that a startup wants to venture into the sustainable energy business and decides to build a green energy trading platform as a web application. They hire a team of requirements engineers, UX experts, and developers. They start the market research, which is a manual step, wherein they find three related platforms offered as Android mobile apps. They have heard that automated means for Persona creation, validation, and evolution are available and decide to use them. With the help of the solution, they start analyzing user comments for those platforms on Google Play. The new Persona Engine, which automatically parses user comments, finds the authors of the comments even on other platforms, scraps publicly available information about those users, aggregates reviews and demographic data and provides results to the requirements engineers on a dashboard to support Persona creation (RC 2.3, Step 4). One of the Personas they identify is a woman in her early thirties who prefers to live a sustainable lifestyle.

Based on the requirements and Personas, UX experts design user interfaces, and the green energy trading platform is developed including built-in feedback mechanisms. Finally, the green energy trading platform is deployed, and several hundred users have registered and are actively using it. The requirements engineers want to see whether the initially envisioned Personas are valid (RC 2.1, Step 1). They use the Persona Engine which, based on the available user feedback and monitoring data, allows them to explore and visualize relevant data. Luckily, the data confirm that the attribute values are correct. For example, the data reveals that most of the users are in their thirties.

User feedback and monitoring data are constantly stored and accessible by the Persona Engine for Persona evolution (RC 2.2, Step 2). About a year after the deployment of the app, the dashboard shows that the Persona evolution mechanism has detected another dominant user group– elderly people. The company is surprised to see these people as users. Looking at the data in more detail, they can also see that this group is struggling with the usability of the app, in particular the energy trading feature. As the company sees potential in supporting this new user group, they create another Persona for elderly people using values for several attributes proposed by Persona Engine (RC 2.3, Step 3). The new Persona is then shared with the development team, together with the feedback of this user group. These insights encourage them to redesign certain user interfaces.

5   Preliminary Results of the Interview Study

We are conducting interviews with experts in requirements engineering and UX design to gather insights on the software development process for Persona creation, validation, and evolution. The aim is to validate the practical relevance of our proposed idea from the practitioners' perspective. We aim to identify which aspects of Persona development (e.g., validation, evolution) are most important to them. So far, we interviewed a total of 3 experts who mainly act as project managers in projects where industry partners and research departments work together. We have selected these experts as they all have several years of experience and a strong background in requirements engineering and in applying user-centered methods. However, as they are working for the same university as the authors, the sampling strategy used so far can be described as convenience sampling. The interview instrument consists of a total of eight questions spread across three sections. The instrument and transcripts of the interviews are provided as additional material.[1] Although the study is still ongoing, we are able to present the first interesting insights.

5.1   Results

To create Personas (*i.e.,* question 2.3), the participants mostly used ethnographic methods, such as interviews. Furthermore, their Personas were also created based on assumptions and through brainstorming with other stakeholders. All participants stated that projects wherein they created Personas followed agile development practices and Personas were created at the beginning of a project. Reflecting on potential improvements to the current Persona creation process, one of the participants even mentioned that he would see value in working more with the actual usage data and using tools, such as Google Analytics, but with current approaches, it might be a lot of work.

When asked about the validation of Personas (*i.e.,* question 2.4), all of them mentioned that Personas were never validated once created. The reason, one of them mentioned, was that there is no specific demand for it from the business side, often due to the limited budget. Two of them mentioned usability testing as some form of validation. One participant argued that involving actual target users is a way to check whether the solution behaves according to user needs as long as these users match the created Persona. Nevertheless, all of them mentioned that validating a Persona makes sense. For example, one of the participants said that she "... would validate a Persona to strengthen or deepen them, or to widen the view on them ...".

The discussion regarding Persona evolution (*i.e.,* question 2.7), revealed that none of them "updated" or "changed" (*i.e.,* evolved) the Personas during implementation, maintenance, or evolution. As such, it did not occur to them to evolve the Personas. However, when we explained our idea of automation in more detail, two of them mentioned that their mental model of the users was evolving, although they did not update the Personas. Nevertheless, all of them were positive about the idea of Persona evolution if data were available.

When asked whether Personas were used during the development phase and the maintenance and evolution phase (*i.e.,* questions 2.5 and 2.6), all of them said that they were used to create UI prototypes only. Developers did not refer to Personas during implementation. They were all uncertain about how Personas were used in the maintenance and evolution of these systems, as they were not involved in it. One of the reasons mentioned for the poor application of the Personas was the lack of resources (*e.g.,* budget). One of the participants reasoned that, often, there is high time pressure to deliver a product, which compromises the use of Personas. Another participant mentioned that having a UX expert in a team can help educate developers and other stakeholders about the applicability of Personas during implementation and maintenance. Nevertheless, all of

---

[1] https://figshare.com/articles/online_resource/Data-driven_Persona_Creation_
Validation_and_Evolution/21552111.

them wished that Personas were used more intensively during development and saw value in using them in the software's maintenance and evolution phase.

Finally, reflecting on our idea (*i.e.,* question 3.1), all of them said that it is meaningful to automate the Persona creation, validation, and evolution process. One of the participants while reflecting on the idea said that it seems meaningful to validate the Personas because user behavior keeps changing on existing software, as well as new users are attracted to software all the time. Further, he compared the idea of Persona evolution to a growing child, "Like a child is growing and getting older or more experienced– Personas evolve similarly, and if you can automate this process over the lifetime of software might definitely be interesting." One of the participants was rather critical and warned us of the risk of "... overfitting the Personas when more data about the potential users is available." Finally, the participants also added their own ideas. For example, one of them mentioned that she would like to visualize the entire customer journey through the use of various Personas.

5.2 Threats to Validity

The ongoing interview study and its outcomes are susceptible to several threats to validity as described by Wohlin *et al.* [12]. *Construct validity*: The variables in this research are measured through interviews, including open-ended questions where participants are asked to express their own opinions. To avoid imposing one's own meaning instead of accurately capturing the viewpoint of the individuals studied, open-ended questions are used to allow participants to elaborate on their answers. *Conclusion validity*: To mitigate incorrect conclusions, each interview was conducted in one session to ensure that answers were not influenced by internal discussions. A pilot study was also conducted to obtain highly reliable measures and avoid poor question-wording. *Internal validity*: In this study, all experts interviewed work for the same university as the authors, potentially introducing bias in the sample selection. To address this, for further interviews, the authors aim to interview practitioners from outside their university. The risk of information loss during the interview is mitigated by audio recording the interviews and using a mobile app to automatically create transcripts. *External validity*: To increase external validity, the observations shared by the interviewees will be validated with the rest of the sample to make the outcomes more generalizable.

6   Conclusion

In this research preview paper, we present the idea of automating the creation, validation, and evolution of Personas. We discuss corresponding research questions, potential research contributions, the envisioned solution, and a fictional usage scenario. The main contribution of our work is the presentation and discussion of initial results from an ongoing expert interview study. Our findings indicate that, in practice, Persona validation and evolution are not yet typically part of the development workflow. However, feedback from practitioners was positive and indicates that our ideas on providing automated support for the creation, validation, and evolution of Personas are welcome.

References

1. Billestrup, J., Stage, J., Nielsen, L., Hansen, K.S.: Persona usage in software development: advantages and obstacles. In: The Seventh International Conference on Advances in Computer-Human Interactions, ACHI, pp. 359–364. Citeseer (2014)
2. Cooper, A.: The inmates are running the asylum. In: Software-Ergonomie'99, p. 17. Springer (1999)
3. Jung, S.G., Salminen, J., Jansen, B.J.: Giving faces to data: Creating data-driven personas from personified big data. In: Proceedings of the 25th International Conference on Intelligent User Interfaces Companion, pp. 132–133 (2020)
4. Miaskiewicz, T., Kozar, K.A.: Personas and user-centered design: How can personas benefit product design processes? Des. Stud. **32**(5), 417–430 (2011)
5. Oriol, M., et al.: Fame: supporting continuous requirements elicitation by combining user feedback and monitoring. In: 2018 IEEE 26th International Requirements Engineering Conference (re), pp. 217–227. IEEE (2018)
6. Park, D., Kang, J.: Constructing data-driven personas through an analysis of mobile application store data. Appl. Sci. **12**(6), 2869 (2022)
7. Salminen, J., Guan, K., Jung, S.G., Jansen, B.J.: A survey of 15 years of data-driven persona development. Int. J. Human-Comput. Interact. **37**(18), 1685–1708 (2021)
8. Salminen, J., Guan, K., Nielsen, L., Jung, S.g., Jansen, B.J.: A template for data-driven personas: analyzing 31 quantitatively oriented persona profiles. In: Human Interface and the Management of Information. Designing Information: Thematic Area, HIMI 2020, Held as Part of the 22nd International Conference, HCII 2020, Copenhagen, Denmark, July 19–24, 2020, Proceedings, Part I 22. pp. 125–144. Springer (2020)
9. Tu, N., et al.: Combine qualitative and quantitative methods to create persona. In: 2010 3rd International Conference on Information Management, Innovation Management and Industrial Engineering. vol. 3, pp. 597–603. IEEE (2010)
10. Walker, D., Myrick, F.: Grounded theory: an exploration of process and procedure. Qual. Health Res. **16**(4), 547–559 (2006)
11. Wieringa, R., Heerkens, H.: Design science, engineering science and requirements engineering. In: 2008 16th IEEE International Requirements Engineering Conference, pp. 310–313. IEEE (2008)
12. Wohlin, C., Runeson, P., Höst, M., Ohlsson, M.C., Regnell, B., Wesslén, A.: Experimentation in software engineering. Springer Science & Business Media (2012)
13. Zhang, X., Brown, H.F., Shankar, A.: Data-driven personas: Constructing archetypal users with clickstreams and user telemetry. In: Proceedings of the 2016 CHI Conference on Human Factors in Computing Systems, pp. 5350–5359 (2016)

Towards a Cross-Country Analysis of Software-Related Tweets

Saliha Tabbassum, Ricarda Anna-Lena Fischer, and Emitza Guzman[✉]

Vrije Universiteit Amsterdam, De Boelelaan 1111, 1081, HV Amsterdam, Netherlands
s.tabbassum@student.vu.nl, {r.a.l.fischer,e.guzmanortega}@vu.nl
https://s2group.cs.vu.nl

Abstract. [**Context and motivation**] Twitter is one of the most widely used micro-blogging platforms. Globally distributed developers and software companies use Twitter to communicate about software updates, bugs and other type of information related to the software. End-users from diverse geographical regions also use Twitter to give feedback about the software they use. Previous research has shown that this feedback is valuable for requirements engineering, containing information such as feature requests and usage scenarios. However, the effect of the country of origin on software-related tweets has not been studied so far. [**Question**] In this paper, we investigate to what extent people from various countries provide distinct feedback regarding certain characteristics on Twitter. [**Principal ideas/results**] We collected 70,759 tweets (Original: 17,940, Replies: 52,819) from popular Twitter support accounts of ten software applications for two months. In the subsequent analysis, we selected the tweets originating from the eight most popular countries and analyzed a sample of 1,813 tweets with the help of automatic and manual content analysis. Results show that out of three characteristics (content, sentiment and text length); content, and sentiment differ significantly at the country level in some cases. These characteristics are used in algorithms automatically processing user feedback. Such algorithms are commonly used for requirements engineering tasks. [**Contributions**] Our findings show the importance of considering software-related user feedback on Twitter from a diverse audience during the design, testing, and validation of feedback processing algorithms to minimize bias concerning different countries of origin.

Keywords: User Feedback · Twitter · Diversity · Countries · Software Evolution · Algorithm Bias

1 Introduction

Due to its ability to feature recent trends and real-time data, microblogging has become a popular method for spreading information. Twitter is the most popular micro-blogging platform among users [18]. More than 500 million tweets are generated by active users daily on Twitter[1].

---

[1] https://www.internetlivestats.com/twitter-statistics/.

© The Author(s), under exclusive license to Springer Nature Switzerland AG 2023
A. Ferrari and B. Penzenstadler (Eds.): REFSQ 2023, LNCS 13975, pp. 272–282, 2023.
https://doi.org/10.1007/978-3-031-29786-1_19

Previous studies [6,13] show that tweets contain valuable software-related information such as bug reports, feature requests and usage scenarios. This user feedback is crucial for eliciting requirements, improving existing applications, and developing future products that are less prone to errors and that address user needs. Software-related user feedback can be submitted by a globally distributed audience on Twitter. Previous studies e.g., [1] on requirements elicitation for software engineering have shown cultural differences among both users and developers either in terms of requirements prioritization for provided services, bug perception, and feature preference. Despite the potential impact of users' country of origin, the majority of approaches for automatically detecting, classifying, summarizing and prioritizing requirements from this user feedback e.g., [2,11,15,21] have not considered a diverse set of country distribution in the data collected for the training and evaluation of their approaches. Ignoring country-related differences in user feedback on Twitter and using tweets from a limited set of countries could introduce algorithm bias—and therefore bias in the requirements engineering process— when using tweets for designing, validating and testing user feedback algorithms. Not taking some countries into account when processing user feedback, could lead to users from these countries limiting or stopping the app usage because their interests and needs are not considered by the software companies. It is, therefore, necessary for app developers and researchers to consider the importance of international diverse user feedback.

When specifically looking at tweets, to our best knowledge, there exists no research that has analyzed country-related differences in tweets giving user feedback about software applications. The goal of this work is to narrow this gap and understand to what extent people from varying countries provide distinct software-related feedback on Twitter regarding characteristics such as content, sentiment, and text length. These characteristics are often used as input in algorithms automatically processing user feedback e.g., [2,11,15,21]. If the data used to create the algorithm is not diverse, this can lead to algorithmic bias.

In this study, we collected 70,759 tweets from Twitter support accounts. Then, we manually annotated and analyzed a representative sample of 1,813 tweets originating in the eight countries that had the most tweets in our original collection. To encourage replication, we make our collected and annotated data, analysis scripts and annotation guide available[2].

This work is the first to study country differences and similarities of tweets directed to software support accounts and present empirical evidence that such tweets contain a considerable amount of user feedback that diverges among several characteristics commonly used for the automatic processing of user feedback, such as content, and sentiment.

2   Related Work

Research on software-related tweets from an end-users' perspective appeared in recent years. Studies have shown that non-technical users tend to give more

---

[2] https://doi.org/10.6084/m9.figshare.18739577.

feedback on social media platforms like Twitter compared to software forums [3]. Prior work has shown that Twitter is an important source for software evolution and crowd-based requirements engineering e.g., [3,6,13]. Our study helps to understand more about the differences in user feedback on Twitter from international non-technical users. Prior work has mined tweets from Twitter support accounts of various applications to extract software-related information e.g., [7,12]. A few studies have used multilingual tweets as the basis for developing models to classify or analyze user feedback e.g., [14,19]. Oehri and Guzman [14] developed an approach for detecting semantically similar feedback among different platforms (one of them being Twitter) and languages.

Regarding user feedback about software, there is some research done on country-level differences in app store reviews [4,9]. Guzman et al. [9] analyzed country and cultural differences in App Store reviews from eight countries and found significant differences among some of the analyzed characteristics of the reviews. Fischer et al. [4] analyzed national cultural differences in app store reviews from eight countries written in five languages and found that some characteristics differed among national cultures. Tizard et al. [20] surveyed end-users and studied their reasons for (not) providing feedback. Although there are already some studies in other domains on country-specific differences in tweets and their characteristics, there is to our best knowledge, no study that addresses country-specific differences in tweets directed to software support accounts that are potentially helpful for requirements engineering and software evolution. With our study we address this research gap.

3   Study Methodology

In this study, we investigate if user feedback on Twitter stemming from different countries differs according to various characteristics. Our main research question is:

RQ: *Does user feedback written by different countries vary?*

In particular, we studied the following three user feedback characteristics of tweets:

- **Content:** Used for classifying the tweets in the context of requirements engineering and software evolution. It consists of five categories: bug report, support request, feature request, noise, and other.
- **Sentiment:** Affect present in the users' tweets. It is based on a five-level Likert scale ranging from very negative to very positive.
- **Character Length:** Refers to the total number of characters in a tweet's text after cleaning it (i.e., removing emojis, extra spaces, ampersands).

We considered the content, sentiment, and character length characteristics since they are frequently considered for automatic prioritization and automatic classification e.g., [7,11,16] of user feedback.

Table 1. Dataset Overview per Country and App

Country	Adobe	Dropbox	Evernote	Google Maps	LinkedIn	Netflix	Slack	Snapchat	Spotify	YouTube	Total
United States	1086	407	306	149	503	560	889	1314	1473	942	7,629
United Kingdom	287	97	85	66	168	555	165	379	554	184	2,540
India	51	23	19	81	167	195	53	61	209	430	1,289
Canada	156	42	30	22	55	101	105	83	184	80	858
Australia	45	26	14	10	24	37	79	30	53	42	360
Netherlands	15	13	6	4	16	16	31	8	30	19	158
Germany	34	7	10	6	14	21	51	10	28	26	207
France	25	11	6	2	12	7	36	28	21	11	159
Total	1,699	626	476	340	959	1,492	1,409	1,913	2,552	1,734	13,200

In the remainder of this section, we explain more in detail the data collection, data sampling, automatic and manual content analysis process, as well as the tests performed in the statistical analysis.

Data Collection. We crawled tweets from the Support Accounts of ten popular software applications stemming from eight different countries, written in English. For this process, we used Twitter's Standard Search API[3] as it was the only API that provided free access to tweets at the time of research. We collected our data between March 15, 2020, and May 3, 2020. We collected tweets directed to Twitter support accounts rather than collecting general tweets mentioning popular software as previous research [7] showed that these have a higher proportion of tweets relevant to requirements engineering and software evolution. We chose popular apps (with more than ten tweets on average per day on their Support Account) stemming from different domains for our study.

In total, we collected 70,759 (Original: 17,940, Replies: 52,819) tweets over the two month period[4]. This dataset also includes tweets from other countries not included in our study, or users who did not specify their location.

We used Google Maps Geocoding API[5] for fetching and saving the country from the profile location attribute provided by Twitter account owners. Then, we selected the eight countries with the most tweets to represent geographically distributed regions. These are, ordered by number of collected tweets, United States of America, United Kingdom, India, Canada, Australia, Netherlands, Germany and France. The tweets from the aforementioned selected countries total 13,200. Table 1 shows the total number of tweets per app for the selected countries.

We selected English as the language of choice for this study because it is a widely spoken language[6], the most used language on Twitter [17] and both annotators (see Section Manual Content Analysis) were fluent in it.

---

[3] https://developer.twitter.com/en/docs/tweets/search/overview.

[4] The term original tweet refers to the actual tweet of a user who posted something on a Twitter Support Account, while a reply thread respond to that original tweet.

[5] https://developers.google.com/maps/documentation/geocoding/start.

[6] https://www.statista.com/statistics/266808/the-most-spoken-languages-worldwide/.

Sample Creation. To create an appropriate sample for our manual analysis and have statistically robust results, we chose a confidence interval of 95% with a minimum sample size per country. For instance, the total tweets from the USA were 7,629 and had a minimum sample size of 366 to ensure a confidence interval of 95%. Furthermore, to equally represent the selected apps in the sample, we took the proportional share of an app's tweets in the respective country and multiplied it by the minimum sample size.

The final sampled dataset consists of 1,813 tweets from eight countries and ten applications.

Automatic Content Analysis. We automatically extracted the character length of the tweets from the final sampled dataset. For the character length, we cleaned the tweets' text, and removed all the emojis, additional spaces, and ampersands. We removed them because users sometimes overused these extra characters without having a meaning for them and this would have skewed the character length of individual tweets.

Manual Content Analysis. We manually annotated the sentiment and content. This decision was taken because the automatic classification of sentiment and content on user feedback have not yet reached an accuracy that is comparable to manual classification e.g., [11,16], and we did not want to introduce noise to our results.

Annotation Process. Two authors of this study performed the manual annotation. Each annotator labeled the content and sentiment of 1,813 tweets separately. To reduce major disagreements between the two annotators, we made use of an annotation guide with definitions and examples for the sentiment levels and content categories. For sentiment, the annotators could label between very negative to very positive, including a neutral scale. The annotation guide specified certain key indicators to determine sentiment such as capitalized words (e.g., "FIRE ALL THE MONKEYS WORKING ON THIS APP" - very negative) or superlatives (e.g., "Best app ever!!!" - very positive). When mixed sentiments were present (e.g., "I love the app, but absolutely hate the last release")—which happened rarely—annotators were instructed to label the sentiment they considered more predominant and which deserved more attention from a requirements engineering and software evolution perspective (the sentiment about the last release in the previous example).

Content categories, could be labelled into five categories (bug report, feature request, support request, other, and noise), used in previous studies [4,9]. While bug reports, feature requests, and support requests are important for requirements engineering and software evolution, the other category refers to topics not relevant for requirements or software evolution, such as a product recommendation or a general complaint. The noise category refers to tweets that are not written in English or contained illegible characters. During the annotation process, annotators could indicate when they were unsure about their labelling; this later helped solve disagreements (see Section Manual Disagreement Handling).

The Cohen's Kappa for the final annotated set was 0.82, indicating a strong agreement among annotators.

Manual Disagreement Handling. The disagreement of the final sampled dataset was handled both manually and automatically. We used a script for the automatic resolution of the disagreement among the two characteristics using the following approach. For the sentiment, we used the average of the two sentiment (transformed numerical) scores and thus followed the same disagreement resolving methodology of previous studies [4]. The main reason for this approach was that most disagreements where in one scale unit, i.e., one annotator labelled a tweet as "positive" and the other as "very positive". Only two reviews out of 1,813 tweets in the whole dataset where marked with more than one scale difference. For the content, we used priority settings (bug report over feature request over support request and other) to select the final annotation.

When one of the annotators was unsure about the labelling of a certain tweet and had marked "unsure" in their labelling (see Section Annotation Process) an additional manual inspection was conducted. In this step, both annotators discussed their disagreement and reached a conclusion.

Hypothesis We tested three hypotheses in our study.

Hypothesis H^1: The proportion of the content differs with statistical significance across selected countries.

Hypothesis H^2: The proportion of the sentiment score differs with statistical significance across selected countries.

Hypothesis H^3: The proportion of the character length differs with statistical significance across selected countries.

Statistical Analysis We analyzed the final sampled dataset using descriptive statistics suitable for each dependent variable. For the content, a categorical dependent variable, we used the Chi-square test of independence to investigate the statistical differences. To validate the results, we performed a posthoc pairwise Chi-square test with Bonferroni correction. For the ordinal variables sentiment and character length, we applied a Kruskal-Wallis test and a Tukey and Kramer's (Nemenyi) posthoc test with Tukey-Dist approximation for independent samples. We consider the findings statistically significant if they have a p-value below 0.05.

4   Results

Content 38% tweets of the tweets in the analyzed dataset are bug reports, 16% feature requests, 27% support requests, and 19% other. The United Kingdom has the highest number of tweets categorized as bug reports, with a proportion of 45%. The Netherlands has the highest number of feature requests, with a proportion of 22%. The number of support requests from India is the highest, with a proportion of 42%. The lowest proportion of bug reports and feature requests are from India (30% and 8% respectively). Figure 1 shows the proportion of content for the analyzed countries.

The proportion of content among the analyzed tweets is statistically different per selected country, χ^2 (21, N=1811) = 92.585, p < 0.001. **Thus, the hypothesis H^1, that the content differs statistically significant can be confirmed.** The posthoc pairwise Chi-square test with Bonferroni correction shows that there is a significant difference between India and every other country except France. The United Kingdom differs significantly from Canada and the United States. The results of the posthoc tests reveal that the distribution of the content is not independent across the countries. This indicates that user feedback content differs significantly across different countries.

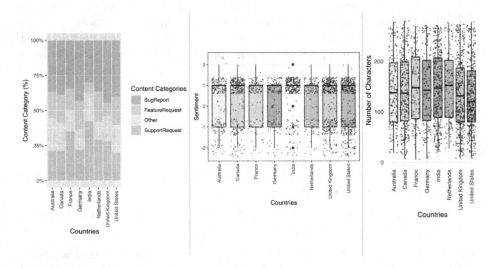

Fig. 1. Descriptive Results for Content, Sentiment and Number of Characters

Sentiment. Most of the tweets' sentiments are positive, leading to a median of 1 for the entire data analyzed. Tweets from each country also have a median of 1, indicating a positive sentiment. Figure 1 shows an overview of the sentiment scores per country. The Kruskal-Wallis test results for the sentiment scores indicate a statistically significant difference among the selected countries with respect to the sentiment, χ^2 (7, N=1811) = 49.291, p < 0.001. **Thus, hypothesis. H^2, that the proportion of sentiment score differs statistically significant across the analyzed countries can be confirmed.** The pairwise comparisons in the posthoc test revealed that Australia (mean sentiment: 0.118), the United Kingdom (mean sentiment: 0.176), and the United States (mean sentiment: -0.008) are significantly different from India (mean sentiment: 0.557) in terms of sentiment expression. This indicates that the sentiment of user feedback differs across some countries.

Character Length. The tweet's character length ranges between 6 and 279 with a median of 137 characters for all analyzed data. The box plot in Fig. 1 shows that the character length does not vary much across countries overall, with medians across countries ranging from 120 to 152. The Kruskal-Wallis test shows that the distribution of the character length differs in at least one of the selected countries, χ^2 (7, N=1811) = 14.896, p = 0.037. The result of the posthoc test shows that no country pair differs significantly. Thus, hypothesis H^3 cannot be confirmed.

5 Discussion

This study analyzes software-related tweets from Twitter support accounts of ten popular software applications, stemming from eight different countries. Its purpose is to investigate whether software-related user feedback available on Twitter differs across various countries. The results show that the answer is partially yes. Two out of three analyzed characteristics (content, sentiment) differ in some cases significantly at the country level. The results of this study have certain implications for both practitioners and academics in the context of requirements engineering and software evolution. Previously, when designing algorithms for processing user feedback, researchers mainly focused on user feedback either from the United States, or from user groups without a clear indication of their country of origin, e.g., [2,11,15,21]. Previous studies have already shown that software engineering practitioners consider the attributes of tweets when prioritizing them [8]. According to the surveyed practitioners, the content of the tweet is one of the most important attributes for prioritization [8]. Our study shows that country differences have in some cases a statistically significant impact on the content of user feedback tweets. Our study shows, for example, that India has a very high proportion of support requests in the sample. This difference could stem from the cultural differences between India and the rest of the countries in the sample. Previous research in other contexts has already found that users from collectivistic cultures (as for example India) have a stronger tendency to ask for support than people from individualistic cultures [5]. These differences are especially important for automatic classification models which could have a better performance among the countries they were trained on or among the countries with the largest amount of specific content categories.

Our results also show that there are statistically significant differences in the tweet sentiment from different countries. The tweets from Australia, UK and USA have significantly different sentiments than tweets from India. This could stem from the cultural differences in the countries. Collectivistic cultures are more hesitant to show emotions to out-group members [5] and are keen to maintain harmony [10], while individualistic cultures (such as Australia, UK and USA are) express their opinion more often [10]. Failing to consider these differences could lead to a misled preference of those countries that tend to express their negative sentiments in a more outwardly manner when prioritizing tweets.

These examples show the potential negative outcomes of not considering user feedback from diverse user groups for applications with a global coverage for creating, training, testing, and using algorithms for processing user feedback. In particular, the large differences in tweets from Western countries and tweets from India already indicate the importance of a diverse sample. However, except for India, all other considered countries are Western, leading to a biased sample. Future studies should investigate the differences in user feedback characteristics when analysing more cultural diverse countries. Additionally, due to the high imbalance of the individual countries (58% USA, 19% UK, 10% India, 13% other) in the sample, the statistical results are prone to errors, as the sample is highly skewed. Subsequent research should confirm these study results with a more balanced sample.

Our study also shows that character length was not significantly different across countries. However, this could also stem from the fact that all analyzed tweets were in English because this was the language in which sufficient tweets from geographically diverse countries could be collected in two months. It could be argued that the users who tweet in English from these countries are not representative for their countries because not all people from non-native speaking English countries are able to post an English tweet. This is another potential bias in the sample. More studies should be conducted to cover tweets in different languages from a broader range of countries.

Our results will hopefully encourage researchers and practitioners to gather and analyze large datasets from various countries so that the algorithms processing user feedback are thoroughly designed, validated, and tested to avoid any algorithm bias. Moreover, the findings provide a first indication of the need to include the users' country of origin as a control variable for ensuring data diversity when designing algorithms for processing user feedback data (from Twitter) that take specific characteristics (e.g., sentiment and content) as input.

6   Conclusion

This study investigated country-level differences in user feedback by analyzing software-related user feedback from Twitter support accounts of ten applications from eight countries. The results show that feedback characteristics such as content and sentiment differ significantly at the country level. These results show that there is a strong need to collect tweets from diverse countries and backgrounds, when designing and testing algorithms for the automatic processing of user feedback, to avoid algorithm bias—as well as when considering feedback as a source of information during requirements engineering and software evolution. Moreover, this study encourages researchers and practitioners to further study the impact of country differences on user feedback from Twitter and other social media platforms (i.e., Facebook) using large and multilingual datasets.

References

1. Alsanoosy, T., Spichkova, M., Harland, J.: Cultural influence on requirements engineering activities: a systematic literature review and analysis. Requirements Engineering, pp. 1–24 (2019)
2. Chen, N., Lin, J., Hoi, S.C., Xiao, X., Zhang, B.: AR-miner: Mining informative reviews for developers from mobile app marketplace. In: International Conference on Software Engineering, pp. 767–778 (2014)
3. El Mezouar, M., Zhang, F., Zou, Y.: Are tweets useful in the bug fixing process? an empirical study on firefox and chrome. Empir. Softw. Eng. **23**(3), 1704–1742 (2018)
4. Fischer, R.A.L., Walczuch, R., Guzman, E.: Does culture matter? impact of individualism and uncertainty avoidance on app reviews. In: International Conference on Software Engineering: Software Engineering in Society (ICSE-SEIS), pp. 67–76. IEEE (2021)
5. Fong, J., Burton, S.: A cross-cultural comparison of electronic word-of-mouth and country-of-origin effects. J. Bus. Res. **61**(3), 233–242 (2008)
6. Guzman, E., Alkadhi, R., Seyff, N.: A needle in a haystack: What do twitter users say about software? In: International Requirements Engineering Conference (RE), pp. 96–105. IEEE (2016)
7. Guzman, E., Ibrahim, M., Glinz, M.: A little bird told me: Mining tweets for requirements and software evolution. In: International Requirements Engineering Conference (RE), pp. 11–20. IEEE (2017)
8. Guzman, E., Ibrahim, M., Glinz, M.: Prioritizing user feedback from twitter: A survey report. In: International Workshop on CrowdSourcing in Software Engineering (CSI-SE), pp. 21–24. IEEE (2017)
9. Guzman, E., Oliveira, L., Steiner, Y., Wagner, L.C., Glinz, M.: User feedback in the app store: a cross-cultural study. In: International Conference on Software Engineering: Software Engineering in Society (ICSE-SEIS), pp. 13–22. IEEE (2018)
10. Hofstede, G.H., Hofstede, G.J., Minkov, M.: Cultures and Organizations: Software of the Mind, Third Edition. McGraw-Hill (2010)
11. Maalej, W., Nabil, H.: Bug report, feature request, or simply praise? On automatically classifying app reviews. In: International Requirements Engineering Conference, pp. 116–125 (2015)
12. Martens, D., Maalej, W.: Extracting and analyzing context information in user-support conversations on twitter. In: International Requirements Engineering Conference (RE), pp. 131–141. IEEE (2019)
13. Nayebi, M., Cho, H., Ruhe, G.: App store mining is not enough for app improvement. Empir. Softw. Eng. **23**(5), 2764–2794 (2018)
14. Oehri, E., Guzman, E.: Same same but different: Finding similar user feedback across multiple platforms and languages. In: International Requirements Engineering Conference (RE), pp. 44–54. IEEE (2020)
15. Panichella, S., Di Sorbo, A., Guzman, E., Visaggio, C.A., Canfora, G., Gall, H.: ARdoc: App reviews development oriented classifier. In: Symposium on the Foundations of Software Engineering, pp. 1023–1027 (2016)
16. Panichella, S., Di Sorbo, A., Guzman, E., Visaggio, C.A., Canfora, G., Gall, H.C.: How can i improve my app? classifying user reviews for software maintenance and evolution. In: International Conference on Software Maintenance and Evolution (ICSME), pp. 281–290. IEEE (2015)

17. Poblete, B., Garcia, R., Mendoza, M., Jaimes, A.: Do all birds tweet the same? characterizing twitter around the world. In: International Conference on Information and Knowledge Management, pp. 1025–1030 (2011)
18. Prasetyo, P.K., Lo, D., Achananuparp, P., Tian, Y., Lim, E.P.: Automatic classification of software related microblogs. In: International Conference on Software Maintenance (ICSM), pp. 596–599. IEEE (2012)
19. Stanik, C., Maalej, W.: Requirements intelligence with openreq analytics. In: International Requirements Engineering Conference (RE), pp. 482–483. IEEE (2019)
20. Tizard, J., Rietz, T., Liu, X., Blincoe, K.: Voice of the users: an extended study of software feedback engagement. Requirements Eng. **27**(3), 293–315 (2022)
21. Villarroel, L., Bavota, G., Russo, B., Oliveto, R., Di Penta, M.: Release planning of mobile apps based on user reviews. In: International Conference on Software Engineering, pp. 14–24 (2016)

Integrating Implicit Feedback into Crowd Requirements Engineering – A Research Preview

Leon Radeck[⊠] and Barbara Paech

Institute for Computer Science, Heidelberg University, 69120 Heidelberg, Germany
{radeck,paech}@informatik.uni-heidelberg.de

Abstract. **[Context/Motivation]** In crowd requirements engineering, users are asked specific questions (explicit pull feedback) to elicit requirements. Existing approaches collect explicit pull feedback by asking the same questions to all users. **[Problem]** Not all questions are meaningful for all users, e.g. regarding a functionality they have not yet used. Furthermore, without knowing the user behaviour giving rise to the feedback, it is difficult to understand the reasons for the feedback. These reasons are important for deriving requirements. **[Principal ideas]** Our idea is to use the user behaviour (implicit feedback) to adapt the collection of explicit pull feedback and the derivation of requirements. We embed this collection of explicit pull feedback into a novel approach that makes use of a rich palette of discussion elements from crowd-based requirements engineering to motivate user participation and to support requirements derivation. **[Contribution]**. To our best knowledge, this is the first approach that combines the collection of implicit feedback and explicit feedback with discussion elements from crowd-based requirements engineering. We sketch our approach and our research and evaluation plan regarding the application of the approach in the context of the interdisciplinary and large-scale research project SMART-AGE with around 500 users.

Keywords: Requirements engineering · Crowd · User feedback · Implicit feedback

1 Introduction

User feedback is essential for the continuous development of software, because it contributes substantially to the elicitation of requirements. Traditional methods of collecting user feedback, such as interviews or workshops, are only feasible with a limited number of users as they are very time-consuming. As the number of users increases, the use of (semi-) automated methods becomes more relevant. These methods do not require the presence of the persons involved, but can be performed remotely and by many stakeholders at the same time [5]. Crowd-based requirements engineering (CrowdRE) is an umbrella term for such approaches to gather and analyse feedback from a large number of users, also called "crowd", to derive validated user requirements [4]. Collecting

© The Author(s), under exclusive license to Springer Nature Switzerland AG 2023

A. Ferrari and B. Penzenstadler (Eds.): REFSQ 2023, LNCS 13975, pp. 283–292, 2023.
https://doi.org/10.1007/978-3-031-29786-1_20

explicit push feedback (feedback intentionally pushed by the user) is one main approach in CrowdRE [15]. Another main approach in CrowdRE is to request users to give explicit pull feedback about the product by asking them questions (e.g. "How satisfied are you with product?", "What are your ideas on how to improve the product?") [15]. The problem is that not every question is always suitable for every user. Users who receive a question about a functionality that they have not used yet, cannot answer the question. In addition, it is difficult for requirements engineers to understand the reasons for the feedback without knowing the user behaviour, which led to the feedback. Our idea is to integrate the capture and analysis of the user behaviour (implicit feedback) into CrowdRE to adapt the collection of explicit feedback and the derivation of requirements.

In this paper, we present our approach CREII (Crowd-based Elicitation with Integrating Implicit Feedback) that employs the tool SMARTFEEDBACK (SF) to tailor the collection of explicit pull feedback to an individual user's usage behaviour. To motivate user participation and to allow for requirements derivation, SF further integrates various established discussion elements from CrowdRE [15]. Users can comment and vote answers in real-time by the use of a discussion platform, they can classify and prioritize explicit feedback, they can indicate a representative sentiment for their feedback and they can discuss about requirements (see Table 1). We also describe the application of CREII in the context of the interdisciplinary and large-scale research project SMART-AGE. In particular, we describe preliminary research in the form of a pilot test, the research plan for the employment of CREII in the main study and our ideas for evaluation. This paper is organized as follows: Sect. 2 gives a brief overview of SMART-AGE and the relevant terminology. Section 3 presents related work. Section 4 describes our approach, and Sect. 5 presents how the approach is applied in SMART-AGE, in particular preliminary research, our research and evaluation plan.

2   Project SMART-AGE and Terminology

SMART-AGE. Recent findings on the role and potential of apps for older adults' quality of life are encouraging. Major examples for these apps are solutions that address social engagement and networking, health and disease prevention, and training and fitness. The 5-year project "Smart Aging in Community Contexts: Testing Intelligent Assistive Systems for Self-regulation and Co-regulation under Real-Life Conditions" (SMART-AGE) is in its core a complex intervention trial aimed at evaluating different constellations of apps. These apps are used by around 500 study participants from two communities in Southwestern Germany (Heidelberg and Mannheim). In this project, we gather feedback for three tablet-based apps, namely (1) an app promoting social networking and social participation (SMARTVERNETZT), (2) an app providing health advice focusing on major areas of older adults' health and functioning (SMARTIMPULS) and, (3) an app to tailor the collection of explicit pull feedback to an individual user's usage behaviour (SMARTFEEDBACK). The gathered user feedback is privacy relevant, as both, voice messages in the explicit feedback and interaction data in the implicit feedback may allow identifying a person. Therefore, we obtain declarations of consent from the study participants. We describe our vision of a CrowdRE process adapted to the needs of older adults and the challenges in implementing our approach in the context of SMART-AGE in more detail in [11].

Terminology. In the following, we distinguish feedback pushed by the user (*push*) or pulled from the user (*pull*), and feedback given with the intent to give feedback (*explicit*) or given unintentionally (*implicit*) [8]. Explicit push feedback are either messages or comments. Explicit pull feedback are answers to questions. Implicit pull feedback are usage data (e.g. clicks on user interface). *Private* feedback are messages and answers from the users, which are only visible to the researchers. *Shared* feedback are answers from the users that are visible to everyone. Implicit push feedback (e.g. comments made by users about the apps during a conversation, that is not happening over SF) is not considered in our approach, as users are instructed to give their feedback via SF only.

3   Related Work

According to the mapping study [13], in approaches up to 2017 mainly explicit feedback is collected and used to derive requirements. Approaches that have collected implicit feedback have used it to estimate service performance or to compute user profiles [13] but not to combine it with explicit feedback. There are more recent approaches that are not included in the study, but combine explicit and implicit feedback. CAFE [2] enriches explicit push feedback with implicit feedback to facilitate the interpretation of the explicit push feedback. CAFE shows the requirements engineer what screens the user visited, what UI elements the user interacted with prior to giving feedback, as well as hardware information (e.g. operation system and device model). CAFE, however, does not give examples on how exactly the interpretation of explicit push feedback is facilitated by the enrichment with implicit feedback. FAME [9] combines implicit and explicit push and pull feedback through an ontology. The ontology links explicit push and pull and implicit feedback by user, timestamp, application and domain concept. The ontology is presented to the requirements engineer, which allows the requirements engineer to get a better understanding of the explicit push and pull feedback and to prioritize requirements derived from the feedback FAME does not use implicit feedback to adapt pull feedback questions. QoE probe [3] and Wuest et al. [16] introduce explicit pull feedback based on implicit feedback. QoE logs the user ID, timestamps of events on feature level (e.g. starting or completing a feature) and user interaction level (e.g. user input or an application output) and then triggers a feedback collection form with the option to answer a question about the users satisfaction and the reasons for the user behaviour. Wuest et al. [16] triggers feedback collection based on user goals in the context of a navigation system. An example for a goal is that the user reached her target destination. This is automatically recognized in the implicit feedback, when the users GPS coordinates match those of the destination, and then explicit pull feedback through a feedback form is triggered. We adapt this in CREII, where we trigger explicit pull feedback when the user does not behave according to the defined ideal usage behaviour. While all of the mentioned approaches combine implicit and explicit feedback, none of them integrates discussion elements from crowd-based requirements engineering (see Table 1) to support user participation and requirements derivation.

4 Crowd-Based Requirements Elicitation Via the CREII Method

In the following, we describe our approach CREII (Crowd-based Elicitation Integrating Implicit Feedback) that employs the tool SF to collect feedback and supports the derivation of requirements. The process is shown in Fig. 1. We first give a brief explanation of Fig. 1 and then we describe the collection of pull feedback based on usage behaviour by using adaptive questions (Sect. 4.1). After that, we describe our plan for bundling similar feedback (Sect. 4.2) and the derivation of requirements (Sect. 4.2) in more detail. An explanation of the steps of CREII with examples is given in Table 1.

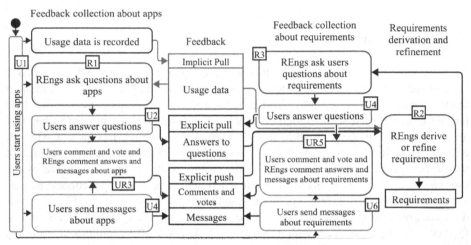

Fig. 1. Diagram representing the process of collecting feedback about apps and requirements, as well as requirements derivation and refinement. Activities in green are executed by users, activities in blue are executed by requirements engineers. Implicit pull feedback and associated arrows are red. Everything else is black.

After the users begin to use the apps (U1), they can provide feedback about the apps by sending messages to the requirements engineers (U4) or they can answer questions about the apps (U2), that the requirements engineers have asked (R1). The requirements engineers ask different types of questions. They ask questions to collect opinions, improvement ideas and problems with the apps and they pose *adaptive questions* that ask for reasons for the observed implicit feedback of the users and corresponding improvement ideas (see Sect. 4.1). The questions can address the app itself, as well as functional and non-functional requirements of the app. After the users have answered questions about the apps, they can comment and vote other answers to the same questions, as far as other users shared their answers (UR3). We stipulate that this discussion possibility enhances the motivation to the user to give feedback. The requirements engineers can also comment the answers to questions and messages about apps to thank the users for their feedback and to ask for clarification (UR3). The users can comment back on messages that they sent (UR3). The requirements engineers do not vote on answers, because they do not want to influence the opinion of the users. Furthermore, we explicitly do not allow users to comment on answers of questions that they did not answer by themselves, because we believe a minimum level of commitment is necessary to have a meaningful discussion. Based on the collected explicit and implicit feedback, the requirements engineers derive and refine requirements (R2). These requirements are presented to the users and users are asked about their opinions (R3). Implicit feedback is used during the derivation, to make sure that users receive only questions about requirements for apps, that they have accessed. To support requirements derivation, users can answer the questions about the requirements (U4), send messages about the requirements to the requirements engineers (U6), and comment and vote on the requirements: The requirements engineers can comment answers and messages about requirements in the same way as answers and messages about apps (UR5). Based on the feedback about the requirements, the requirements engineers derive new requirements and refine existing requirements (R2). Whenever users give feedback, they have the opportunity to assign a priority (low, medium, high) to the feedback and a sentiment (very dissatisfied, dissatisfied, neutral, satisfied, very satisfied). These two attributes can be used by the requirements engineers to bundle feedback (see Sect. 4.2). The users can also indicate whether to share the feedback with other users or whether to give the feedback privately to the researchers. This is important because certain users place high value on privacy [12].

4.1 Collecting Pull Feedback by Using Adaptive Questions

Adaptive questions ask for reasons for the observed implicit feedback of the users and corresponding improvement ideas. The process of asking adaptive questions for the app SMARTIMPULS is illustrated in Fig. 2.

The user with user id "User1" starts interacting with the app (A). The resulting implicit feedback is sent to SF (B). The implicit feedback consists of the ID of the user (*UserID*), the app that was used (*App*), the event that happened (*Event* – e.g. CLICK for clicking on a user interface element or START for starting the app), the context of the event (*Context* – e.g. which user interface element was clicked on), a foreign ID referencing an entity of the app that was used (*FID* – e.g. the ID of the answer) and the data to which the event was created (*Created*). SF receives the implicit feedback and saves it to the database (C). SF now periodically loads the history of the implicit feedback (D) and checks whether it does not represent the ideal usage behaviour of SMARTIMPULS (E). The ideal usage behaviour is configured initially by the requirements engineers. It consists of different metrics, e.g. the ideal usage frequency of the app, the ideal usage duration of the app and the ideal answer rate to questions, as in SMARTIMPULS the user has to answer certain questions about his or her health. Checking for ideal usage behaviour means calculating the metrics based on the history of implicit feedback (e.g. accumulating the time difference between START and STOP events per day, to get the usage duration per day) and then comparing it to the expected ideal behaviour (10 min per day). The check happens once per day. If the implicit feedback does not represent ideal usage behaviour, the user receives an adaptive question (F), which asks for the reason and for improvement ideas. Adaptive questions are only asked again after some time has passed, so that the user does not feel disturbed. When collecting pull feedback with adaptive questions (for an example, see Q2 in Table 1), we combine asking for the reason of a users' usage behaviour with asking for an improvement idea, because knowing the

Table 1. Explanation of steps of CREII with examples

Step	Step description (discussion elements are underlined)	Action (Q = Question, A = Answer, C = Comment, M = Message)
R1	Requirements engineers ask questions	*Q1*: How could SMARTIMPULS be improved in your opinion? *Q2* (adaptive): Why do you not use SMARTIMPULS every day"
U2	Users answer questions	*Q1A1*: It would help to be reminded to answer questions *Q2A1*: I find the questions not suitable for me
UR3	Users comment and vote	Some users vote for *Q1A1*, some users vote for *Q2A1* *Q1A1C1*: "I'd like to be reminded every day."
UR3	Requirements engineers comment	*Q1A1C2*: "Thank you very much for your input." *Q2A1C1*: "Why are the questions not suitable for you?
U4	Users send messages	*M1*: The letters of the app are too small for me to read
R2	Requirements engineers derive and refine requirements	From *Q1A1* a new system function "Remind user to answer questions" is extracted. By *Q1A1C1* the SF is detailed by adding a rule about the frequency of reminding. *M1* details the NFR Accessibility
R3	Requirements engineers ask questions about requirements	To validate the SF, the question *Q3* "How would you like a new functionality in SMARTIMPULS that reminds you every day to answer questions? Please explain your judgement?" is asked To validate *M1*, the question *Q4* "How would you like an increased font size in SMARTIMPULS? Please explain your judgement?" is asked
U4	Users answer questions	*Q3A1*: "I would love that, because I am a bit forgetful." *Q3A2*: "I am a bit sceptic." *Q4A1*: "That's a good idea, then I use the app without my glasses."
UR5	Users comment and vote	Some users vote for *Q3A1*, some user vote for *Q3A2* *Q3A2C1*: "Me too, I don't know if that helps"
UR5	Requirements engineers comment	*Q3A2C2*: "Thanks for giving feedback. Why are you sceptic?"
U6	Users send messages	*M2*: I don't want to tell it publicly, but I think reminders about answering questions would stress me."

...

reason alone might not be enough to derive a requirement. Adaptive questions can avoid asking users for feedback before the users have gained minimal experience with the app

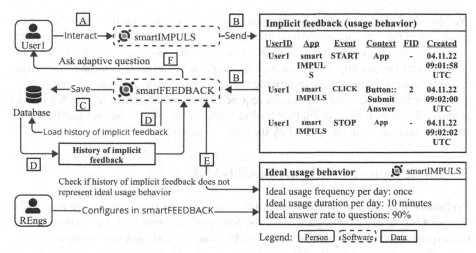

Fig. 2. Diagram representing the process of asking adaptive questions.

or functionality, which would be disturbing [15]. In SMART-AGE, each user joins the study at a different time. We give the users a few days to get familiar with the apps and then start asking the questions relative to the users' start date.

4.2 Bundling of Explicit Feedback

The more explicit feedback is collected, the greater the likelihood that feedback will be similar. To save effort, we bundle similar feedback on the basis of its attributes before deriving requirements. Table 2 shows attributes that we deem useful for bundling.

Table 2. Attributes of feedback used for bundling (R = Specified by requirements engineer, U = Specified by user, A = Automatically recorded by SF)

R	U	A	Attributes
X			Task Oriented Requirements Engineering (TORE) Category [10] (Goal & Task, Domain, Interaction, System Level)
X			Degree of readability (measured by different common readability formulas)
	X		Category (improvement, problem, opinion, neutral)
	X		Sentiment (very happy, happy, neutral, sad, very sad)
	X		Priority (low, normal, high)
		X	Implicit feedback (usage behaviour)

We think that TORE [10] could help with bundling, because feedback can be grouped by different levels. We think that the degree of readability of feedback is helpful for bundling, because unreadable feedback can contribute less to requirements derivation.

We also think that the category of feedback plays a role for requirements derivation. Feedback of the category improvement could have more potential for deriving requirements than feedback which is classified as an opinion. Feedback can also be bundled by sentiment and priority. Feedback with low happiness and an indication of the reason could be especially useful for improving the apps. Feedback that was assigned a high priority by the user is more relevant for requirements derivation than feedback with low priority. We also think that implicit feedback can be helpful to bundle feedback. For example, feedback can be bundled by the users' usage time.

5   Application of CREII in SMART-AGE

Preliminary Research. We performed a convincing proof-of-concept version of CREII in a pilot test with 20 participants over the period of one week. Overall, 208 responses to 24 questions and 33 messages were sent. Feedback for SF was very positive.

Research Plan. We follow the design science methodology proposed by Wieringa [14]. We plan to deploy CREII in SMART-AGE to collect feedback and derive requirements for the apps from around 500 users in the first five months of the study and then follow up with one month of requirement validation and refinement. For the evaluation of CREII, we also ask questions about the acceptance of SF by using the System Usability Scale [1]. In order not to overwhelm the users, we plan to ask them no more than five questions on the same day. As a reminder and for motivation, we also plan to remind users after a week of inactivity to participate in the feedback collection again.

Evaluation. Table 3 shows our research questions.

Table 3. Research questions

Usage behaviour	
RQ1	Does the usage behaviour of a user influence the quality or quantity of the provided feedback?
Feasibility	
RQ2	Is it feasible to collect high quality feedback with CREII and SF?
RQ2.1	What is the quantity and quality of the collected feedback?
RQ2.2	Do adaptive questions and discussion elements influence the quality or quantity of the provided feedback?
RQ2.3	Does reminding to give feedback influence the quality or quantity of feedback?
RQ3	Is it feasible to collect high quality requirements with CREII and SF?
RQ3.1	What is the quantity and quality of the derived requirements?
Acceptance	
RQ4	What is the acceptance of SF?

RQ1 investigates whether the usage behaviour of a user influences the quality or quantity of feedback. For example, high usage duration could lead to feedback of higher quality. RQ2 and RQ3 investigate the feasibility of CREII and SF to collect high quality feedback and high quality requirements. RQ2.1 investigates the quantity and quality of the collected feedback and RQ3.1 investigates the quantity and quality of the derived requirements. Investigating the quantity and quality of derived requirements allows us to make a comparison to similar approaches [15]. RQ2.2 investigates the influence of adaptive questions or discussion elements on the quality and quantity of feedback. If adaptive questions or discussion elements have a positive influence on the quality and quantity of feedback, then it would make sense for others to implement this practice as well. RQ2.3 investigates whether reminding to give feedback affects its quality of quantity. Reminding is easy to implement and would be an implementable practice for others. RQ4 evaluates the acceptance of SF through the questions of the System Usability Scale [1].

Quality of Requirements. We assess the quality of requirements manually. We use established quality criteria from the International Requirements Engineering Board (IREB) manual [7], such as Adequacy, Necessity, Unambiguity, Completeness, Understandability, Verifiability, Consistency, Redundancy.

Quality of Feedback. We assess the quality of feedback manually. To our best knowledge there does not exist an established set of quality aspects for user feedback about software. We therefore derive quality aspects from established standards. We plan to adapt the characteristics for data quality of ISO 25012 [6] and the quality criteria for requirements (IREB) on feedback and to establish metrics that let us quantify each quality aspect manually.

Limitations and Risks. Eliciting pull feedback requires the users to answer the questions of us researchers. This can be time-consuming and strenuous, especially for elderly people. In [11] we discuss how CREII is tailored to the individual needs of older adults.

Acknowledgement. We thank the Carl Zeiss Foundation for the generous 5-year funding of SMART-AGE (P2019-01-003; 2021–2026).

References

1. Brooke, J.: SUS - a quick and dirty usability scale. In: Jordan, P.W., Thomas, B., Ian Lyall, M., Bernard, W. (eds.) Usability Evaluation in Industry, pp. 207–212 (1996)
2. Dzvonyar, D., Krusche, S., Alkadhi, R., Bruegge, B.: Context-aware user feedback in continuous software evolution. In: Proceedings of the International Workshop on Continuous Software Evolution and Delivery, CSED 2016, pp. 12–18 (2016).https://doi.org/10.1145/2896941.2896952
3. Fotrousi, F., Fricker, S.A.: QoE probe: a requirement-monitoring tool. CEUR Workshop Proc. **1564**, 7–8 (2016)
4. Groen, E.C., et al.: The crowd in requirements engineering: the landscape and challenges. IEEE Softw. **34**, 44–52 (2017). https://doi.org/10.1109/MS.2017.33

5. Groen, E.C.: How Requirements Engineering can benefit from crowds. Requirements Eng. Mag., 1–13 (2016)
6. International Organization for Standardization/International Electrotechnical Commission: Software engineering—Software product Quality Requirements and Evaluation (SQuaRE)—Data quality model ISO/IEC 25012:2008(E) (2008)
7. IREB: Certified Professional for Requirements Engineering – Foundation Level. Karlsruhe, Germany (2015)
8. Maalej, W., Happel, H.-J., Rashid, A.: When users become collaborators: towards continuous and context-aware user input. In: International Conference OOPSLA, pp. 981–990. ACM (2009). https://doi.org/10.1145/1639950.1640068
9. Oriol, M., et al.: FAME: supporting continuous requirements elicitation by combining user feedback and monitoring. In: IEEE Requirements Engineering Conference (RE), pp. 217–227. IEEE (2018). https://doi.org/10.1109/RE.2018.00030
10. Paech, B., Kohler, K.: Task-driven requirements in object-oriented development. In: do Prado Leite, J.C.S., Doorn, J.H. (eds.) Perspectives on Software Requirements. The Springer International Series in Engineering and Computer Science, vol. 753. Springer, Boston, MA (2004). https://doi.org/10.1007/978-1-4615-0465-8_3
11. Radeck, L., et al.: Understanding IT-related well-being, aging and health needs of older adults with crowd- requirements engineering. In: Workshop on Requirements Engineering for Well-Being, Aging, and Health of the International Requirements Engineering Conference, pp. 57–64. IEEE (2022). https://doi.org/10.1109/REW56159.2022.00018
12. Tizard, J., Rietz, T., Blincoe, K.: Voice of the users: a demographic study of software feedback behaviour. In: IEEE International Conference on Requirements Engineering, pp. 55–65 (2020). https://doi.org/10.1109/RE48521.2020.00018
13. Wang, C., Daneva, M., van Sinderen, M., Liang, P.: A systematic mapping study on crowdsourced requirements engineering using user feedback. J. Softw. Evol. Process. (2019). https://doi.org/10.1002/smr.2199
14. Wieringa, R.J.: Design Science Methodology for Information Systems and Software Engineering. Springer, Heidelberg (2014). https://doi.org/10.1007/978-3-662-43839-8
15. Wouters, J., Menkveld, A., Brinkkemper, S., Dalpiaz, F.: Crowd-based requirements elicitation via pull feedback: method and case studies. Requirements Eng. 27, 429–455 (2022). https://doi.org/10.1007/s00766-022-00384-6
16. Wüest, D., Fotrousi, F., Fricker, S.: Combining monitoring and autonomous feedback requests to elicit actionable knowledge of system use. In: Knauss, E., Goedicke, M. (eds.) REFSQ 2019. LNCS, vol. 11412, pp. 209–225. Springer, Cham (2019). https://doi.org/10.1007/978-3-030-15538-4_16

RE in Practice

Authoring, Analyzing, and Monitoring Requirements for a Lift-Plus-Cruise Aircraft

Tom Pressburger[1]([✉]), Andreas Katis[2]([✉]), Aaron Dutle[3]([✉]),
and Anastasia Mavridou[2]([✉])

[1] NASA Ames Research Center, Moffett Field, CA, USA
tom.pressburger@nasa.gov
[2] KBR, NASA Ames Research Center, Moffett Field, CA, USA
{andreas.katis,anastasia.mavridou}@nasa.gov
[3] NASA Langley Research Center, Hampton, VA, USA
aaron.m.dutle@nasa.gov

Abstract. **[Context & Motivation]** Requirements specification and analysis is widely applied to ensure the correctness of industrial systems in safety critical domains. Requirements are often initially written in natural language, which is highly ambiguous, and as a second step transformed into a language with rigorous semantics for formal analysis. **[Question/problem]** In this paper, we report on our experience in requirements creation and analysis, as well as run-time monitor generation using the Formal Requirement Elicitation Tool (FRET), on an industrial case study for a Lift-Plus-Cruise concept aircraft. **[Principal ideas/results]** We study the creation of requirements directly in the structured language of FRET without a prior definition of the same requirements in natural language. We focus on requirements describing state machines and discuss the challenges that we faced, in terms of creating requirements and generating monitors. We demonstrate how realizability, i.e., checking whether a requirements specification can be implemented, is crucial for understanding temporal interdependencies among requirements. **[Contribution]** Our study is the first complete attempt at using FRET to create industrial, realizable requirements and generate run-time monitors. Insight from lessons learned was materialized into new features in the FRET and JKIND analysis frameworks.

1 Introduction

The process of writing requirements for safety critical systems can be an arduous task, as engineers need to avoid ambiguous semantics and ensure that the resulting specification excludes unsafe system behavior. Formal specification can help engineers overcome both obstacles, as requirements are translated into unambiguous constructs using mathematical logic. Still, writing requirements using a formal language is not straightforward, especially when the author lacks a solid background in logical concepts. Furthermore, the analysis of such requirements can often leave engineers in a state of confusion, as they struggle with the interpretation of both positive and negative results.

This is a U.S. government work and not under copyright protection in the U.S.; foreign copyright protection may apply 2023

A. Ferrari and B. Penzenstadler (Eds.): REFSQ 2023, LNCS 13975, pp. 295–308, 2023.
https://doi.org/10.1007/978-3-031-29786-1_21

The Formal Requirements Elicitation Tool (FRET) [9] is an active, open-source research project [1] developed at NASA Ames, providing a highly accessible requirements engineering and analysis framework. It is designed so that engineers with varying levels of experience in formal methods can express requirements using structured natural language, observe their behavior through interactive simulation, and analyze their correctness with respect to their realizability; i.e., answer whether a system implementation exists that is guaranteed to conform to the given specification no matter the inputs received from its environment. The requirements can then be used to generate run-time monitors, which are programs that detect the violation of a particular requirement during the execution of the system. A tool chain beginning with FRET allows for partially automatic generation of such monitors.

In this paper, we briefly present FRET (Sect. 2) and report our experience using it to author and analyze requirements (Sect. 4) for an industrial case study on a Lift-Plus-Cruise concept aircraft (Sect. 3) as well as generate run-time monitors (Sect. 5). There is a focus on formulating state machines using FRET. We showcase challenges that we encountered, corresponding to common problems in requirements engineering, from expressing said requirements, to checking their realizability and actually interpreting the analysis results. We furthermore discuss how we were able to address issues, not only through the process of refining the specification, but also in terms of improving existing features of FRET to improve explainability of analysis results. This top-to-bottom study provided us with valuable insights, which we describe through lessons learned (Sect. 6).

Related Work: Previous work explored using FRET for industrial-level case studies, in cases where natural requirements specification already existed [6,20]. In contrast, in this paper we create requirements from informal diagrams and incorporate realizability checking as part of the workflow. Similar studies have been conducted in the past for other requirements specification tools. Previous works in the RAT [24] SPECTRA [19] and LTSA [17] tools has showed how requirements expressed in Linear Temporal Logic could be evolved, guided by the results of consistency and realizability analysis in a Boolean setting. The EARS-CTRL [18] tool provides a natural language, and its analysis for synthesizing controllers is also in a Boolean setting. In comparison, requirements written in FRET's language can deal with linear arithmetic expressions over unbounded integer and real numbers. A study analyzing control software in the AGREE framework identified errors in specification using realizability checking [3]. Notably, the checking algorithm used is known to be unsound w.r.t. unrealizable results [8], whereas FRET employs sound procedures [12,16]. The CLEAR [4] tool also uses a constrained natural language to formalize requirements. It can check completeness and consistency, but not realizability.

2   Background

FRET provides a collection of features for the creation, management and analysis of requirements. We next present the features that were used in this paper, namely the FRETISH language, the realizability checking component and finally the requirements export functionality to achieve synthesis of run-time monitors.

Requirement Specification and Formalization. Users write requirements in FRETISH, i.e., a restricted natural language with standard mathematical expressions [9]. A FRETISH requirement is described using up to six sequential fields (the * symbol designates mandatory fields): 1) `scope` specifies the time intervals where the requirement is enforced, 2) `condition` is a Boolean expression that triggers the `response` to occur at the time the expression's value becomes true from false, or is true at the beginning of the scope interval, 3) `component*` is the system component that the requirement is levied upon, 4) `shall*` is used to express that the component's behavior must conform to the requirement, 5) `timing` specifies when the response shall happen, subject to the constraints defined in `scope` and `condition` and 6) `response*` is the Boolean expression that the component's behavior must satisfy.

FRETISH provides 8 scopes: *global, in, before, after, notin, only in, only before,* and *only after.* The scope *global* means *always*; the others are with respect to when the system is in a mode or satisfies a Boolean expression. The optional condition field is introduced by any of the words *upon, when,* or *if,* which are synonymous. FRETISH provides 10 timings: *immediately* (meaning: at the same time point), *at the next timepoint, always, eventually, never, for N* time steps, *within N* time steps, *after N* time steps, *until bool_expr,* and *before bool_expr.* When the scope is omitted it is taken as *global*; when the condition is omitted, it is taken as `true`; when the timing is omitted, it is taken as *eventually.*

The Boolean expressions use the standard logical symbols, as well as standard arithmetic symbols and relations. FRETISH also supports several predefined predicate symbols: *preBool(init,x)* (resp., *preReal(init,x)*) denotes, at the first time point, the value of the Boolean (resp., real) expression *init,* and subsequently the value of the Boolean (resp., real) expression *x* at the previous time point; *absReal(x)* denotes the absolute value of the real-valued expression *x*; and *FTP* is true at the first time point in the execution, and is false otherwise.

FRETISH requirements are based on rigorous semantics and thus, have a precise, unambiguous meaning. Once the requirements are specified, FRET produces formalizations in several logics. In this study, we make use of past-time metric linear temporal logic (pmLTL).

To capture commonly occurring requirement patterns, FRET provides a *template* facility. This allows the user to construct FRETISH requirements by instantiating placeholders in a template.

Realizability Checking. Informally, a specification is realizable if we can implement a system, such that it always conforms to the given requirements, while considering inputs from an uncontrollable environment (sensors, user input, etc.). A proof of realizability not only establishes the truth that a system can be implemented for the given requirements, but also the fact that, given proper care in the system implementation, the requirements themselves are free of conflicts that would translate into unsafe behavior.

The analysis portal in FRET provides means to examine specifications in terms of their realizability, as well as generate artifacts that help engineers further understand the analysis results. The following features are available [13]:

— *Compositional analysis:* As a preprocessing step, FRET decomposes the specification, if possible, into a set of connected components, based on the outputs exercised in each requirement. This decomposition is sound w.r.t. realizability, allowing the independent analysis of each one of the computed components [21].

— *A portfolio of engines and algorithms that support infinite theories.* FRET uses both the KIND 2 [5,16] and JKIND [7,12] model checkers for realizability analysis. Both engines are SMT-based, supporting unbounded theories of integer and real arithmetic, while also providing means to compute counterexamples from unrealizable specifications, in the form of deadlocking execution traces.

— *Diagnosis of unrealizable specifications.* FRET employs diagnostic algorithms to provide further feedback in unrealizable requirements. This is achieved through the computation and simulation of minimal sets of unrealizable requirements (known as minimal conflicts) [14,15]. Counterexamples that demonstrate unrealizability can be graphically displayed in the interactive simulator in FRET.

Exporting Requirements for Monitor Synthesis. Having created a set of realizable requirements, we can now generate runtime monitors. To this end, FRET generates and exports a specification that can be digested by the OGMA tool [23] for the generation of COPILOT monitors [22]. This specification contains formalized requirements and information about the variable types referenced in the requirements. OGMA then produces an input specification for COPILOT, and finally COPILOT generates C code suitable for use in hard real-time systems, running without dynamic memory allocation in predictable space and time. The C code accepts inputs to be monitored and invokes user-provided handlers when the requirements are violated. The creation and integration of these monitors is intended to be as seamless as possible; the properties to be monitored are written in FRETISH, and little to no code is required to be written by hand.

3   The Lift Plus Cruise Case Study

Because of its ability to be used at many stages of the development lifecycle, and the familiarity of the researchers with the tool, FRET was chosen as a main component in a NASA project studying safety assurance for a novel Lift-Plus-Cruise (LPC) electric Vertical Takeoff and Landing (eVTOL) aircraft. There are several different concepts for VToL aircraft being investigated by the aviation community, including NASA [26]. One such design has a number of lifting rotors

Fig. 1. The LPC vehicle.

attached to the wings and a forward pushing propeller on the rear of the aircraft (Fig. 1). NASA is developing models of the flight characteristics of this LPC concept, as well as simulation capabilities, and control schemes [11].

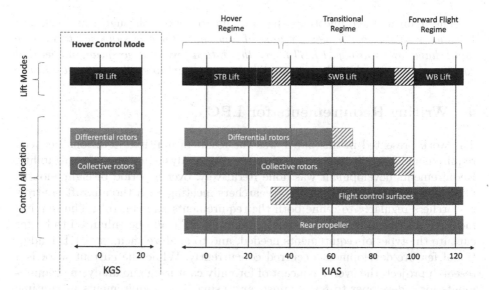

Fig. 2. Control Allocation Schedule. KGS/KIAS: ground/indicated air speed (knots).

The project investigated aspects of safety assurance of the aircraft including hazard analysis, requirements capture, formal modeling, and runtime monitoring. FRET was used to capture requirements for the vehicle, and the collection of requirements served as a model of how the vehicle was expected to behave. Some of these requirements were then used to generate runtime monitors for use in the simulation environment.

Due to the design of the aircraft, several distinct control regimes may apply at different phases of flight. For example, during takeoff and landing, the aircraft motion is controlled by the lifting rotors only, and the flight surfaces (wings, ailerons, etc.) have no effect (thrust-borne mode, TB). On the other hand, during the higher speeds of the en-route phase, the wings provide lift, the rear propeller provides thrust, and the lifting rotors are inactive (wing-borne mode, WB). *Collective* control means that all of the rotors are commanded to increase or decrease torque, leading to more or less "heave" (vertical climb). *Differential* control means that the rotors are commanded to have differing amount of torques, enabling control of pitch, yaw and roll.

Figure 2 shows the ranges of air/ground speeds for the control regimes. The hashed areas indicate regions of hysteresis; i.e., control lag. For example, if the vehicle is slowing down from the wing-borne mode (WB), the transition to semi-wing-borne (SWB) kicks in at an indicated airspeed of 90 knots (kias <= 90.0), whereas if the vehicle is speeding up from a SWB mode, the transition to WB mode occurs at kias > 100.0 knots; similarly for the transitions between semi-thrust-borne (STB) mode and SWB mode. The vehicle remains in the thrust-borne mode (TB) as long as kgs <= 20.0 knots and Hover Control (HC) mode is selected.

The main research questions that we aim to answer through this work are: 1. *Can we take informal descriptions of how the vehicle is supposed to operate and behave, and (through FRET) turn this into a formal description/model that can be analyzed?* and 2. *Can we use this formal model to easily create monitors?*

4   Writing Requirements for LPC

The work presented in this paper was the result of multiple iterations between requirements formalization and their respective analysis in terms of realizability. Requirements development was done iteratively, over a period of eight months part-time, with the requirements researchers meeting with the aircraft controls researchers regularly to refine both the requirements and controls. The requirements development revealed some ways that FRET could be enhanced to better capture the types of requirements needed, and to analyze them, so FRET additional feature development occurred concurrently. While the current work is a research project, the overall concept of formally capturing and analyzing requirements for a developer to test against, and using these requirements as runtime monitors, is envisioned as a method to help assure safety of future aircraft.

4.1   Initial Formalization

To validate the control scheme concept, and facilitate use in further development and refinement of control software, we undertook the formal modeling in FRET of the control allocation of the LPC concept during the landing transition phase. This phase transitions from fully wing-borne flight to fully thrust-borne, with intermediate phases semi-wing-borne and semi-thrust-borne.

Our task is to develop realizable requirements for the control schedule (Fig. 2). The complete set of requirements is in the technical report [25]. The variables used in these requirements, as well as their types are shown in Table 1. For the purposes of realizability checking and monitor generation, we need to declare each variable as either an *input* or *output*. An output is a variable that the system controls. An input is a monitored variable, one whose value is set by the environment that the system has no control over.

We start with a requirement that the vehicle be in one of the lift modes at each time point. Note that integer constants in Table 1 are used to simulate a lift-mode enumerated type.

Table 1. LPC Variables.

cr	boolean	output
dr	boolean	output
fcs	boolean	output
HC	boolean	output
rearprop	boolean	output
kgs	double	output
kias	double	output
wind_speed	double	input
lift_mode	integer	output
TB	integer	constant 0
STB	integer	constant 1
SWB	integer	constant 2
WB	integer	constant 3

[LIFT_MODE]: The vehicle **shall** always satisfy lift_mode = TB | lift_mode = STB | lift_mode = SWB | lift_mode = WB

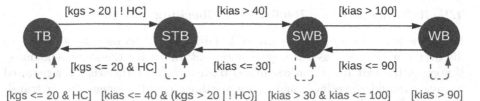

Fig. 3. Lift Mode State Machine derived from Fig. 2. The acronyms are: HC = hover control, B = borne, T = thrust (rotors), W = wing, S = semi-, kgs = ground speed (knots), kias = indicated air speed (knots).

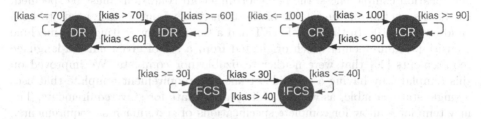

Fig. 4. Differential Rotors (DR), Collective Rotors (CR), and Flight Control Surface (FCS) state machines derived from Fig. 2.

We also require that the rear propeller be always used, except in HC mode: **[REARPROP]:** The vehicle **shall** always satisfy rearprop xor HC

To specify the control schedule requirements, we chose the clear and succinct way that state machines provide, and expressed those state machines in FRETISH. Initially, we transformed what is shown in Fig. 2 into state machines. E.g., for the required behavior of the lift modes, we created Fig. 3: the four states correspond to the lift modes and the black, solid-line guarded transitions define when a mode change may happen. For instance, when in STB mode and the ground speed is less than or equal to 20 knots (kgs <= 20), the pilot, or an automated control system, can switch to HC mode, allowing the aircraft to enter the TB lift mode. Similarly, the control allocation state machines are represented in Fig. 4. The guards on the transitions refer to the conditions on the indicated airspeed in knots (kias) and ground speed (kgs). Initially, we designed Figs. 3 and 4 without the red, dashed-line loop transitions.

To capture *transition* requirements, we created the following FRET template to express transitions from state s_0 to a state s_1 under condition p:

Upon state = s_0 & p the vehicle **shall** at the next timepoint satisfy state = s_1

E.g., the transition originating from state WB to state SWB in Fig. 3 can be written as follows: **[WB_TO_SWB]:** Upon lift_mode = WB & kias <= 90.0 the vehicle **shall** at the next timepoint satisfy lift_mode = SWB

4.2   Refinement Using Realizability Checking

Using the realizability analyzer over this initial set of requirements, led us to discover that we also need a *stay* requirement that says the state remains s_0 if none of the exit transition conditions from s_0 hold. Otherwise, the required behavior is under-specified, and hence anything could happen after a transition to a particular state when no transition condition applies. In particular, realizability analysis, as shown in Table 2, reported a realizable trace where the aircraft state transitions from wing-borne mode directly to thrust-borne mode without visiting intermediate modes. The stay requirements are necessary for specification completeness: the behavior under all conditions must be specified, so the disjunction of the guards of the transitions from a state needs to be a valid formula [10]. In the past, FRET had a template for writing state-machine transition requirements, which originated from a set of given natural-language requirements [20] that were neither realizable nor complete. We improved on this template, by having a simplified transition requirement template that uses a single state variable, as well as adding a template for stay requirements. The new templates allow for complete specifications of state machine requirements.

One could express the stay requirement in FRETISH as: When state $= s_0 \& P$, the vehicle **shall** at the next timepoint satisfy state $= s_0$, where $P = !p_1 \& \ldots \& !p_n$ is the conjunction of negated guards that belong to outgoing transitions of s_0. However, this would only constrain the value of state when the condition transitioned from false to true, not whenever the condition held. Instead, the stay requirement can be expressed with the following FRET template (see [25]):

> Vehicle **shall** always satisfy if preBool(false, $state = s_0 \& P$) then state $= s_0$

Currently, this is formalizable by FRET only in pmLTL. This was adequate for this case study, since both realizability analysis and monitor generation rely on the past-time formalization. If, in a different situation, a future-time formula is needed, it can be expressed in FRETISH without preBool as Upon state $= s_0 \& P$ the vehicle **shall** until state $= s_0 \& !P$ satisfy state $= s_0$. This says that the system, upon entering state s_0 when no transition condition applies, will remain in state s_0 until *and including* the time point where a transition condition holds. The two formulations were shown, using the NuSMV model-checker, to be equivalent. Although equivalent logically, realizability analysis using the second formulation was 15 to 100 times slower; we are investigating the cause. We show below stay transition requirements from the wing-borne mode (Fig. 3) and flight control surfaces (Fig. 4). Other stay and transition requirements were written in a similar manner. These requirements correspond to the dashed-line loop transitions (Figs. 3 and 4).

[WB_STAY_ON_pre]: The vehicle **shall** always satisfy if preBool(false, lift_mode = WB & kias > 90.0) then lift_mode = WB

[WB_STAY_ON_until]: Upon lift_mode = WB & kias > 90.0 the vehicle **shall** until lift_mode = WB & kias <= 90.0 satisfy lift_mode = WB

Table 2. Example trace from incomplete specification.

Variable \ Step	0	1	2	3	4	5	6
HC	false	false	false	false	false	false	false
kgs		120	120.25	110.25	111.5	103.5	100	90.25
kias		120	120.25	110.25	111.5	103.5	100	90.25
lift_mode		WB	TB	STB	SWB	WB	SWB	SWB

Table 3. Example trace from the final specification.

Variable \ Step	0	1	2	3	4	5	6	7	8	9	10	11
HC	false	false	false	false	false	false	false	false	false	false	true	true
kgs	120	110	100	90	80	70	60	50	40	30	20	20
kias	120	110	100	90	80	70	60	50	40	30	20	20
lift_mode	WB	WB	WB	WB	SWB	SWB	SWB	SWB	SWB	SWB	STB	TB

[FCS_STAY_OFF]: The vehicle **shall** always satisfy if preBool(false, !fcs & kias <= 40.0) then !fcs

[FCS_TURN_ON]: Upon !fcs & kias > 40.0 the vehicle **shall** at the next timepoint satisfy fcs

So far, we have specified the required behavior for transitioning from wing-borne lift mode to thrust-borne mode. Still, we are not done: we need to specify initial conditions, as well as a time target before which the transition should complete. We try the scenario where the initial airspeed is 120 knots, the initial lift mode is wing-borne, the ground speed always equals the airspeed, and the airspeed changes by no more than 10 knots in consecutive time points:

[INIT_KIAS]: The vehicle **shall** immediately satisfy kias = 120.0

[INIT_LIFT_MODE]:
The vehicle **shall** immediately satisfy lift_mode = WB <=> kias >= 90.0

[KIAS_KGS]: The vehicle **shall** always satisfy kias = kgs

[KIAS_DERIVATIVE]: The vehicle **shall** always satisfy
FTP | absReal(preReal(0.0, kias) − kias) <= 10.0

All that is now left is to define a possible goal about *lift_mode*:

[REACH_HOVER]: The vehicle **shall** within 10 ticks satisfy lift_mode = TB

We now claim that we have a complete formalization. Is it realizable, though? Careful inspection should result in a "no" answer, as 10 ticks is not enough time to complete the transition. The realizability analysis supported this claim: the requirements are unrealizable for 10 ticks and realizable for 11 ticks. Table 3 shows a positive trace from the latter result, where the system is able to complete the transition from wing-borne to thrust-borne in a proper manner, exercising the intended intermediate mode transitions.

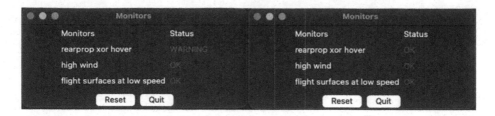

Fig. 5. Runtime monitor displays: monitor violation (left), no violation (right).

4.3 Reasoning About the System's Environment

Notably, the requirements presented thus far do not constrain the system's input. We experimented with additional requirements involving wind, changing **[KIAS_KGS]** to specify that kgs is the sum of kias and the wind_speed input variable (hence uncontrollable). Furthermore, we added the following assumption on the environment: **[WIND_SPEED_assumption]:** The vehicle **shall** always satisfy absReal(wind_speed) <= 30.0

We expected that this assignment for kgs would prohibit entering TB mode because the wind would prevent kgs <= 20. However, in about a minute, realizability checking said that it was realizable. Examination of a positive trace revealed that this was due to kias becoming negative; i.e., the vehicle flying backwards. The diagram we were initially given (Fig 2) is misleading: the vehicle can only maneuver backwards slowly, while in TB mode, to make small corrections while landing. When a requirement was included that said kias >= 0, the requirements were shown to be unrealizable, even when increasing the time limit in **[REACH_HOVER]** to 16 ticks. Strengthening the assumption to |wind_speed| <= 20 fixed the issue, as the requirements were shown to be realizable within 13 ticks, which makes sense as kias needs to be reduced to zero for kgs to be <= 20, for any valid wind speed.

5 Generation of Run-Time Monitors

We integrated the C code generated by COPILOT into the FlightDeckZ Vehicle Simulation Environment [2], monitoring three requirements described earlier: **[REARPROP]**, **[FCS_STAY_OFF]**, and **[WIND_SPEED_assumption]**. FlightDeckZ is a system that incorporates physics models of the LPC concept with flight controllers, and a visualization system, to allow for fairly realistic flight simulation of the LPC vehicle model with experimental controllers.

The first two monitors express requirements that we expect from the control system of the LPC model, while the last monitor expresses an environmental property that may be of interest to a pilot during an actual flight (as most eVToL systems are not designed to take off in high winds). This difference here is intentional. The first two monitors are likely more useful to a system developer, and so can likely be removed from use once a stable and trustworthy control system is in place. The last monitor is something that may be integrated into a system display on a real aircraft. The status of the monitors is displayed to the users with a simple window frame, with descriptions of the monitors and their current status displayed side-by-side (Fig. 5).

6   Lessons Learned

We list below lessons and FRET needs and improvements resulting from the experience of using FRET in this case study.

Expressiveness of FRETish: In this effort, requirements were written directly in FRETish based on informal diagrams describing desired behavior, rather than being translated from an initial natural language description. Thus, we were interested in understanding whether FRETish provides adequate expressiveness and clarity and whether we are able to capture requirements that observe complex interaction behavior for generating meaningful runtime monitors.

We were able to express in FRETish all the requirements of the control allocation schedule. Writing these requirements directly in FRETish made them more detailed while avoiding ambiguities; a lot of attention was given to understanding their semantics and how small changes in their syntax affect it.

We also found limitations: FRET lacks an enumerated type facility; a workaround with integer constants was used instead for the lift modes in Table 1. Also, a condition that enforces the response whenever the condition is true, not just triggering the response upon the condition becoming true from false, would have been useful, as discussed in Sect. 4.2.

Usefulness of Tool Assistance in Writing Requirements: Crucial to the requirement formulation process were the interactive simulator of FRET and the realizability checking mechanism that guided the discussion to corner cases, important sanity checks, and complete requirement sets (see Sect. 4.2).

Formulating correctly the FRETish for state transitions involved some subtlety, but once the FRET templates were devised, they were used to specify 26 out of the 53 LPC requirements. Since state machines are frequently used in requirements development, we expect that the FRET templates could be useful to others who wish to formulate complete and realizable state machine requirements. On the other hand, instead of formulating such requirements in FRETish, the ability to express requirements in a state-machine notation directly could be a useful addition to FRET.

Usability of Feedback from Realizability Analysis in the Form of Positive and Negative Traces, and Minimal Conflicts: In several cases (e.g., the cases mentioned in Sect. 4.3 that revealed negative air speed, and in Sect. 4.2 the need for "stay" requirements for completeness), we needed evidence to understand why a specification was realizable. This motivated a new feature in JKIND and the FRET analysis portal that computes and displays a satisfying, i.e., positive, trace showing how the requirements are realizable. We achieved this by using the proof produced by realizability checking. More specifically, when a specification is proved realizable by the underlying tools, a symbolic fixpoint of "good system states" is computed. We reuse this fixpoint to compute and present to the users valid system execution traces of bounded length, that can be seen as indicative runs of a system that is, by definition, guaranteed to always comply with the specification. Examples of such generated positive traces were shown in Tables 2 and 3. Furthermore, we enabled the use of the simulator to interact with the requirements in context, starting from the satisfying trace (see [25]).

Unrealizable results also contributed to the refinement of the specification. E.g., note how we allow the vehicle to control the HC variable (i.e., declared as an output). The fact that it needed to be an output was pointed out by the realizability analysis: the specification was unrealizable when the variable was originally declared as an input, because the environment could decide to never switch to hover control mode. When a set of requirements is unrealizable, it is left up to the FRET user to puzzle out from a negative trace and experimentation why the requirements don't allow a positive trace. In particular, minimal conflicts can be subsets of requirements that discard necessary requirements, for example, **LIFT_MODE**. Further research is needed in the area of providing helpful counterexamples. We sometimes found it sufficient to find the cause just by examining which requirements were in the conflict set.

Dealing with Requirement Versions: During realizability analysis, we refined our requirements multiple times. Thus, we ended up with several different versions of the same requirements that can be used within different subsets of requirements. This motivated a new feature in the FRET analysis portal that allows the user to easily select which requirement versions should be included in each realizability check (see [25]). In certain cases, we ended up with logically equivalent requirement versions. We thus think that there should be a capability of the FRET interface to test the equivalence of requirements, without the user needing to escape to other tools.

Easy Monitoring: FRETISH allows for the easy specification of many complex and time-based interactions inside a system. For example, in the LPC model, if one of the lifting rotors fails, the mirrored rotor on the other wing should be turned off, so that a thrust imbalance does not occur. In FRETISH, one could easily specify a property that says "Upon rotor_1_fail, the vehicle shall within 5 s satisfy rotor_4_power_off". Such a monitor could then be automatically generated, and requirements violations could be detected without post-simulation analysis, or even without the need for manual writing of code that collects and assesses the state of the system over periods of time.

Monitor Semantics Mismatch: We discovered an issue with FRET-generated COPILOT monitors during the integration and testing process. Due to the fact that the requirements are turned into pmLTL, the interpretation of each requirement is the statement "always in the past, requirement x holds." What this means is that at each time step, the monitor is determining if there has ever been a violation. Hence even if the system returns to a state that is determined to be safe, the monitor is still considered violated. For example, if the wind ever goes above 30 knots, then even after the wind calms, the statement "The wind shall always be below 30 knots" is false, so the monitor stays on. Currently, a workaround "reset" button restarts the monitors, effectively erasing all past history, mitigating the issue.

7 Conclusion

This experience report paper showed how certain aspects of a concept Lift-Plus-Cruise aircraft were captured in requirements written in FRETISH and how realizability analysis was crucial for guiding the evolution of the requirements. The main requirements engineering challenges that we encountered stemmed from the iterative process of refining requirements with respect to realizability. These challenges were not apparent until after the step of analysis was reached.

To answer our main research questions: we were successful in turning informal descriptions into an analyzable formal model through FRET and subsequently using this formal model to easily create monitors. To this end, the FRET model did fulfill its purpose. Additionally, experience with this case study led us to improve FRET as well as to point to future work such as adding to the expressiveness of FRETISH and revisiting the semantics of run-time monitoring.

Acknowledgements. We acknowledge Michael Feary, John Kanishige, and Kimberlee Shish who explained the vehicle used in this study and provided Fig. 2, and Dimitra Giannakopoulou who did an early requirements development. Thanks also to the anonymous reviewers who provided detailed improvement suggestions. This work was supported by the Advanced Air Mobility and System Wide Safety projects in the NASA Aeronautics Mission Directorate's Airspace Operations and Safety Program. Andreas Katis and Anastasia Mavridou were supported by contract NASA 80ARC020D0010.

References

1. FRET. https://github.com/NASA-SW-VnV/fret.git
2. Archdeacon, J., Iwai, N., Feary, M.: Aerospace cognitive engineering laboratory (ACELAB) simulator for electric vertical takeoff and landing (eVTOL) research and development. In: AIAA Aviation Forum (2020)
3. Backes, J., Cofer, D., Miller, S., Whalen, M.W.: Requirements analysis of a quad-redundant flight control system. In: NFM 2015 (2015)
4. Bhatt, D., Ren, H., Murugesan, A., Biatek, J., Varadarajan, S., Shankar, N.: Requirements-driven model checking and test generation for comprehensive verification. In: NFM 2022 (2022)
5. Champion, A., Mebsout, A., Sticksel, C., Tinelli, C.: The Kind 2 model checker. In: CAV 2016 (2016)
6. Farrell, M., Luckcuck, M., Sheridan, O., Monahan, R.: Fretting about requirements: formalised requirements for an aircraft engine controller. In: REFSQ 2022 (2022)
7. Gacek, A., Backes, J., Whalen, M., Wagner, L., Ghassabani, E.: The JKind model checker. In: CAV 2018 (2018)

8. Gacek, A., Katis, A., Whalen, M.W., Backes, J., Cofer, D.: Towards realizability checking of contracts using theories. In: Havelund, K., Holzmann, G., Joshi, R. (eds.) NFM 2015. LNCS, vol. 9058, pp. 173–187. Springer, Cham (2015). https://doi.org/10.1007/978-3-319-17524-9_13

9. Giannakopoulou, D., Pressburger, T., Mavridou, A., Schumann, J.: Automated formalization of structured natural language requirements. Inf. Softw. Technol. **137**, 106590 (2021)

10. Heitmeyer, C.L., Archer, M., Bharadwaj, R., Jeffords, R.D.: Tools for constructing requirements specifications: the SCR toolset at the age of ten. Int. J. Comput. Syst. Sci. **20**(1), 19–35 (2005)

11. Kanishege, J., Lombaerts, T., Shish, K., Feary, M.: Command and control concepts for a lift plus cruise electrical vertical takeoff and landing vehicle. In: AIAA Aviation Forum and Exposition, San Diego, CA, June 2023

12. Katis, A., et al.: Validity-guided synthesis of reactive systems from assume-guarantee contracts. In: TACAS (2018)

13. Katis, A., Mavridou, A., Giannakopoulou, D., Pressburger, T., Schumann, J.: Capture, analyze, diagnose: realizability checking of requirements in FRET. In: CAV 2022 (2022)

14. Könighofer, R., Hofferek, G., Bloem, R.: Debugging unrealizable specifications with model-based diagnosis. In: Haifa Verification Conference (2010)

15. Könighofer, R., Hofferek, G., Bloem, R.: Debugging formal specifications: a practical approach using model-based diagnosis and counterstrategies. Int. J. Softw. Tools Technol. Transfer **15**(5–6), 563–583 (2013)

16. Larraz, D., Tinelli, C.: Realizability checking of contracts with Kind 2 (2022)

17. Letier, E., Heaven, W.: Requirements modelling by synthesis of deontic input-output automata. In: 2013 35th International Conference on Software Engineering (ICSE), pp. 592–601. IEEE (2013)

18. Lúcio, L., Rahman, S., bin Abid, S., Mavin, A.: EARS-CTRL: generating controllers for dummies. In: MODELS (Satellite Events), pp. 566–570 (2017)

19. Maoz, S., Ringert, J.O.: Synthesizing a lego forklift controller in gr (1): a case study. arXiv preprint arXiv:1602.01172 (2016)

20. Mavridou, A., et al.: The ten Lockheed Martin cyber-physical challenges: formalized, analyzed, and explained. In: Proceedings of the 28th IEEE International Requirements Engineering Conference (2020)

21. Mavridou, A., Katis, A., Giannakopoulou, D., Kooi, D., Pressburger, T., Whalen, M.W.: From partial to global assume-guarantee contracts: compositional realizability analysis in FRET. In: Formal Methods (2021)

22. Perez, I., Dedden, F., Goodloe, A.: Copilot 3. Technical report NASA/TM 2020-220587, April 2020

23. Perez, I., Mavridou, A., Pressburger, T., Goodloe, A., Giannakopoulou, D.: Automated translation of natural language requirements to runtime monitors. In: TACAS 2022 (2022)

24. Pill, I., et al.: Formal analysis of hardware requirements. In: DAC 2006 (2006)

25. Pressburger, T., Katis, A., Dutle, A., Mavridou, A.: Using FRET to create, analyze and monitor requirements for a lift plus cruise case study. Technical report NASA/TM 20220017032 (2023)

26. Silva, C., Johnson, W.R., Solis, E., Patterson, M.D., Antcliff, K.R.: VTOL urban air mobility concept vehicles for technology development. In: AIAA 2018 (2018)

Knowns and Unknowns: An Experience Report on Discovering Tacit Knowledge of Maritime Surveyors

Tor Sporsem[1]([⊠])(iD), Morten Hatling[1], Anastasiia Tkalich[1](iD), and Klaas-Jan Stol[1,2](iD)

[1] SINTEF Digital, 7034 Trondheim, Norway
tor.sporsem@sintef.no
[2] University College Cork, Cork, Ireland

Abstract. **[Context]** Requirements elicitation is an essential activity to ensure that systems provide the necessary functionality to users, and that they are fit for purpose. In addition to traditional 'reductionist' techniques, the use of observations and ethnography-style techniques have been proposed to identify requirements. **[Research Problem]** One frequently heard issue with observational techniques is that they are costly to use, as developers who would partake, would lose considerable development time. Observation also does not guarantee that all essential requirements are identified, and so luck plays a role. Very few experience reports exist to evaluate observational techniques in practice, and for organizations it is difficult to assess whether observation is a worthwhile activity, given its associated cost. **[Results]** This report presents experiences from DNV, a global leader providing maritime services who are renewing an information system to support its expert users. We draw on several data sources, covering insights from both developers and users. The data were collected through 9 interviews with users and developers, and over 80 h of observation of prospective users in the maritime domain. We capture 'knowns' and 'unknowns' from both developers and users, and highlight the importance of observational studies. **[Contribution]** While observational techniques are costly to use, we conclude that essential information is uncovered, which is key for developers to understand system users and their concerns.

Keywords: User involvement · Expert Knowledge · Requirements engineering · Tacit Knowledge · Ethnographic Techniques

1 Introduction

"We can know more than we can tell"
—Polanyi [17]

"We're not good at knowing what we know"
—Ken Jennings, *Jeopardy!* champion[1] [14]

---

[1] Jennings holds the record of the longest streak of wins of the popular TV game show 'Jeopardy!'

© The Author(s), under exclusive license to Springer Nature Switzerland AG 2023
A. Ferrari and B. Penzenstadler (Eds.): REFSQ 2023, LNCS 13975, pp. 309–323, 2023.
https://doi.org/10.1007/978-3-031-29786-1_22

Organizations in all domains rely on software solutions to support their employees in achieving their core business goals. Automation to support professionals and experts in their manual jobs can be traced back to the early days of computing [12, 21]. Initially, development processes for software were modeled after traditional engineering cycles, focusing on problem formulation and analysis, systematically developing systems based on requirements that could be identified.

Many of the early software development methods sought to provide support to systems developers in structurally and systematically design and implement systems—the so-called structured approaches; specific practices included structured analysis, structured design, data-driven design, and structured coding [5, 12]. A key characteristic of these practices is that they all focus on that what can be observed and articulated; further, they take a 'reductionist' approach in that these methods consider only technical entities such as components, control structures, and data flows [5], ignoring what we could label "system-in-action" requirements that capture the subtleties of how users actually use systems.

In recognition of the importance of capturing and managing the right system requirements, the Requirements Engineering (RE) discipline emerged [15], and has been concerned with identifying, modeling, communicating, and documenting requirements [16]. Despite a very rich and mature RE literature today, RE remains a major challenge because systems are usually developed by people other than the intended users. Software systems that fail to meet the needs of expert users may threaten those users' ability to do their job and, indirectly, the core business activities of the organization. A key problem is the knowledge gap that exists between the analyst/developer, and the expert user who possesses a high level of expertise. Understanding how this gap can be closed has been a longstanding goal of the RE discipline.

As one of the opening quotes suggests, Polanyi [17] argued that much human knowledge acquired by highly skilled experts through experience is impossible to articulate. In seeking to understand whether and how expertise can be articulated, several scholars have invoked the term 'tacit knowledge' [6]. Gervasi et al. [7] drew on a notable 2002 press briefing of the late Donald Rumsfeld, Secretary of Defense during the U.S. invasion of Afghanistan and Iraq. Rumsfeld argued there are different types of knowledge, or 'knowns': known knowns, known unknowns, and unknown unknowns.[2] Gervasi et al. [7] suggested that there is a fourth type of knowledge:

"An unknown known is knowledge that a customer holds but which they withhold from the analyst."

We note that this 'withholding' is likely to be unintentional. Tacit knowledge, then, fits that definition, i.e. tacit knowledge is an unknown known [7]. Table 1 presents the four types of knowns, considering two important roles in RE, as Gervasi et al. [7] suggested: the system analyst/developer, and the user. Some knowledge is held by both developers and users (known knowns), whereas other knowledge is known to only one but not the

---

[2] These different types of 'knowns' map very well to Phillip Armour's "Orders of Ignorance" published two years prior, in 2000 [1]. This might be a rare unintended instance where SE research has had an impact on global political rhetoric.

other, e.g. unknown knowns represent knowledge held by users, whether they are aware of it or not, but unknown to developers.

Table 1. Developers and users' knowns and unknowns (based on Gervasi et al. [7] and Sutcliffe and Sawyer [20])

		Analysts and Developers	
		Known to developers	Unknown to developers
Users	Known to users	**Known Knowns**: relevant knowledge that users know and that can be articulated for software developers	**Unknown Knowns**: relevant knowledge that users know (whether consciously or without realizing it), but which is not yet articulated and thus not known yet to software developers
	Unknown to users	**Known Unknowns**: relevant information that developers are aware of (know), but which they don't know yet. Users may be unaware of this knowledge, or have forgotten it	**Unknown Unknowns**: potentially relevant information, but both developer and user are unaware that it is missing. Developers lack relevant domain knowledge, and users are unaware of the knowledge that they rely on

The RE field has discussed different requirements elicitation techniques at length [4, 8, 20] and it lies beyond the scope of this paper to present a full discussion. Commonly discussed techniques are interviews, workshops, scenarios, and observation [4, 8, 20, 22]. Observation is often mentioned as a part of conducting ethnographic studies; several papers have discussed ethnography or observational approaches to support requirements elicitation and design [9, 11, 19]. Early studies proposing to integrate ethnography for RE recognized that traditional techniques *do not take into account actual work practices* [19]. While there has been some fruitful discussion and analysis of observational methods to uncover 'unknowns,' a few issues seem to remain. For example, ethnographically-informed or observational methods for RE have been suggested to require considerable resources, time in particular, making them less attractive. Further, other issues associated with ethnographic research is that it may suffer from ambiguous interpretation [20], a lack of technical competence of ethnographers [3], and the serendipitous nature of identifying new requirements through ethnography [20], i.e. the reliance on luck. Finally, the number of studies that evaluate the use of observational approaches including ethnography for requirements engineering has remained limited, despite several important contributions in the 1990s and early 2000s [3, 9–11]. One might wonder, given the drawbacks listed above, whether organizations should bother with observational approaches. A lack of experience reports on the use of observational or ethnographic techniques hinders organizations in deciding whether this approach is worth the considerable cost. There seems to be an acceptance that there is no advantage in any specific technique over the use of structured interviews [8, 20], and so an open question is: what value does observation offer in a requirements engineering context?

Thus, the goal of this experience report is to highlight the importance of observation to identify unknown knowns and unknowns, and report lessons learned from the field, in a domain that hitherto has not been studied in this context. Several of the seminal papers in the software engineering and requirements engineering literature reported on an air traffic control system, which represents a very specific setting whereby its users operate in a fixed location. This experience report focuses on surveyors who inspect ships for certification, necessary to allow them to operate in international waters. Surveyors, unlike air traffic controllers, operate in a different setting *every single day*. This makes characterizing these actors' work environment more challenging as each ship is unique and thus it is important to recognize the varying work settings of these experts. We illustrate how developers, who had used traditional requirements elicitation techniques such as interviews, were struggling to understand these expert users, and indeed had not gained important insights that we classified as unknown knowns and unknown unknowns (see Table 1). We conclude by juxtaposing our findings with prior literature, adding clarifications and commentary, and identify some implications for practice.

2 Methods

This experience report draws on data collected from different sources at DNV, a major service provider in the maritime sector. As researchers of the SINTEF Digital Process Innovation group, we are involved in an ongoing project with DNV focused on Digital Transformation, which provided the backdrop of this investigation. In the remainder of this section we describe DNV and procedures for data collection and analysis.

2.1 Description of DNV

DNV is a leading service provider in the maritime sector, with about 3,700 employees operating globally. DNV's core business is compliance verification of vessels (ships of any size); successful verification leads to issuing of necessary certificates that vessels require in order to secure marine insurance and sail and operate in international waters. Certificates are normally issued annually, with a more thorough five-year survey. Vessels are costly to run; therefore, they are continuously in operation. Surveys are typically conducted during visits to ports or shipyards, when vessels load or unload cargo, or undergo maintenance. Every survey job is tailored to the unique characteristics of a vessel, and its operation plan in order to reduce the interruption to normal operations. This means that survey procedures are often broken down into parts, with each part of the survey potentially being conducted in a different port.

DNV is currently modernizing its survey support system, which surveyors use to conduct and manage surveys. The system is used for planning survey jobs, document compliance, reporting of 'findings,' (that is, issues that require fixing before compliance can be signed off), looking up a vessel's history, and issuing of certificates. The current desktop version for Microsoft Windows was released in 2004, and at the time of data collection, DNV was developing a new web-based solution to allow continuous development of new features. Development is organized as an in-house project with a release date when the new solution goes live and the old system shuts down.

DNV employs approximately 1,000 surveyors globally who are the primary users. This group of users tend to dislike new digital tools – or in the words of one surveyor, *"we don't like change."* DNV management was concerned that if the new system gained a bad reputation, the cost would rise dramatically, possibly outweighing the benefits of the new system, requiring significant resources to overcome resistance in adoption. In other words, management put a premium on developing a system that pleases its intended users.

2.2 Data Collection and Analysis Procedures

We collected data during a nine-month period; data collection and analysis were interleaved, and followed procedures described by Seaman [18]. The data collection activities included semi-structured interviews and several site visits for observation of surveyors at work. The first site visit for observation was treated as a pilot study, and from this we gained valuable insights into how this group of users interacts with software technology and a general understanding of their role; based on this we designed observation guides and semi-structured interview guides.

Interviews can be a valuable source of information as it allows in-depth conversation with experts, but it is only one of many potential methods to collect data in field studies [13]. Interviews fall in a category of methods that Lethbridge et al. have labeled *inquisitive* techniques [13], in that a researcher must actively engage with interviewees to get information from them. A second category is *observational* techniques, which includes observation of professionals. Both types of techniques have benefits and drawbacks; interview data may be less accurate than observational data, but observational techniques may introduce the Hawthorne effect, whereby professionals' processes change when they are observed [13].

A total of nine interviews were conducted: four software developers, one manager, one implementation manager, and three surveyors. The focus of these interviews was to develop an understanding of the purpose of the new system and how developers elicited user requirements. All interviews were transcribed, resulting in 105 pages of text. Following the interviews, we conducted a total of seven observations. We observed three more surveyors for two days each, and three surveyors for one day each. Three of the surveyors were situated in Norway and four in the Netherlands.

The first two authors conducted observations of seven surveyors. These onsite activities were organized in collaboration with DNV's central scheduler who assigns jobs to surveyors. The site visits involved shadowing the surveyors for the full day; this included accompanying surveyors during inspection of vessels, including crawling through narrow storage tanks, climbing crane towers, as well as driving for hours to reach remote ports during which surveyors could also have phone calls with colleagues, and having lunch together. Our impression is that the surveyors appreciated the opportunity to show their work practices and expressed themselves freely. The researchers conducting observations were dressed similarly to crew and surveyors, including all the required Personal Protective Equipment (PPE) (including safety helmet, ear muffs, safety shoes, etc.). In a way, we were more like apprentices than researchers. Research notes and pictures were constantly captured over the course of data collection, and reflections were written immediately afterwards, resulting in 59 pages of notes produced and about 100 photographs of surveyors in action (see Fig. 1).

The first and second authors jointly analyzed the data and immersed themselves in the material. A word processor was used for both open-ended coding and memoing [18]. Examples of labels include:

- "use of phone calls, not chat, to maximize bandwidth of communication and realtime feedback"
- "surroundings force surveyors to take breaks during their work day"

These two labels were grouped in a theme "adapting to surroundings." After we completed the data analysis, we used member checking, a procedure to assess the validity of our findings by presenting them in a workshop involving surveyors and DNV management, and adjust any misapprehensions. Overall, their response was confirmative.

3   Findings

This section presents the key findings. We first discuss requirements elicitation practices at DNV; the remainder of the section is organized using the framework presented in Table 1; a summary of findings is presented in Table 2.

It should be clear that the 'users' in this context are domain experts, namely, surveyors of the DNV organization, who have very extensive experience; the term 'users' does not therefore apply to other types of users.

We do not discuss Known Knowns in further detail, because that is knowledge shared among developers and surveyors alike. One example of this is that surveyors preferred to keep the new system as similar to the old one, and developers were aware of this and tried to accommodate this.

3.1   Requirements Elicitation at DNV

The team we interacted with did not have any dedicated requirement engineers; requirements therefore were elicited by the developers. Development teams relied mainly on two methods to capture the surveyors' requirements.

1. **Workshop.** A three-day workshop with a user representatives group of 50 surveyors face to face in the early phases to gather as much information as possible about how surveyors work and interact with technology. One developer explained their focus on the gap between current features and what surveyors need and expect: *"We were discussing current solutions and what they [the surveyors] miss."* Following the workshop, a UX-designer created user stories based on the results.
2. **User tests.** Variants of one-on-one test sessions virtually on Microsoft Teams to test usability and functionality developed from the user stories gathered in the workshop. One surveyor in the user representative group explained:
   > *"I share my screen and they sit and take notes along the way or ask [questions]. We did tests where I kind of got instructions on what to do (...) and tests to check if I intuitively could find what to do."*

Fig. 1. From left to right: planning the survey, photographing issues, surveying in difficult environments, paper artifacts remain important, and one of the researchers on-site. No photographs could be taken on tankers as we had no explosive-proof camera.

These tests were facilitated by the UX-designer and ranged from strictly orchestrated, to tests where the user was given a task and encouraged to explore the system on their own to solve it. Tests were conducted every 2–3 months with the same selection of surveyors as the one attending the first workshop.

Table 2. Summary of Findings

	Known to developers	Unknown to developers
Known to users	**Known Knowns:** – Surveyors preferred to keep the new software as similar as possible to the old one, which developers knew and tried to accommodate. (Not discussed in the findings section, because this is unproblematic knowledge.)	**Unknown Knowns:** – Surveyors rely on "gut feeling" and "on the go" decision making. – Surveyors prefer to discuss issues over the phone with colleagues, instead of in writing (e.g. chat or email), ignoring the 'chat' function that was designed for this. – Surveyors spend much time interacting with people onboard of vessels, which is essential for a successful survey, but is hard if not impossible to capture.
Unknown to users	**Known Unknowns:** – Developers frustrated with an inability to test their assumptions. – Developers lacked domain knowledge and realized they were missing information when merely talking to surveyors. – Developers envisioned the user according to how they understood them from the workshop and interviews, filling in blanks using their own logic rather than by asking the surveyor, and their experience might be limited.	**Unknown Unknowns:** – Divergence between observational and interview data: surveyors simplified, generalized, and abstracted when talking about their job, but reality is different. – Surveyors must constantly adapt their workflow to changing circumstances; the survey process is not straightforward and is tailored to the context. – The application to report findings (issues that require fixing) technically works, but is not used as originally conceived by designers due to inconvenient menu navigation and illogical object naming.

3.2 Known Unknowns: What Developers Know They Don't Know

Developers were aware that they did not know certain aspects. To better develop and gain insights into these known gaps in their knowledge ('Known unknowns') developers had continued access to the user representative group during development. This proved useful when developers sought to ask for clarifications while reading user stories and developing features. One of the developers explained that: *"Some assumptions that we've had before proved slightly different."*

Developers described this combination of observation during user tests on the one hand, and the ability to contact surveyors directly for follow-up questions on the other hand, as 'ground-breaking' because they had not had such close contact with users before. At the same time, they sought to acquire an even deeper insight into the user's context and observe surveyors use their software in the real world, to understand *"How does this [software] relate to how they actually work?"* Developers were aware they lacked this understanding, and expressed a preference to visit the world of surveyors and conduct observations themselves. One developer shared that: *"I have not been on a tanker before, so I have no idea what things really look like."*

Developers argued they could not obtain all the essential information about user needs due to a lack of basic domain knowledge of surveying. One of the developers recognized that context information was crucial to gaining a deeper understanding of the survey process in practice:

> *"I would like to get on a boat and see what the actual work process looks like [...] there is a lot they [surveyors] cannot include when they explain it in the office, versus when you are actually out physically with them."*

At the same time, observing a surveyor in real life is not without challenges, not least of which include the cost associated with site visits as well as the purchasing of prerequisite PPE, and the cost associated with lost developer productivity for the time they travel. One developer highlighted:

> *"We have always requested that we visit a ship so that we can actually connect the dots between our domain knowledge versus what we actually see in practice."*

Developers were frustrated that some of their work had to rely on assumptions that were impossible to test through traditional methods such as workshops and interviews. It seemed there was a shared recognition among the developers that if they could not test these assumptions, the software would not fulfill its potential.

3.3 Unknown Knowns: What Developers Don't Know, but Users Do

There was also a category of knowledge that developers were not aware of, but the surveyors were. All surveyors we shadowed were highly experienced in their role. When boarding a vessel, they would quickly gain a 'feeling' of the vessel's condition, as they had learned to recognize subtle clues of technical problems ("findings") through observations and conversations after years of experience. Subtle clues include the freshness of the paint, the general tidiness of the deck, the condition of the lights, and signs of stress among the vessel's captain and crew. By piecing together these clues, surveyors adapted the survey job to uncover the most crucial findings. For example, one of the surveyors we followed decided to check all so-called ex-lights (lights certified for explosive environments) after having been only minutes on board a vessel. The surveyor then continued to make numerous other findings. When we asked why he decided to go straight for the lights, he said that it was a combination of experience and pieces of information he gathered when boarding the vessel, concluding: *"I get this gut feeling."* He did not need to spend time analyzing what to focus on in his survey but intuitively made this decision *"on the go"*—the checklist or survey support system did not prompt the surveyor. This was knowledge he found impossible to describe to developers, arguing: *"You can only learn it through years of experience."* We made similar observations with other surveyors, all of whom shared the same explanation.

Although surveyors tried to explain this type of knowledge during interviews, it was not until we observed this ourselves, while attending surveys on vessels, that we gained an adequate understanding of how such knowledge impacted the way the survey was conducted. It had remained unknown to us as researchers and to the developers.

Because the surveyors struggled to communicate many critical aspects of their work to others (non-surveyors), they felt they were unable to articulate their needs clearly to developers. Previously, DNV had digitalized the old paper-based checklists to relieve surveyors of hours at the printer. Despite knowing most checklists by heart and rarely printing them, surveyors welcomed this improved accessibility. However, whereas the old checklists automatically marked newly added check items in red, this feature disappeared with the introduction of a new system; this forced surveyors to search for this information that was now 'hidden' within hierarchies of application menus. Surveyors agreed that a lack of understanding of surveyor work among software developers was the reason for this shortcoming.

Another crucial part of surveying, that remained completely unknown to developers, is how surveyors establish relations with the captain and crew upon entering a vessel. This is important because both captain and crew act as 'gatekeepers' as well as facilitators for the survey. Good working relations results in a smoother survey because the captain and crew willingly support the surveyor in accessing the vessel's different parts, i.e. by opening hatches, stopping maintenance work, clearing gas tanks, etc. They are powerful allies because they can provide flexibility to the surveyor. Surveyors and crew constantly negotiate which survey activities are convenient based on the current situation on the vessel. For instance, when a surveyor planned to inspect tanks, a cooperative relationship with the Chief Officer gained him increased access and guidance during the survey.

Establishing good relations happens in social situations. Typically, when surveyors board a vessel they go straight to the bridge and meet the captain and First Officer. Coffee is offered, sometimes cakes, and usually a polite conversation ensues about for example the vessel's history and mutual acquaintances. They then move to planning the day and negotiating what surveyor activities are possible considering the vessel's operations that day. Drawings of the vessel and survey-checklists are commonly central artifacts for achieving a shared vision amongst them, because they are able to show and tell, pointing to details as a basis for discussions (see Fig. 1). These artefacts are found either on the surveyor's computer screen or printed by the captain to make it easier for them to gather around.

Social situations, like these, remain unknown to developers because surveyors perceive this to be an informal part of the survey process, and as irrelevant to developers:

"I spend half my time going around the vessel talking to people. That human part is not very well captured [in work instructions, procedures, etc.]."

This becomes a challenge because developers do not know that the software they develop is a critical artefact in establishing good relations between surveyors and captains.

Another example of unknown social interactions occurs when surveyors talk on the phone. Surveyors are dealing with problem solving in complex surroundings and often need to discuss their situation with colleague surveyors. In a time-pressed world they reach out for colleagues because they are able to explain their problem and surroundings within minutes, let their colleagues ask questions, and interpret the situation together to agree on how to proceed. Such conversations contain valuable and almost-instant information about problem-solving that DNV attempted to capture by introducing a chat function. By capturing these conversations through chats in the application, knowledge could be saved and processed to benefit other surveyors experiencing similar problems. However, surveyors only used the chat for straightforward and simple issues, leaving the more complicated issues for phone conversations, meaning the most valuable and interesting conversations were never captured. One surveyor argued that:

"It [a phone call] creates fewer misunderstandings, less dissatisfaction, more understanding and makes it easier to reach an agreement. You don't waste time writing e-mails"

We observed such conversations between surveyors and actors like captains, engineers, managers, superintendents, ship owners, authorities, and colleague surveyors. Characteristics of such phone calls remained unknown and indeed invisible to developers.

These were some examples of critical pieces of expert knowledge about surveyors' daily jobs, constraints and requirements that developers simply were not aware of, and that surveyors struggled to articulate during requirements elicitation.

3.4   Unknown Unknowns: What Neither Developers nor Users Know

Finally, there was a category of knowledge that neither developers nor the surveyors were aware of, the unknown unknowns. The surveyors' workflow was highly dependent on, and tailored to a complex context with constantly changing circumstances onboard the vessels. This was an important issue for developers to understand, yet they were not aware of this, and nor were surveyors readily aware of this as they described their work in interviews. For example, surveyors were forced to take short spontaneous breaks caused by changing circumstances on deck. On one occasion, they had to wait for a crew performing gas measurements before the surveyor could enter a tank—strict protocols are in place before people may enter certain hazardous areas of a vessel. The loading and unloading of cargo also affected the survey activities that could be performed; bunkering (refuelling of a vessel) could also limit certain activities, such as a black-out test that tests back-up generators. Surveyors would use these breaks to capture findings. Although surveyors were expected to use the application to do this, we observed that this involved large numbers of clicks and taps when navigating menus to register findings. Surveyors found this too cumbersome, because breaks were usually short and unpredictable. In addition, several surveyors found it challenging to locate the correct 'object' in the application because their naming did not always make sense to them. Therefore, they avoided using the application during inspections and instead preferred to register their findings manually, for example by taking photographs of issues, or handwritten notes.

Although the application seemed to be an excellent digitalization effort in theory, it failed to support the experts in practice because it was incompatible with how they performed their job in a complex and unpredictable environment. The surveyors perceived that the system failed because their context was too complex to describe all of its aspects to software developers. Interestingly, when we compared our observational data and interview data, we discovered significant divergences in how surveyors *said* they worked and how they *actually* worked. During interviews, surveyors would generalize and simplify how they worked, leaving out details and abstracting away aspects that are hard to comprehend without field experience. During observation, however, they referred to situations they participated in as a starting point for more detailed explanations. They were not able to translate their expert needs to software requirements. They acknowledged that these requirements remained unknown to them and that developers would benefit by making observations in the real world. One of the surveyors proposed that *"They [developers] have to come out here and see for themselves."*

In sum, both developers and surveyors recognized the need for developers to observe the world of surveyors to better understand the needs of this group of expert users.

4   Discussion and Conclusion

Most prior work in the software engineering domain was conducted by a relatively limited number of authors, detailed many insights from a select number of case studies, in particular studies of air traffic controllers [10]. Several other studies are situated in the HCI and CSCW communities in the 1990s and early 2000s, recording lessons for designing interactive systems that became increasingly popular in the 1990s [2].

In this paper we report a number of insights (see Table 3). For example, whereas prior literature has suggested the high cost of ethnographic or observational studies might preclude organizations from using these strategies, we would argue that in some domains the gap between developers and users is simply too large to bridge. Even a few days of observing surveyors at work provided extensive insights that could have saved many hours of developer time. It is important at this point to distinguish between 'observation' and 'ethnography'; the former being a technique that we focused on primarily, whereas the latter is a more encompassing strategy that would take far more time.

A second issue with observation is that it depends on serendipity and 'lucky' circumstances that would lead to identifying new requirements. While we agree this is an issue in *identifying* requirements, it is less of an issue to *understanding* the nature, context, and constraints of an expert's daily job. We argue this is an important distinction to keep in mind when planning site visits for developers or requirement engineers; the goal of observation then becomes one of "walking in the user's shoes," to understand the system-in-action context before attempting to capture requirements.

Analysts and developers are trained professionals who look at systems through a lens of the *affordances* that technologies offer, but this too has a reductionist 'smell': looking at technologies and how they "map" to possibilities, rather than user preferences. It is possible, of course, to add a chat function that seeks to capture potentially valuable conversations—but this ignores users' *preference* to *not* use system features, but use simple means such as a phone for a direct and possibly private conversation.

Table 3. Contributions and Implications

Insights from prior literature	Findings of this Study	Implications for Practice
Observation and ethnography are resource-intensive activities. Early scholars proposed 'quick and dirty' approaches" [9]	The gap between developers and users may be too large to bridge; observing surveyors even for a few days provided valuable insights that are difficult to articulate	The cost associated with letting developers go into the field for even a week may well be worth it, potentially improving the quality of requirements analysis
Observation and ethnography depend on 'luck' and serendipity to identify requirements [20].	Observing expert surveyors at work for even a few days provided deep insights into the constraints and work practices	Observation is not only about identifying requirements and features, rather, it can be valuable for developers to understand the real-world context and constraints (see Fig. 1)
Integrating ethnographic observations into structured methods of requirements analysis is very challenging [19]. The reductionist character of RE (focusing on components, data flows, processes) is not compatible with ethnographic inquiry [3].	Surveyors prefer ad hoc communication *outside* the system's functionality (rather than built-in features e.g. chat). The social interactions and goodwill between surveyor and captain/crew are essential for a successful survey, which cannot be captured in system features.	Rather than seeking ways to integrate observational findings into requirements analysis, consider system boundaries and develop a good understanding of the social context where systems are implemented
Experts find it difficult to articulate their expertise during requirements analysis [19]. Ethnography recognizes work activities as they are actually conducted, rather than some idealized version of it [10]	Surveyors frequently rely on 'gut' instincts developed over years of experience. Surveyors are often assigned jobs on short notice, and have to adjust surveys based on continuously changing circumstances	Not all work practices can be digitalized. The valuable experience and gut instincts cannot be replaced with a rigid workflow. Develop systems that empower experts, rather than aiming to digitalize existing workflows

Finally, experts are known to experience difficulty articulating their knowledge, certainly when this includes dependency on their gut instincts that developed over years. No guidelines, handbook, or IT system can replace this. This leads us to argue that, perhaps, not all work practices can (and should) be digitalized. In a way, we are touching upon the boundary of a current popular trend in the IT industry, namely a search to digitalize everything. While we do not deny that software systems can greatly improve our lives and productivity, care should be taken to understand how software solutions can *support* our work practices rather than replace them.

In conclusion, software systems are designed for expert users, but in some domains these experts have very extensive tacit knowledge. Drawing on previous insights on tacit knowledge, we reported experiences with some of the issues in a domain where changing circumstances, users' 'gut instincts' and extensive experience, and physically challenging environments are important issues when seeking to identify unknown knowns and unknowns.

Acknowledgements. We thank the participants of this research for sharing their insights, and the surveyors whom we observed for their patience and insights. We are grateful to DNV for funding this research together with The Norwegian Research Council (grant number: 309631). For the purpose of Open Access, the authors have applied a CC BY public copyright licence to any Author Accepted Manuscript version arising from this submission.

References

1. Armour, P.G.: The five orders of ignorance. Commun. ACM **43**(10), 17–20 (2000)
2. Coad, P., Yourdon, E.: Object-Oriented Analysis. Prentice-Hall, Inc. (1990)
3. Crabtree, A., Rodden, T.: Ethnography and design? In: International Workshop on Interpretive Approaches to Information Systems and Computing Research (2002)
4. Davis, A., Dieste, O., Hickey, A., Juristo, N., Moreno, A.M.: Effectiveness of requirements elicitation techniques: empirical results derived from a systematic review. In: 14th IEEE International Requirements Engineering Conference, RE 2006, pp. 179–188. IEEE (2006)
5. De Marco, T.: Concise notes on software engineering. Yourdon Inc. (1979)
6. Gacitúa, R., et al.: Making tacit requirements explicit. In: 2nd International Workshop on Managing Requirements Knowledge, pp. 40–44 (2009)
7. Gervasi, V., et al.: Unpacking tacit knowledge for requirements engineering. In: Maalej, W., Thurimella, A. (eds.) Managing Requirements Knowledge. Springer, Heidelberg (2013). https://doi.org/10.1007/978-3-642-34419-0_2
8. Hickey, A.M., Davis, A.M.: Elicitation technique selection: how do experts do it? In: Proceedings of the 11th IEEE International Requirements Engineering Conference, pp. 169–178 (2003)
9. Hughes, J., O'Brien, J., Rodden, T., Rouncefield, M., Sommerville, I.: Presenting ethnography in the requirements process. In: Proceedings of 1995 IEEE International Symposium on Requirements Engineering, RE 1995, pp. 27–34. IEEE (1995)
10. Hughes, J.A.: Ethnography, plans and software engineering. In: IEE Colloquium on CSCW and the Software Process. IET (1995)
11. Hughes, J.A., Randall, D., Shapiro, D.: Faltering from ethnography to design. In: Proceedings of the 1992 ACM Conference on Computer-Supported Cooperative Work, pp. 115–122 (1992)
12. Jensen, R.W., Tonies, C.C.: Software Engineering. Prentice-Hall (1979)
13. Lethbridge, T.C., Sim, S.E., Singer, J.: Studying software engineers: data collection techniques for software field studies. Empir. Softw. Eng. **10**, 311–341 (2005)
14. Levitt, S.: Episode 4. Ken Jennings: "Don't neglect the thing that makes you weird". People I (mostly) Admire. Podcast, 3 October 2020
15. Mead, N.R.: A history of the international requirements engineering conference (RE) RE@21. In: 21st IEEE International Requirements Engineering Conference, pp. 21–221 (2013)
16. Paetsch, F., Eberlein, A., Maurer, F.: Requirements engineering and agile software development. In: 2003 Proceedings of the 12th IEEE International Workshops on Enabling Technologies: Infrastructure for Collaborative Enterprises. IEEE Computer Society (2003)
17. Polanyi, M.: The tacit dimension. In: Knowledge in Organizations, pp. 135–146. Routledge (2009)
18. Seaman, C.B.: Qualitative methods in empirical studies of software engineering. IEEE Trans. Softw. Eng. **25**(4), 557–572 (1999)
19. Sommerville, I., Rodden, T., Sawyer, P., Bentley, R., Twidale, M.: Integrating ethnography into the requirements engineering process. In: 1993 Proceedings of the IEEE International Symposium on Requirements Engineering, pp. 165–173. IEEE (1993)

20. Sutcliffe, A., Sawyer, P.: Requirements elicitation: towards the unknown unknowns. In: 21st IEEE International Requirements Engineering Conference (RE), pp. 92–104 (2013)
21. Tracy, K.W.: Software: A Technical History. ACM (2021)
22. Zachos, K., Maiden, N., Tosar, A.: Rich-media scenarios for discovering requirements. IEEE Softw. **22**(5), 89–97 (2005)

Feel It, Code It: Emotional Goal Modelling for Gender-Inclusive Design

Diane Hassett[1]([✉]), Amel Bennaceur[2], and Bashar Nuseibeh[1,2]

[1] Lero, University of Limerick, Limerick, Ireland
Diane.Hassett@ul.ie
[2] The Open University, Milton Keynes, UK

Abstract. Context and motivation: Organisational values such as inclusion are often explicit, providing a common language to guide behaviour and motivate employees. Personal values are often less explicit but do guide individuals' decisions, and when challenged they generate an emotional response. However, understanding organisational values and linking them to implicit personal values of employees can be challenging.

Question/problem: In this paper, we investigate the use of emotional goal models to act as a link between organisational and personal values.

Principal ideas/result: We argue that when designing processes and systems for enacting organisational values, requirements engineers must consider the diverse personal values of the employees. We completed a case study within a multi-national organisation and identified pain points on career journeys which amplify the disparity of experience between men and women. We applied emotional goal modelling to elicit requirements for inclusive processes. We suggest that emotional goals can serve as a proxy for personal values and can support the formulation of requirements for designing processes cognizant of the organisational value of inclusion.

Contribution: Our empirical evaluation suggests that the modelling of emotional goals can support the operationalisation of values as requirements for gender-inclusive organisational processes and systems.

Keywords: Emotional Goal Modelling · Inclusive Design · Values

1 Introduction

Organisational values *(such as honesty, respect and diversity)* are usually explicit and documented. Personal values are often implicit and represent individuals' beliefs, reflecting what is important and serving as a guide for life choices [23]. When organisational and employees' personal values are aligned they make for positive work attitudes [22]. When activated, values become infused with feelings, which can be seen in emotional responses [23]. Lack of alignment between the values of organisations and the values of their employees can result in conflicts [12].

Many organisations, especially in male-dominated professions such as engineering and software development, advocate for greater gender diversity in the

© The Author(s), under exclusive license to Springer Nature Switzerland AG 2023
A. Ferrari and B. Penzenstadler (Eds.): REFSQ 2023, LNCS 13975, pp. 324–336, 2023.
https://doi.org/10.1007/978-3-031-29786-1_23

workplace. There is an increasing emphasis on methods that identify gender-inclusivity biases within software (such as GenderMag [29]), but the socio-technical environment in which software is created is equally important [14]. In this paper we propose that understanding employees' values can help organisations design processes which reflect their values of *gender-inclusivity*.

Peoples' emotions play a significant role in the user acceptance of innovation [5]. Value-based Requirements Engineering (VBRE) uses emotions as cues to stakeholders' reactions arising from personal values to support communication processes [28]. However there is a lack of methods which use emotions to represent personal values as requirements in the design of processes and systems [17].

Emotional Goal Models (EGMs) build on requirements modelling to construct people-oriented models which show how users want to feel [15]. They have been successfully used to guide design in domains such as health [15] and homelessness [3]. In this paper, we focus on organisational processes and build on EGMs for identifying emotional goals as requirements for gender-inclusive processes. We investigate the following research questions:

RQ1: Can emotional goal models serve as requirements for design of a (software) system that promotes inclusion within the organisation?

RQ2: Can emotional goals contribute to a method for representing personal values as requirements?

To address those questions, we conducted an empirical study to understand the emotional experience of men and women in the organisation and extract the broad patterns that are important and relevant for the conception, design, and development of gender-inclusive processes. Mapping the experience of men and women during their career highlighted disparities at onboarding, development, and promotion. To translate employees' experience into tangible goal-driven requirements, we used EGMs to represent the functional, quality and emotional goals of employees during onboarding processes. We then evaluated the emotional goals generated and their link to employees' personal values. The contributions of this paper are twofold:

- *Applying emotional goal modelling to design gender-inclusive processes.* We propose a novel application of EGMs to the design of organisational processes that reflect emotional goals of men and women who experience those organisational processes in differing ways.
- *Evaluation of emotional goals for their representation of personal values within goal modelling.* We extend EGM to propose a method of eliciting and reflecting on emotional goals relationship to personal values. We reflect on the use of emotional goals to operationalise value requirements for the design and development of processes and systems that align organisational values and the personal values of its employees in the context of gender-inclusivity.

The remainder of the paper is structured as follows. Section 2 reviews related work. Section 3 presents the empirical study for mapping employees experience. Section 4 explains how we leverage EGMs to identify gender-inclusive requirements. Section 5 discusses the findings, lessons learned, and implications for requirements engineering (RE). Section 6 concludes the paper.

2   Background and Related Work

"Values are individual representation of societal goals. As elusive societal goals change, individuals' values will sometimes lead and sometimes reflect this change" [10]. Schwartz's theory of basic values is the most commonly applied within software engineering [24]. It identifies 10 value categories described by conceptual definitions in terms of motivational goals.

Goal modelling has emerged as a promising technique to operationalise values in software engineering [18,21]. 'Soft goals' have been used to represent non functional requirements such as trust in agent-oriented models [31]. Goal models can also be used to understand value trade-offs based on software features [21].

However, these approaches do not address the values of the processes surrounding the creation of software. In this paper, we build on Schwartz's theory of values [24] to represent personal values within organisational processes for onboarding.

To operationalise values such as *inclusion*, organisations implement support structures such as mentoring, networking and senior leadership sponsorship [8]. However, there is a lack of systematic processes to link organisational and personal values and reflect these as requirements to design inclusive organisational processes. This paper seeks to define such a systematic process.

Emotional goal modelling builds on requirements modelling to capture the emotional needs of users and construct people-orientated models that show how users want to feel [15]. They serve as high-level representations of the functional, quality and emotional goals of stakeholders, making them suitable for informing technical design. EGMs can be used as communication tools to assist in developing a shared understanding of a problem. They have also been applied within 'living labs' which support co-creation between stakeholders and development teams during the design process [20]. Emotional-led domain modelling has been considered to represent needs of diverse users (e.g., age, culture, personality, emotions) with adaptive software [7]. However the method lacks a way of ensuring emotions are linked to design and implementation decisions and features. In platforms for enhancing social interaction, successful applications have been found to be driven by user emotional engagement. A key challenge is to capture and understand users' emotional requirements so that they can be incorporated into interaction design [25].

Psychologically-driven requirements engineering assigned value goals based on peoples' roles in specific contexts, and emotional goals based on responses toward the occurrence of values within a system [1].

The study of affective computing and affective states in requirements engineering recognises the role emotions play in acceptance and negotiation activities [6]. However, there is a lack of modelling techniques that consider emotional goals as a means of representing personal value requirements within organisational processes. In this paper, we build on existing EGM methodology to evaluate a link between emotions and personal values within organisational processes.

3   Capturing Disparities in Employees Experience

This section presents the empirical study we conducted to understand employees' experiences of joining and working within the organisation. We start by describing the multi-national organisation in which the study took place. We then report on how we identified the key disparity between men and women's experience.

3.1   Study Setup

We identified participants from a large manufacturing organisation. The organisation involved is highly automated, heavily reliant on its software technology and committed to equality and inclusion for its workforce. The study was set up to look at the experiences of employees and investigate respective viewpoints of their career journey. We interviewed 11 employees, 6 women and 5 men, from a cross section of Engineering (7), Quality (1), and Operations (3) functions. The engineers' experience ranged from graduate entry level through to senior leadership. They included 4 engineers (1–4 yrs experience), 4 mid-career middle managers (5–10 yrs experience) and 3 senior managers (greater than 10 yrs experience). In this paper we will refer to the participants as employees. Interviews were semi-structured, and lasting between 50–70 min and were conducted over a 3 month period. While prior studies have focused on uncovering bias within software teams [30], we focused on broader organisational representation to capture a diverse range of viewpoints.

We used thematic analysis [2] as a systematic method to analyse and code the responses of participants. It is an iterative method of analysis and synthesis allowing patterns and themes of the employees' experience to be identified. It consists of three main phases: *descriptive, interpretative and thematic.* In the descriptive phase, we transcribed all interviews and established initial codes based on the quotations and continued to develop these as we worked through the interview transcripts. In the interpretative phase, we categorised the data with similar codes into subthemes, we identified 42 subthemes, such as *building confidence, building trust, role models and self induced pressures.* We grouped and regrouped different codes to identify patterns in the data and create meaningful themes describing how interrelationships between people and contexts fit together. In the thematic phase, we created a descriptive story using excerpts of interviews to describe the employee experience. Further details about the analysis can be found in Hassett (2018) [8,9]. We structure the findings as employee experience maps [27] which allow us to visualise and contrast experiences.

3.2   Experience Mapping

We used positive and negative sentiment [11] to reflect employees' experience as shown in Fig. 1. We identified three key themes in the employee experience: *onboarding, development,* and *promotion.* Within these, subthemes were identified in the iterative coding process and plotted along the x-axis. Onboarding (when new to the organisation) includes three subthemes: introduction, getting experience, and developing networks. Development includes the following subthemes: identifying career paths, developing expertise, utilising networks and opportunity. Onboarding (when promoted to a new role) includes four subthemes: onboarding for promotion, supports, personal change, and being an ambassador.

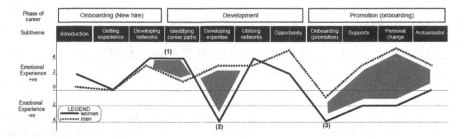

Fig. 1. Mapping Employee Experience

An example of a positive quotation is *"having someone you can talk to, in kind of an informal everyday basis"*. An example of a negative quotation is *"Moving into the team leader role. I think we could do with a bit more support definitely, and guidance"*. The y-axis scale is the scaled relative quantity of positive or negative quotations associated with each group for that subtheme rather than graded on magnitude of emotion. For example, in *development: developing expertise*, there were 3 positive quotations (from men) and 4 negative quotations (from women), this corresponded to a disparity in experience illustrated by the greyed area in Fig. 1. When the experiences were categorised, we observed disparity between men and womens' experience at three points.

(1) During *onboarding*, women actively network, seek out mentors, and contacts to discuss technical and career areas. Men are less worried about networking, with more informal "organic" networks.

(2) During *development*, men were assigned more technical tasks whereas female employees took more organisational/procedural tasks. For example, there were two employees with the same engineering qualification; the man was assigned to Operational Support (fixing, designing, changing), the woman was assigned to Quality Engineering (assessing risks, authorising changes). In this case, the woman described her experience *"I needed to know more, won't speak up even if know answer is correct"*. While the man indicated more confidence *"Because even the shot in the dark, I have more information about the shot in dark"*.

(3) The *promotion* from engineer to manager reflects the most significant career transition. We found that the support of a close network of like-minded peers to provide informal advice and discuss challenges, helped employees adapt to the transition. The availability and benefits of informal support to men at promotion was evident, e.g., *"Without an ounce of doubt, my peers and colleagues have been a major support for me, the fact that other peers may have gone through the same experience, the fact that I am able to sit down and have a chat and go through"*. This compares starkly with the experience of one woman, who describes *"I would have reached out for help sooner"*.

We will now look to translate these (emotional) experiences (such as inconsistent onboarding processes) into goals for gender-inclusive processes.

4   Applying Emotional Goal Modelling to Design Inclusive Processes

The data from thematic analysis is rich and detailed but lacks interpretative power [4]. We used EGMs to translate employees emotional experience into desired functional, quality and emotional goals. The emotional goals capture the desired feelings of stakeholders within a socio-technical system and how they relate to one another [15].

Emotional goals are defined as *"the desired reflective level emotion of a role"* [15]. This definition relates to Normans' three levels of emotions which are visceral, behavioural and reflective [19]. Emotional goals differ from functional goals and non-functional quality goals which represent intended properties of the system that affect the behavioural level [15]. The employee experience map identified disparity between men and womens' emotional experience during onboarding, development and promotion. We now develop and propose EGMs to address these disparities, and design the organisational process of onboarding to create more gender-inclusive experiences for employees. We will now describe the modelling phase in which we identify and capture functional, quality and emotional goals to create EGMs (RQ1). We then describe how we evaluated emotional goals as a link to personal values (RQ2).

4.1   Modelling

The first stage in creating EGMs was to analyse the coded interview responses from the thematic analysis, extract key themes and identify them as functional, quality or emotional statements. These are coded as *do* (functional), *be* (quality) and *feel* (emotional) goals. The second step of the EGM analysis was to create future-based "should" statements and assign these to the relevant quote [9]. The following is an example of how an employee quote was transformed into functional, quality and emotional goals for the theme of *onboarding*:

– Quote: *"Sometimes success can be a failure. You can fail but you can learn something out of it which will be the next big success. We don't seem to manage that very well."*

– Subtheme: learn from failure/experience approach
– Goal type: feel (emotional)
– Goal (should statement): acceptance to make mistakes

The final stage was to create the model to represent the functional, quality, and emotional goals associated with stakeholder roles (such as manager, employee, engineer). We followed three steps: (i) moving from left to right of the figure, place stakeholders, functional goals on model, (ii) add the quality and emotional goals and connect related functional goals, and add the process outcome. Finally, (iii) assess and iterate the model to ensure goals and interactions make sense and the model can be easily understood by stakeholders.

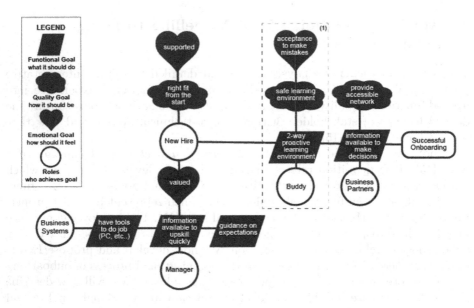

Fig. 2. Emotional Goal Model for Onboarding

Figure 2 depicts the EGM describing the set of functional, quality and emotional goals for stakeholders in the onboarding process. It makes explicit those goals to support the organisational value of *inclusion*. We use notation for motivational goal modelling [13,26]. Relationships between goals and stakeholders are indicated by non-directional connector lines. This is to visually represent the multiple interactions within the EGM and allow the model to be understood by non-technical stakeholders. The stakeholder (or "role") shape represents an individual/organisation. The functional goals are represented by a parallelogram. Quality goals are represented by a cloud. Emotional goals are represented by a heart. There were six functional goals, three quality goals and three emotional goals. The outcome is *'successful onboarding'*. Stakeholders include *new/promoted employee, manager, buddy, business partners and systems*. In this

EGM, the functional goals describe what the process **should do** (e.g.,'*correct tools to do the job*' and '*access to information*'). The quality goals describe how the process **should be** (e.g., '*to provide accessible network*' and '*safe learning environment.*') The emotional goals describe how the process **should feel** and include '*supported*', '*valued*' and '*acceptance to make mistakes*'.

From the EGM, we generated goal statements such as '*Buddies need to create a safe learning environment for the new hire, while letting them have access to a proactive learning environment where it is acceptable to make mistakes*' (refer to (1) on Fig. 2). Another EGM was also generated for development [8].

4.2   Mapping Emotional Goals and Personal Values

We then performed an evaluation of emotional goals generated within the EGM to link with personal values definitions [24] using value terms [23]. We proposed an initial mapping of emotional and value goals for inclusive processes. We aligned functional, quality and emotional goals from the EGM as shown in Fig. 3. Functional and quality goals supported fulfilling the emotional goals for onboarding and development. For example, the functional goal of '*2-way proactive learning environment*' and quality goal of '*safe learning environment*' aligned with the emotional goal of '*acceptance to make mistakes*'. Within the context of the stated functional and quality goals, this emotional goal aligned with the value goal '*self-direction - freedom to determine ones own actions*'. To support the organisational value of *gender-inclusion* we assigned the value goals based on alleviating the negative experience of women.

Organisational process outcome	Functional goal	Quality goal	Emotional goal	Value goal
successful onboarding	information available to upskill quickly	right fit from the start	valued, supported	personal security - feeling others care about me, a sense of belonging
successful onboarding	2-way proactive learning environment	safe learning environment	acceptance to make mistakes	self-direction - freedom to determine ones own actions
successful continuous development	enable supportive team environment	ability to develop networks	inclusive	universalism - concern, equality for all
successful continuous development	opportunity to display own work	bringing new/ diverse opinion	you can make a difference	self-direction - freedom to determine ones own actions

Fig. 3. Mapping Emotional and Value Goals for Inclusive Processes

To represent *gender-inclusion* as an organisational value goal, we were able to successfully align emotional goals '*supported*', '*valued*', '*inclusive*', '*acceptance to make mistakes*' to personal values goals of '*personal security*', '*self-direction*', '*universalism*' and '*self-direction*' respectively. The linkage between emotional goals and value definitions provides process-related requirements that the organisation can use to realise the desired value of inclusion.

We propose and demonstrate a systematic process for linking an organisation value of *gender-inclusion* to employees' personal values via emotional goals.

4.3 Threats to Validity

We now discuss some threats that might challenge our findings.

- *Sample size and selection criteria.* The selection was limited to 11 participants as no new subthemes and categories were emerging, representing theoretical saturation. As this study focused on gender inclusive representation of men and women, we would recommend future studies incorporate GenderMag assessment as well as hybrid working models [29].
- *Coding and thematic analysis.* The interviews were transcribed and coded by one researcher. To minimise reliability bias, sample transcript was cross checked with a second researcher to ensure factual interpretation.
- *Generalisability.* While these results are from one organisation, we are unable to say if they can be generalised and further studies would be required.

5 Findings and Lessons Learned

In this paper, we used emotional responses to identify disparity in the experience of men and women employed in a multinational organisation. We suggested that this emotional response is in part due to misalignment between the values of the organisation and those of employees impacted by the current processes [23]. This section discusses findings, lessons learned, and their implications for requirements engineering.

5.1 Findings

We now reflect on how our work addresses the research questions posed in Sect. 1.

RQ1: Can emotional goal models serve as requirements for design of a (software) system that promotes inclusion within the organisation? Using the organisational value of *gender-inclusion* we used emotional goal models to design gender-inclusive processes for onboarding (and development). We found that employees wanted to feel *'valued'*, *'supported'*, and that it was *'acceptable to make mistakes'*. We propose that EGMs can represent a high level goal model which make explicit the functional, quality and emotional goals within organisational processes.

RQ2: Can emotional goals contribute to a method for representing personal values in organisational processes? We evaluated emotional goals as a link to personal values and provided linkage between emotional goals and personal values for inclusive design. The emotional goals *'valued'*, *'supported'*, *and that 'it was acceptable to make mistakes'* aligned with values of *'personal security'* and *'self-direction'* respectively. While our work was limited to one organisation, and limited to men and women gender participants, the results are encouraging. We suggest that emotional goals can serve as a proxy for personal values and can support the formulation of requirements for designing processes cognizant of the organisational value of *gender-inclusion*.

5.2   Lessons Learned

We now discuss in more detail some challenges identified during the process of designing EGM and mapping them to values.

Conflicting Values. We identified two sources of conflicting values. Firstly, to meet the organisational value of *gender-inclusion* we linked the emotional goal of feeling *'valued'* to the personal value *'personal security'*. However, when we consider the positive experience of men, the same emotional goal of feeling *'valued'* could have been associated with the value of *'benevolence - being a reliable and trustworthy member of the ingroup'*. This values decision poses a challenge for RE during design of organisational processes and systems. However, we propose that maintaining alignment with the higher level organisational value of *gender-inclusion* can ensure that under-represented groups are considered explicitly. Secondly, we found the emotional goals of *'valued'* and *'acceptance to make mistakes'* represented values of *'personal-security'* and *'self-direction'* respectively. When placed within Schwartz's value continuum, these would be in conflict [24]. We believe the inclusion of emotional goals in requirements gathering offers an opportunity to mediate values conflicts and trade-offs.

Emerging Values. When evaluating the EGM for onboarding, there was an absence of 'system' level goals such as *privacy and autonomy*. As with any RE process, designing with values and emotion is an iterative process. As we are developing from organisational processes to a specific software system context, additional values and emotions may emerge and will need to be considered and integrated into the goal models. Therefore we propose future work (i) using EGM-led prototypes, and (ii) iterating EGMs with stakeholders, to understand how new emotional goals and personal values surface. Existing RE tools and methods will need to be reviewed to integrate and maintain these emergent requirements.

In summary, the evaluation of emotional goals as representation of personal values within organisational processes identified two key challenges, those of conflicting values and emerging values. We propose future work to include use of emotional goals to mediate value trade-offs and supporting ongoing design iteration using EGMs and EGM-led prototypes early in requirements gathering. In other words, emotional goals offer an approach to translating employees' lived experience during career development into values requirements for software systems.

5.3   Implications for RE

As society evolves, so do its values and priorities. To adapt, RE needs to consider not only what values to prioritise, but how these values are reflected in practice [16]. Emotions provide one representation of personal values. Modelling them as requirements necessitates reflection on existing RE methods and techniques to investigate how they can be extended to be more explicit in representing and reasoning about different values. *Inclusion* represents an example of an organisational value that reflects societal values. The redesign of organisational processes informed by emotional responses of stakeholders will benefit society. However, it can also lead to challenges. For example, prioritising some values, such as *inclusion* above immediate economic value, can be disruptive to traditional approaches of prioritising requirements. Indeed, expanding to consider other organisational values such as *sustainability* will require existing methods to adapt in order to incorporate an additional set of complex value requirements, whose costs and benefits can then be more systematically considered.

6   Conclusion

This paper presented a systematic approach to design gender-inclusive processes using emotional goals to link organisational and personal values. It takes a step towards making requirements of personal values explicit through emotions. It thus provides an approach to operationalise employees' personal values.

The results can be extended in a number of ways. These include evaluating if emotional goals can be used to link other organisational values such as *sustainability* with personal values. We plan to extend the work to consider conflicting and emerging values within EGM by (i) prioritising emotional goals when personal values are conflicting, and (ii) analysing the use of EGMs and EGM-led prototypes to surface emotions and values early in the requirements engineering process. Our ambition is that by designing inclusive organisational processes, then the products, the services, and ultimately the software that these organisations create, will reflect values of the society that we want to live in.

Acknowledgements. This work was supported, in part, by Science Foundation Ireland grants 16/RC/3918 (Confirm), and 13/RC/2094 P2 (Lero), EPSRC grant (EP/R013144/1) (SAUSE), and UKRI Trustworthy Autonomous Systems Node in Resilience (EP/V026747/1). Thanks to Patrick Slevin for his encouragement and support during the empirical study, and to Helen Sharp and Andrea Zisman for feedback on early revisions.

References

1. Alatawi, E., Mendoza, A., Miller, T.: Psychologically-driven requirements engineering: a case study in depression care. In: ASWEC. IEEE (2018)
2. Braun, V., Clarke, V.: Using thematic analysis in psychology. Qual. Res. Psychol. **3**(2), 77–101 (2006)

3. Burrows, R., Lopez-Lorca, A., Sterling, L., Miller, T., Mendoza, A., Pedell, S.: Motivational modelling in software for homelessness: lessons from an industrial study. In: RE (2019)
4. Cruzes, D.S., Dybå, T.: Synthesizing evidence in software engineering research. In: ACM/IEEE International Symposium on Empirical Software Engineering and Measurement (2010)
5. Eskelinen, J., Robles, A.G., Lindy, I., Marsh, J.B., Muente Kunigami, A.: Citizen-driven innovation: A guidebook for city mayors and public administrators. Technical report, The World Bank (2015)
6. Fucci, D., Kuhn, S., Maalej, W.: The second international workshop on affective computing for requirements engineering (affectre2019). In: RE Workshops (2019)
7. Grundy, J., Khalajzadeh, H., Mcintosh, J.: Towards human-centric model-driven software engineering. In: ENASE, pp. 229–238 (2020)
8. Hassett, D.: Finding a Voice: Using Design Ethnography And Emotional Goal Modelling To Empower Employees In STEM Careers. Master's thesis, Department of Design Innovation (2018)
9. Hassett, D.: Finding a voice: Codebook (2018). http://bit.ly/3AzLxTt
10. Kahle, L.R., Beatty, S.E., Homer, P.: Alternative measurement approaches to consumer values: the list of values (lov) and values and life style (vals). J. Consum. Res. 13(3), 405–409 (1986)
11. Kim, S.M., Hovy, E.: Determining the sentiment of opinions. In: 20th International Conference on Computational Linguistics, pp. 1367–1373 (2004)
12. Liedtka, J.M.: Value congruence: the interplay of individual and organizational value systems. J. Bus. Ethics 8(10), 805–815 (1989)
13. Marshall, J.: Agent-based modelling of emotional goals in digital media design projects. In: Innovative Methods, User-Friendly Tools, Coding, and Design Approaches in People-Oriented Programming, pp. 262–284. IGI Global (2018)
14. Mens, T., Cataldo, M., Damian, D.E.: The social developer: the future of software development [guest editors' introduction]. IEEE Softw. 36(1) (2019)
15. Miller, T., Pedell, S., Lopez-Lorca, A.A., Mendoza, A., Sterling, L., Keirnan, A.: Emotion-led modelling for people-oriented requirements engineering: the case study of emergency systems. J. Syst. Softw. 105, 54–71 (2015)
16. Morley, J., Floridi, L., Kinsey, L., Elhalal, A.: From what to how: an initial review of publicly available ai ethics tools, methods and research to translate principles into practices. Sci. Eng. Ethics 26(4), 2141–2168 (2020)
17. Mougouei, D., Perera, H., Hussain, W., Shams, R., Whittle, J.: Operationalizing human values in software: a research roadmap. In: ESEC/FSE (2018)
18. Mussbacher, G., Hussain, W., Whittle, J.: Is there a need to address human values in domain modelling? In: MoDRE (2020)
19. Norman, D.: Emotional Design: Why We Love (or Hate) Everyday Things. Basic Books (2007)
20. Pedell, S., et al.: Methods for supporting older users in communicating their emotions at different phases of a living lab project. Technol. Innov. Manage. Rev. (2017)
21. Perera, H., Mussbacher, G., Hussain, W., Shams, R.A., Nurwidyantoro, A., Whittle, J.: Continual human value analysis in software development: a goal model based approach. In: RE (2020)
22. Posner, B.Z., Schmidt, W.H.: Values congruence and differences between the interplay of personal and organizational value systems. J. Bus. Ethics 12(5) (1993)
23. Schwartz, S.H.: An overview of the schwartz theory of basic values. Online Read. Psychol. Cult. 2(1) (2012)

24. Schwartz, S.H., Cieciuch, J., Vecchione, M., Davidov, E., et al.: Refining the theory of basic individual values. J. Pers. Soc. Psychol. **103**(4) (2012)
25. Sherkat, M., Mendoza, A., Miller, T., Burrows, R.: Emotional attachment framework for people-oriented software. arXiv preprint arXiv:1803.08171 (2018)
26. Sterling, L., Taveter, K.: The Art of Agent-Oriented Modeling. MIT Press (2009)
27. Stickdorn, M., Hormess, M.E., Lawrence, A., Schneider, J.: This is Service Design Doing: Applying Service Design Thinking in the Real World. O'Reilly (2018)
28. Thew, S., Sutcliffe, A.: Value-based requirements engineering: method and experience. Requirements Eng. **23**(4), 443–464 (2018)
29. Vorvoreanu, M., Zhang, L., Huang, Y.H., Hilderbrand, C., Steine-Hanson, Z., Burnett, M.: From gender biases to gender-inclusive design: an empirical investigation. In: CHI (2019)
30. Wang, Y., Redmiles, D.F.: Implicit gender biases in professional software development: an empirical study. In: Kazman, R., Pasquale, L. (eds.) ICSE-SEIS (2019)
31. Yu, E., Liu, L.: Modelling trust for system design using the i^* strategic actors framework. In: Falcone, R., Singh, M., Tan, Y.-H. (eds.) Trust in Cyber-societies. LNCS (LNAI), vol. 2246, pp. 175–194. Springer, Heidelberg (2001). https://doi.org/10.1007/3-540-45547-7_11

A Product Owner's Navigation in Power Imbalance Between Business and IT: An Experience Report

Lotte Mygind[1] , Jens Bæk Jørgensen[1(✉)] , and Lutz Prechelt[2]

[1] Mjølner Informatics A/S, Aarhus, Denmark
{lmy,jbj}@mjolner.dk
[2] Freie Universität, Berlin, Germany
prechelt@inf.fu-berlin.de

Abstract. [**Context and motivation**] We consider a company where software development was previously a minor activity and today is a major activity with high priority and attention. Software is now developed according to Scrum, and the company can be seen as being in an *agile transition*. [**Question/problem**] What are the relevant specifics of this organization and which product owner behaviors appear to be valuable or problematic, respectively? [**Principal ideas/results**] A fear of disruption put the development teams under pressures that led to low efficiency in an interesting way. The introduction of an *IT product owner* to assist a *business product owner* reduced this effect, but a problematic power imbalance still remains. [**Contribution**] Not only do agile technical teams need feedback from their product owner, the product owner also needs meaningful, effective feedback from the teams. Our experience report shows how this can be improved by the introduction of an IT product owner when the organization otherwise has insufficient focus on certain important dynamics of software engineering.

Keywords: Requirements Engineering · Product Ownership · Project Organization · Cooperation between Business and IT · Software Quality

1 Introduction

We consider a company where one of the authors of this paper has been a consultant as product owner. The product owner, an integral part of Scrum [1], is a key role in modern agile software projects in the industry, and the role is crucial for the quality of the developed solutions. A product owner should have knowledge of requirements engineering and should facilitate the interplay between business stakeholders and software development teams. The product owner must contribute significantly to the elicitation, specification, validation and prioritization of requirements in a given organizational context.

In [2], Lauesen's taxonomy for functional requirements includes goal-level, domain-level, product-level, and design-level. In most software development projects, functional requirements on all four levels are relevant and exist with different degrees of explicit

© The Author(s), under exclusive license to Springer Nature Switzerland AG 2023
A. Ferrari and B. Penzenstadler (Eds.): REFSQ 2023, LNCS 13975, pp. 337–350, 2023.
https://doi.org/10.1007/978-3-031-29786-1_24

or implicit representation. A product owner must be able to work at all four levels and must facilitate bridging between the levels with assistance from experts, such as an IT architect or a user experience expert.

Product ownership in software projects is described in numerous papers, for example, [3–6]. These papers are case studies by researchers, while this paper is an experience report based on a practitioner's participation in a project for two years.

This paper introduces the company, project and product owner role in Sects. 2, 3 and 4. Section 5 presents a quantitative analysis of the backlog. Section 6 describes key observations about the product owner's work and the lesson learned: there is a power imbalance between business and IT which is too big. Related work is discussed in Sect. 7, which is followed by the conclusions in Sect. 8.

2   The Company

The company[1] is a traditional company in the fossil fuel industry. The main products are threatened by disruption in the form of electrification. In response to this, the company has introduced a new business area with a number of electrical products, which are highly interactive and digital. Looking back 5–10 years, the company had fewer than 20 employees working with IT, and today this has grown to around 150 people.

The company has a business division and an IT department, including a large IT development department headed by an IT development manager. A user experience department is part of the business division, see Fig. 1.

The product owner role is shared between a business product owner and an IT product owner, employed in the new business area and IT respectively. The IT product owner is one of the authors of this paper.

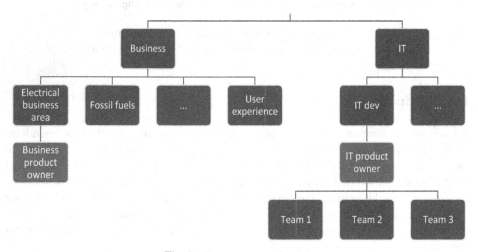

Fig. 1. Company organization.

---

[1] The company is kept anonymous.

3 The Project

3.1 History

After the new business area was introduced in late 2019, a software development project to support it was initiated in early 2020. An IT architect, several software developers and the business product owner joined the project within a few months.

The staffing has grown significantly since then. As of now, there are three Scrum teams, each with a Scrum master. In addition, there is an IT architect, a technical supporter, two testers, and the IT and business product owners[2], and these roles support all teams. The total number of team members in the IT department is 24 plus eight at a subcontractor. Moreover, the company plans to hire more developers for an additional team in the near future. Lastly, several user experience specialists work with the product. Figure 2 illustrates the timeline.

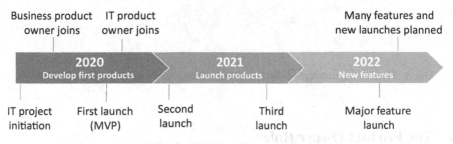

Fig. 2. Timeline of the project.

The company expects significant IT development for a decade ahead across several product suites in the new business area. For this reason, the activities can be seen as numerous, concurrent projects, or a *program* in the sense of project management literature. We use the term *project*, because this is how the company describes it.

3.2 Project Organization

The IT product owner works with all three Scrum teams. Furthermore, there is an IT project manager whose main responsibility is to create realistic plans for major releases. He is also a member of a steering group along with the IT development manager.

The chart in Fig. 3 illustrates the main flow of development tasks. The business product owner describes new domain-level requirements and communicates these to the IT product owner for analysis and feedback. When needed, the IT product owner consults the architect for an early, deeper analysis of the most complex requirements. For most requirements, the architect will not see the user stories until sprint start, and this works well most of the time. If the IT product owner has overlooked any non-trivial technical issues, the user story is taken out of the sprint to be analyzed further.

---

[2] Two additional IT product owners were hired shortly before the author left the project. This article describes the time with only one IT product owner.

When applicable, the user experience team cooperate with the IT product owner and describe a detailed solution for the user interface, i.e., design-level requirements. There are numerous alternate flows and backward loops, in addition to the main flow depicted in Fig. 3.

The IT product owner writes user stories (in a broad interpretation of this term) for specification to the development teams and defines the verification criteria. When the developers have implemented a user story, the testers verify it according to the verification criteria. Finally, the functionality is deployed to the production environment.

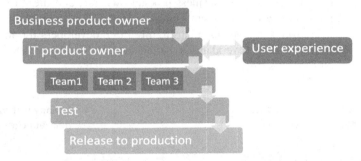

Fig. 3. Flow of requirements-related and development tasks in the project organization.

4   The Product Owner Role

Before this project, the company used product owners from the business division, with the mindset that the business would order IT deliveries from the IT department.

When the project started in early 2020, the business area was new, and a product owner was hired as the second employee in the business area. He has a background in business administration with limited experience of IT development projects and no domain knowledge. As the project grew, it became clear that he needed assistance in requirements engineering, including elicitation, specification, validation and prioritization of requirements.

The company hired one of the authors as a consultant who could teach the user story technique to the business product owner. She is an experienced product owner and has extensive domain knowledge in the relevant industry. She joined the project in October 2020. During the following months it became clear that the business product owner only had the time for and the interest in goal- and domain-level requirements, so the author was given a long-term contract, and people gradually started to describe her as the "the IT product owner", a new role in the organization (Fig. 4).

In June 2021 the management of the IT development asked the IT product owner and the IT project manager to describe the IT product owner role and a list of qualifications. After this, IT management decided that all IT projects should have such an IT product owner, so most projects in the organization now have an IT product owner.

The Scrum Guide [1] does not state whether the product owner is a part of the business organization or a part of the IT organization. The role of the product owner

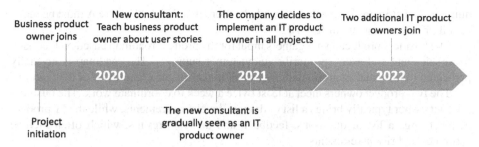

Fig. 4. The product owner role has evolved during the lifetime of the project.

is to "maximize the value of the product resulting from the work of the Scrum Team." The product owner must "define and communicate product goals, maintain the product backlog and ensure the backlog is transparent, visible and understood." In the project considered here, the product owner role is implemented by the joint work of the business product owner and the IT product owner.

4.1 Division of Work with Requirements

In most cases, the requirements elicitation, specification/description and prioritization are shared between the steering group, product owners and user experience as shown in Fig. 5.

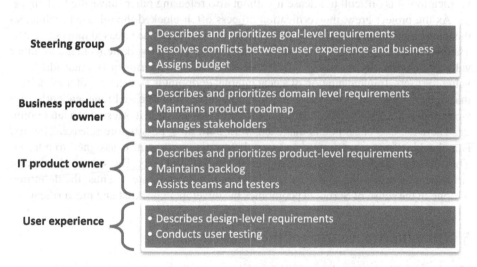

Fig. 5. The work with requirements is shared between various roles.

The goal-level requirements of the steering group are based on input from the product roadmap. The steering group study the product roadmap in order to find the most important areas and define goal-level requirements, such as "At the end of 2023 the number of customers for product X should be more than Y" or "Feature A must be ready for all private customers in July 2023".

The business product owner gathers input for the product roadmap based on feedback from sales and end users (primarily via customer support). The roadmap is normally presented to the steering group without any estimates of the complexity of the features.

The two product owners meet at least twice a week to coordinate work. The business product owner typically brings a list of domain-level requirements, while the IT product owner brings a list of questions, feedback and analysis results, which often include estimates and risk assessments.

When the IT product owner believes there is a good breakdown of a requirement with sufficient implementation details, the architect makes a rough estimate. Whenever the estimate is higher than expected by the business or high compared to the relative value of the feature, the IT product owner discusses the situation with the business product owner. This often leads to re-prioritization.

4.2 User Stories, Verification and Definition of Done

When the IT product owner joined the project, she intended that all work items should be user stories describing the direct value for a user role. We explain below that the IT product owner applies a modified approach.

The teams use a continuous integration strategy, where automated tests are implemented by the developers as part of the coding, and these tests are supplemented with work item verification by a tester. If a work item has been active for a long time before verification, it is difficult to release it without also releasing other (unverified) changes.

As the project grew, the verification process often blocked the release pipeline, so the teams now prefer small, independent work items that can be released independently. Therefore, the IT product owner creates small, independent work items that add either value or functionality, and these are not necessarily traditional *user stories* that add direct value for users. Implementation of a new internal endpoint is an example of a work item that does not add value for users, but it can be tested and released independently. If a work item is not testable, the IT product owner makes sure that it is tested at a later point.

There are two checkpoints that each work item must pass before release. The first is a checklist for the software developers before the work item is assigned to a tester. This includes checks of code quality, peer review and unit tests. The second checkpoint is verification according to the criteria described in the work item. Thus, the definition of done in the sense of Scrum [1] combines the developer checklist and the verification.

5   Quantitative Analysis of the Product Backlog

The status and history of the project's work items are analyzed in this section in order to see if the data may serve to illustrate the state of the project and the IT product owner's work. We discuss the data in further details in Sect. 6.

The project has a history of a growing backlog[3], see Fig. 6. The number of bugs is low due to an explicit prioritization: At each sprint planning, resolution of bugs is prioritized highest.

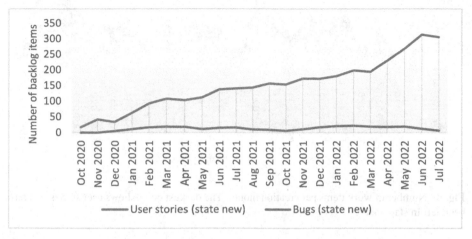

Fig. 6. The size of the backlog (new work items) is growing, but the number of bugs is stable.

The work items are grouped in features, and there are 93 features in state new or active[4] (65% see Fig. 7). In Sect. 6.1, we explain how a high business pressure may simultaneously result in many features in state new. Furthermore, Sect. 6.2 shows how the pressure at the same time may create a situation with many active, incomplete features.

Fig. 7. 35% of features are closed. The majority of features (51%) are in state *new*.

To understand the features that have not yet been fully implemented (closed), we have investigated the lifecycle of the individual backlog items, see Fig. 8. In total 22% of the backlog items created during this time period are still in state new. This will be discussed further in Sect. 6.

---

[3] All data in this section is for work items created before August 1st, 2022. The data has been extracted from the historical user stories in Azure Devops.

[4] The state of the work items on August 1st, 2022.

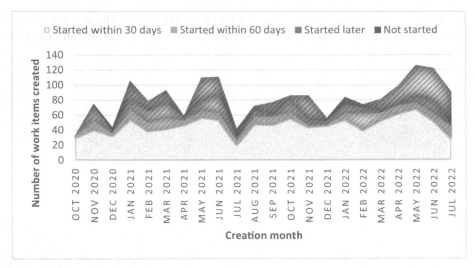

Fig. 8. Number of work items per creation month. The darkest color shows user stories that have been left in state *new*[5].

6   IT Product Owner Observations and Lesson Learned

Based both on the description of the product owner role and the quantitative analysis of the preceding sections, we will now describe two observations and a main lesson learned, which have their roots in the power imbalance between business and IT. This imbalance occurs in many companies, since their main focus is to manufacture and sell their products. Software development is just one activity among many others, and sometimes not a high-attention activity seen from the perspective of a company's business side and senior management. The imbalance is predictable, and it is our responsibility as software professionals to handle it properly to ensure good working conditions for software teams for the benefit of the company as a whole.

6.1   Mismatch Between Organizational Pressure and Product Owner Capacity

In early 2021, the business pressure increased. In fact, the business product owner extended the IT roadmap of the project to include 81 domain-level requirements.

From mid-2021 it became difficult for the IT product owner to find time for all the work. The number of customers grew rapidly, and since the IT product owner was the point of contact to the business, she constantly had to delegate production errors and support issues to team members. At the same time, the company gradually added more developers, so the task of supporting the work of all developers and testers grew. In the autumn of 2021, the project went from two to three Scrum teams.

---

[5] The state of the work items was extracted from Azure Devops on September 30th, 2022. Only work items created before August 1st, 2022 have been included.

The early estimates made by the IT product owner have only had limited impact on the priorities of the 81 domain-level requirements. There was, for example, one domain-level requirement that the business expected to be completed in a few sprints, whereas the IT product owner estimated it to be a complex subproject with a time horizon of many months. The business product owner did not accept this as a fact, so the IT product owner spent several months breaking the domain-level requirement into work items for more detailed estimation. In the end, the project was postponed because of the high estimates, and this was escalated to the steering group. The requirement was, however, immediately replaced by another subproject that had not been estimated at all. In this way, nobody evaluated whether the business value would match the effort for the new requirements, and implementation started before the full analysis had been done.

In this way, postponed subprojects are part of the reason behind the many features in state new on Fig. 7. Furthermore, the corresponding user stories are represented as a spike of unimplemented work items in Fig. 8. At the end of 2021 we estimate that the product owner spent 75% of her time on day-to-day support of the development teams. At this point, there were still approximately 80 domain-level requirements. Some of these were themselves full-sized projects. If we estimate an average of two workdays of product owner work for each domain-level requirement, this adds up to eight months of full-time work, which can be completed over 32 months (at 25% capacity for this work). This amount of time is far beyond reasonable and is an impediment for agile development.

On Fig. 8 the darkest color illustrates requirements that are dropped. This reflects the high number of ideas on the roadmap that the IT product owner has spent time on, but the team has not. Thus, there is a mismatch between the business' requests for analysis and the capacities of both the IT product owner and the development teams.

At the same time Fig. 8 shows that many work items are implemented after more than 60 days, and for these, the IT product owner must often update the descriptions before the backlog item is handed over to the development teams. After two months the software has often had changes to relevant data structures or interfaces invalidating the old description. Thus, the IT product owner must find time to analyze and describe more items than the teams implement, and the items that do get implemented must often go through an extra revision, because implementation is delayed.

The data illustrates a high-pressure environment where the IT product owner must analyze a high number of requirements. The effect was that she could not always provide qualified input to the prioritizations of the business product owner and steering group in time.

There are two obvious solutions. The first is to add more resources to the IT product owner function, which the company has done in mid-2022. The new resources are strong profiles who will be able to work as two cooperating IT product owners. It is unclear whether the two new IT product owners have enough capacity to make sure that steering group prioritizations are consistently based on estimates.

The second solution is to add a mechanism for prioritization and long-term planning of the IT product owner's work. This could, for example, be in the form of a board where the work of the IT product owner is tracked (for example, in status new, active and done), and this board could be used for communication with the business product owner in order to prioritize the work and limit the number of parallel tracks. This has not yet been implemented. Such planning of the product owner's work is difficult to accept for the business, because the high pressure creates an environment where the business always wants more analysis in the hope that the product owner can find an easy solution for the next problem.

6.2 Insufficient Trade-Offs Between Business Demands and Software Quality

In this project, we have often seen that when the most central functionality supporting a domain-level requirement has been implemented, the business wants to prioritize something else. Under the pressure, the business wants new features quickly, and as a result, the product has a lot of barely-viable solutions, which are reflected as active features in Fig. 7.

While the number of customers has grown significantly, the number of people working in administrative functions to perform workarounds for missing functionality has also grown significantly. At the same time the performance of the IT development teams is negatively affected by an increasing number of database updates and support for the administrative workers.

In Scrum, one of the product owner's most important tasks is to make prioritization continuously, sprint after sprint. As we see here, domain-level requirements with high priority might get lower priority at some point, and many features are never completed. In the opinion of the IT product owner, architect and, in fact, all development team members, the business product owner's prioritization is too volatile, resulting in incomplete features that create extra work for the development teams in the long run.

While a lot of bad decisions have been avoided because of good communication between the business and IT product owners, it has been difficult to find the right balance; this has sometimes been subject to intense debates between the business product owner and the IT product owner. When everything is broken into small, independent work items for prioritization (see Sect. 4.2), the business product owner can pinpoint elements that seem minor from a functional point of view.

In the aftermath of all these minimal implementations, the IT product owner has learned that due to the high business pressure, it is difficult for the business product owner to prioritize features that are not directly linked to the product roadmap. After a lot of discussions between the two product owners, the business product owner has extended the roadmap with some features that ensure continued scalability, operability and performance – more generally, software quality. In this way, the IT product owner has pushed technical requirements to be recognized and prioritized.

6.3 Lesson Learned: Power Imbalance Between Business and IT is Too Big and Must Be Addressed

The observations above illustrate the power imbalance between the business and IT. The company is in a quick transition from a traditional, industrial company to an IT service company, but the company has not yet reached maturity as to how to prioritize larger business roadmaps.

The company has taken one important step in the transition, namely hiring IT product owners to work with all development teams, but the IT product owners have not been given a proper mandate. In the current set-up, the IT product owners can argue their cases and deliver estimates for early prioritization, but they are often overruled.

The power imbalance between business and IT must be addressed to ensure continued and efficient progress in the company's agile transition. Looking back, the IT product owner could have attempted to do more to communicate this observation to upper management at an earlier point in time in an attempt to contribute to a resolution. This is difficult, however, as the root causes lie in the company's organization and its inherent culture, so it is likely that the necessary changes will not happen in a short time frame.

One approach for the communication could be to show some of the data in Sect. 5 to higher-level management. The figures illustrate the ever-changing directions of the project with many features that are not *closed* (Fig. 7) and many work items where implementation starts late or never starts (Fig. 8). Furthermore, the IT product owner could have kept a record of goal-level requirements where the teams were told to start implementation immediately without further analysis (see an example in Sect. 6.1) and situations where the recommendations of the IT product owner were overruled by the business (see examples in Sect. 6.2).

7 Related Work

The Scrum guide [1] insists that "The Product Owner is one person, not a committee." In the project considered in this paper, it was vital that the product owner became a committee consisting of business product owner and IT product owner. The reason was threefold: scale - the business product owner being too busy to fully do all the things the teams needed; maturity - the business product owner not understanding the implications of his actions well enough; and power - there was no one who could successfully oppose the business product owner when needed.

Our experience, that the product owner gets too busy when the organization around her grows, is of course a common problem, and there are several approaches to scale agile development, including the scaled agile framework (SAFe) and Scrum@scale. For a recent overview of approaches, see Edison et al.'s survey paper [7].

Scalability is obviously an important issue, but there are also other issues which are relevant to discuss. We use Bass and Haxby's framework of roles in product owner teams as a theoretical lens for obtaining a useful alternative view of our case [8]. They consider three dimensions they call Scale, Distance, and Governance. Distance effects are absent in our case; scale effects have already been discussed; regarding governance, Bass and Haxby suggest three roles for solving Governance issues: Technical Architect, Governor (who watches over adherence to corporate technical policies), and Risk Assessor. The latter was required for solving the problem in our case: Evaluate and highlight risks involved in a particular course of action, so that problematic routes can be avoided. Bass and Haxby also recommend making product owner teams explicit, which happened naturally in our case.

For understanding the maturity dimension, the mapping study by Unger-Windeler et al. [3] is helpful. It maps out topics and findings from 30 studies of product owners in industrial settings. Like [8], they find that product owners for larger-scale settings will have to be teams. In terms of roles, they list no fewer than 14 distinct roles for product owners that have been discussed in multiple (as opposed to only one) publications. Two of them help understanding what went wrong in our organization initially, before the IT product owner was introduced: Prioritizing the backlog [4, 9, 10] and managing expectations [11, 12]. Both are rather basic aspects of product owner work. Yet our business product owner was unable to manage expectations. The reason was that he did not know what to expect because he was relatively inexperienced with large-scale software development. For the same reason, his prioritization used inappropriate criteria, and resulted in the negative side effects we have described.

As for the power imbalance problem, Scrum has two roles responsible for solving that: The product owner, by prioritizing work appropriately, and the Scrum Master, by fighting against distractions (such as the support work arising from the incomplete features). Boehm and Turner called these tasks the "protector" role already in 2005 [13]. Yet in our case, coach roles had barely developed in the organization and the product owner was the very source of the problem. Therefore, it required the introduction of the IT product owner to get to effective software development.

8 Conclusions

In this paper, we have described the work of the IT product owner and her navigation in an organization where there is an imbalance between business and IT. We have supplemented this with a quantitative analysis to illustrate and discuss the state of the project and the product owner's work. While the analysis has the obvious weakness that it may show the product owner's work methods more clearly than the general state of the project, it would be interesting to compare such data for a larger number of projects.

In general, a product owner builds a bridge between business and IT, but this requires an organization that is in better balance than the one described in this paper. As IT professionals, we should work on strengthening our communication with stakeholders with no or low interest in software development in the organizations we work in. An example of this is described in [14].

We have been partially successful in doing that: Upper management has recognized the benefits of strong product ownership, and for this reason, as mentioned in Sect. 4, it has been decided to apply a set-up with a business product owner and an IT product owner to all other IT projects in the organization.

A reason for this is that the product ownership in the considered project has been strong in many respects, and it has been visible for upper management that this has been the case. There is a well-structured, well-described backlog, strong communication between business and IT and good access to support for developers and testers. The organization saves development time, because it is ensured that focus is on features that bring the right amount of value compared to the effort. The developers are, in general, very satisfied with this approach. They get well-defined user stories in sprints, so they can focus on their favorite activity, namely coding. They have immediate access to clarification from the IT product owner when needed and they are always able to discuss alternative implementation ideas with the architect or IT product owner. Furthermore, they experience only few interruptions from analysis of future features.

On the other hand, the company's agile transition is certainly not yet completed. It is a problem that many business stakeholders in the company still see IT as a supporting function where they "order" IT systems. The terminology in the business is: "We tell you what we want and when, and you tell us the cost".

We believe that the company could get more value by involving IT more and earlier in identifying where to get the best value compared to the effort. The company has to some degree done this by introducing the IT product owner role, but the IT product owner has only had the power of argumentation - and has often been overruled.

References

1. Schwaber, K., Sutherland, J.: The Scrum Guide – The Definitive Guide to Scrum: The Rules of the Game. https://scrumguides.org/docs/scrumguide/v2020/2020-Scrum-Guide-US.pdf
2. Lauesen, S.: Software Requirements – Styles and Techniques. Addison Wesley (2004)
3. Unger-Windeler, C., Klünder, J., Schneider, K.: A mapping study on product owners in industry: identifying future research directions. In: 2019 IEEE/ACM International Conference on Software and System Processes (ICSSP), Montreal, Quebec, Canada. https://doi.org/10.1109/icssp.2019.00026
4. Bass, J.M., Beecham, S., Razzak, M.A., Canna, C.M., Noll, J.: An empirical study of the product owner role in scrum. In: ICSE 2018: Proceedings of the 40th International Conference on Software Engineering: Companion Proceedings, Gothenburg, Sweden. IEEE (2018). https://doi.org/10.1145/3183440.3195066
5. Matturro, G., Cordoves, F., Solari, M.: The role of product owner from the practitioner's perspective. An exploratory study. In: 16th International Conference on Software Engineering Research and Practice (SERP 2018), Las Vegas, Nevada, USA, pp. 113–118. CSREA Press (2018)
6. Kristinsdottir, S., Larusdottir, M., Cajander, Å.: Responsibilities and challenges of product owners at spotify - an exploratory case study. In: Bogdan, C., et al. (eds.) HCSE/HESSD -2016. LNCS, vol. 9856, pp. 3–16. Springer, Cham (2016). https://doi.org/10.1007/978-3-319-44902-9_1
7. Edison, H., Wang, X., Conboy, K.: Comparing methods for large-scale agile software development: a systematic literature review. IEEE Trans. Softw. Eng. 48(8), 2709–31 (2022). IEEE. https://doi.org/10.1109/tse.2021.3069039
8. Bass, J.M., Haxby, A.: Tailoring product ownership in large-scale agile projects: managing scale, distance, and governance. IEEE Softw. 36(2), 58–63 (2019). https://doi.org/10.1109/MS.2018.2885524
9. Bass, J.M.: How product owner teams scale agile methods to large distributed enterprises. Empir. Softw. Eng. 20(6), 1525–1557 (2014). https://doi.org/10.1007/s10664-014-9322-z
10. Bass, J. M., Beecham, S., Razzak, M. A., Canna, C. N., Noll, J.: An empirical study of the product owner role in scrum. In: Proceedings of the 40th International Conference on Software Engineering: Companion Proceedings (2018). https://doi.org/10.1145/3183440.3195066
11. Finsterwalder, M.: Does XP need a professional Customer? In: Proceedings of the XP2001 Workshop on Customer Involvement (2001)
12. Sverrisdottir, H.S., Ingason, H.T., Jonasson, H.I.: The role of the product owner in scrum: comparison between theory and practices. Procedia Soc. Behav. Sci. 119, 257–267 (2014). https://doi.org/10.1016/j.sbspro.2014.03.030

13. Boehm, B., Turner, R.: Management challenges to implementing agile processes in traditional industrial organizations. In: IEEE Softw. **22**(5), 30–39 (2005). IEEE. https://doi.org/10.1109/ms.2005.129
14. Jørgensen, J.B., Christensen, H.L., Hansen, S.T., Nyeng, B.B.: Effective communication about software in a traditional industrial company. In: 2022 IEEE 44th International Conference on Software Engineering (ICSE), 5th International Workshop on Software-Intensive Business, Pittsburg, Pennsylvania, USA. IEEE (2022). https://doi.org/10.1145/3524614.3528625

Eliciting Security Requirements
– An Experience Report

Roman Trentinaglia^(✉) (iD), Sven Merschjohann(iD), Markus Fockel(iD),
and Hendrik Eikerling(iD)

Safe & Secure IoT Systems, Fraunhofer IEM, Paderborn, Germany
{roman.trentinaglia,sven.merschjohann,markus.fockel,
hendrik.eikerling}@iem.fraunhofer.de

Abstract. [**Context and motivation**] Cyber-physical systems, like
modern cars and industrial automation systems, are highly connected
and complex. [**Question/problem**] Their various interconnections open
interfaces for attackers, and their complexity increases the risk of undetected security vulnerabilities. Hence, an important part of requirements
engineering is threat modeling. It is a means to elicit security assets,
goals, and assumptions, and to derive required security controls. Effective threat modeling needs a systematic workshop setup. [**Principal
ideas/results**] In this paper, we report our experiences and lessons
learned from threat modeling workshops that we conducted with industry partners from the domains of industrial automation, health care,
smart home, and automotive. [**Contribution**] In conclusion, we derive
a set of open challenges.

Keywords: threat modeling · STRIDE · requirements elicitation

1 Introduction

Cyber-physical systems are highly connected and complex. Automotive systems
for autonomous driving use a large variety of sensors, car-to-X communication,
and high-performance control units to determine the current driving situation
and plan next actions in real-time. In industrial automation, systems become
more and more sophisticated and connected as part of Industry 4.0. Industrial
controls are connected to the cloud for data analytics (e.g., for condition monitoring) and can be updated and extended via app stores.

The increasing number of communication interfaces open new doors for cyber
attacks, and the growing complexity increases the risk of undetected security
vulnerabilities. Hence, an important part of requirements engineering is *threat
modeling* [7]. It is a means to elicit security assets and goals, determine threats
to the system and assess their risk, and to derive security assumptions and
requirements (i.e., required security controls).

Xiong and Lagerström published a systematic literature review on threat modeling [13]. They concluded that there is no common definition of threat modeling

© The Author(s), under exclusive license to Springer Nature Switzerland AG 2023
A. Ferrari and B. Penzenstadler (Eds.): REFSQ 2023, LNCS 13975, pp. 351–365, 2023.
https://doi.org/10.1007/978-3-031-29786-1_25

and most literature they found used a manual threat modeling approach. Yskout et al. assessed the state of practice in the field of threat modeling by interviewing threat modeling practitioners [14]. They conclude that threat modeling, as an engineering discipline, is currently at a low level of maturity, mostly performing "whiteboard hacking".

We believe that effective threat modeling needs a systematic workshop setup. In this paper, we report our observations and lessons learned from systematic threat modeling workshops that we conducted with industry partners from the domains of industrial automation, health care, smart home, and automotive.

Our threat modeling workshops are based on STRIDE [7]. It is a mnemonic representing its six categories of threats, and a method to find threats from these categories in an architectural diagram. We explain the details of our workshop setup in Sect. 2. An extended version of our method with development process integration was certified for compliance with the industrial automation security standard IEC 62443 and presented in [2].

Considering STRIDE, Scandariato et al. performed a descriptive study and Tuma and Scandariato did a comparative study [5,9]. Both papers used university students as study participants that were provided with documentation about a system that was used for threat modeling. In contrast to that, we applied our STRIDE-based method in industry-funded workshops to identify threats and security requirements for commercially used products. All participants (except us) were stakeholders for the product (e.g., product owners, software architects, developers). Examples of the analyzed products are an industrial control system, a system assisting in surgeries, and a cloud backend used by smart home appliances. All in all, this report is based on the observations from threat modeling six products from the four aforementioned domains.

This paper is structured as follows. In Sect. 2, we explain the setup of our threat modeling workshops. Section 3 describes central observations that we made when conducting the workshops. In Sect. 4, we describe the lessons we learned from those observations. Section 5 describes threats to validity of this experience report. Section 6 concludes the paper with a set of open challenges that we derived from the lessons learned.

2   Setup of Our Threat Modeling Workshops

An introduction to our method can be found in [1]. In this section, we explain the details needed to understand our observations and lessons learned.

Our threat modeling workshops are based on STRIDE [7]. It is a mnemonic representing six categories of threats *spoofing, tampering, repudiation, information disclosure, denial of service,* and *elevation of privilege*. It is also a method to find threats from these categories in an architectural description of the system of interest (SoI), typically a *data flow diagram* (DFD). A matrix defines which STRIDE category is of prior applicability to which element of a DFD.

There are tools that support in the application of STRIDE. We typically use the Microsoft Threat Modeling Tool (TMT)[1] in our workshops. For a list

---

[1] https://aka.ms/threatmodelingtool.

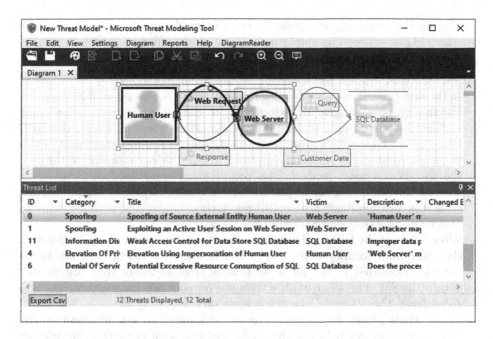

Fig. 1. Microsoft Threat Modeling Tool

of threat modeling tools see [6]. Figure 1 shows a screenshot of the tool. The diagram shows a DFD representing a web server that interacts with an SQL database and a human user. Below the diagram, possible threats are listed. The first threat is relevant to the highlighted elements in the diagram, specifically, to the data flow from the human user to the web server. It specifies that the human user could be spoofed, i.e., an attacker could try to access the web server pretending to be a valid user.

In the following, our workshop setup is described in more detail. For preparation, the customer has to select the participants for the workshop. We recommend not more than six participants, to enable discussions but nevertheless keep focused enough within the given time frame. Typically, our customers do not have their own security experts. Hence, they hire us to fulfill that role. The customer's participants should cover relevant stakeholders of the SoI, e.g., developers, system and software architects, product owners, quality assurance, risk management and technical support. From our side, the workshops are moderated by two security requirements engineers to share the work of giving hints on possible attacks and countermeasures, moderating discussions, operating the tool, taking notes, and time-keeping. The workshops are typically constrained to eight hours, divided in two four-hour sessions on separate days. The workshops can be performed offline, on-site at the customer's premises, or online using conferencing tools like Microsoft Teams.

During the workshop we guide, teach, and moderate the participants. We use the TMT, start with an empty model, and enter information step by step as provided by the participants. The workshop is structured in three phases:

1. *Specifying the System and Determining its Security Context*
   In the first phase, we start modeling the customer's SoI as a data flow diagram. For this purpose, we give an introduction to data flow diagrams beforehand, so that the participants understand the notation. Then we create the DFD based on the information given by the participants. An important part of this is the elicitation of security assumptions about the environment of the SoI, the so-called *security context*. This comprises assumptions/requirements of security measures that are in place where the SoI is deployed (e.g., operating system security-updates are applied regularly).

2. *Determining Assets and Security Objectives*
   After finishing the DFD and security context, in the second phase, we identify the assets and security objectives of the SoI and its assets. The assets are directly annotated in the DFD, so that it is clear where they are located within the SoI. The security objectives are noted down for the complete SoI.

3. *Identifying Threats and Countermeasures and Assessing Risk*
   In the third phase, we identify threats and possible countermeasures and assess each threat's risk. For the initial identification of threats, we use the threat generation feature of the TMT which follows the basic STRIDE rules (i.e., matrix mentioned above) to generate a list of potential threats for each element of the DFD. Typically, the list of generated potential threats is quite long (more than 100 threats), so that we first explain the different types of generated threats and then pick one threat for further examination. For the selected threat, we discuss whether it is truly applicable to the SoI, and, if so, identify possible countermeasures and note them down. These counter-measures may either already exist or form security requirements yet to be realized. An additional important step is manually adding further threats to the SoI that come up during the discussions.

   Finally, the participants determine the risk of the selected threat by playing *protection poker* [11]. Protection poker is a variant of planning poker, in which every participant is asked to play simultaneously a numbered card for assessing the risk. If differing card numbers are played among the participants, the ones with the highest and lowest numbers played explain their reasoning. Then a next round is played in which again everybody plays a card. This procedure is repeated until consensus for the risk value is achieved. In this way, first, the ease of an attack is determined, and afterward, the impact of a successful attack. Both values are then multiplied to a single risk value.

   Afterward, the next threat is selected and the steps repeated until either all threats were considered or the allotted time for the workshop is used up.

After the workshop is finished, we provide the customer with the threat model created during the workshop. In addition, we create a report listing the results, especially to highlight top priority (high risk) threats, security assumptions, and security requirements. These materials can then be used by the customer to continue with their secure development of the SoI. Furthermore, they also give a head start on performing threat analysis for another system on their own.

3   Observations

In the following, we describe our observations from conducting the threat modeling workshops as described in Sect. 2. These observations serve as the basis for lessons learned, which we present in Sect. 4.

In general, we have found that threat modeling is a lightweight activity and can be applied in any development process, be it classic V-model or agile. We have noticed positively that our participants have always been highly motivated, worked well together, and had many valuable discussions concerning the SoI's security. The discussions were often sparked by the DFD or by the generated potential threats. Especially for the generated threats, as intended, the participants carefully evaluated whether or not they were relevant to their SoI. If a threat was deemed relevant, the participants quickly got into brainstorming potential countermeasures. Accordingly, the workshops led to fruitful results although most participants were no security experts. In fact, some participants told us afterward, that they regularly think in STRIDE categories, because the workshop raised their security awareness and was learning-by-doing.

Despite these positive observations, we have also made the following eight observations that we want to improve upon.

OB1: STRIDE results in many fine-grained threats
The STRIDE approach works iteratively along the DFD. That means, for each element of the diagram a set of possible threats is analyzed. According to our observations, this local focus results in the threats (and corresponding countermeasures) being rather fine-grained. Correspondingly, depending on the size of the underlying data flow diagram, STRIDE often leads to a huge amount of threats (also known as *threat explosion* [10]). For example, for an extensive DFD (with more than 20 elements), the TMT generated more than 400 threat candidates during one of our workshops.

OB2: Threat models can get large and cluttered
During our workshops we observed that, depending on the analyzed SoI, threat models can get quite large. It can be difficult to set the right scope and level of detail for a workshop so that the outcome remains manageable. We further noticed that the diagram canvas in the TMT is constrained in size. Therefore, one is limited in the number of diagram elements that can be specified in a well-arranged way. This is especially problematic when larger DFDs (exceeding more than 20 elements) are to be specified, as one has to scroll between different parts of the diagram. Therefore, when the models get big, it might be necessary to split the content. While the TMT actually does offer the user the possibility to create additional diagrams, these are displayed in a separate pane. The DFD elements are thereby not linked in any way (except by name) and also the generated threat candidates are completely

separate. In summary, we did not find a good use case for having different diagrams in one model in the TMT.

OB3: Finding suitable countermeasures is not a trivial task

The generated potential threats often only include some general advice on possible countermeasures, e.g., using HTTPS instead of HTTP. However, typically, threats can be mitigated in several different ways. The effects that the introduction of a new countermeasure has on the system have to be figured out by the users of the TMT on their own. Especially when countermeasures require architectural changes of the system, this is difficult for the users to foresee. As an example, to introduce an authentication scheme, typically at least one new component is required, that handles the authentication and management of the valid users. Apart from these architectural effects of countermeasures, often trade-offs with other requirements of the system need to be made, e.g., between usability and security objectives. In this regard, the TMT only gives hints at which technologies should be looked at for possible countermeasures, while leaving the users with the difficult decisions mostly on their own.

OB4: The modeling of chosen countermeasures, assets, security objectives, and assumptions is not explicitly supported by the TMT

As part of our workshops, we additionally determine aspects like the security context, assets and security objectives for the SoI, as well as countermeasures for the identified threats (cf. Sect. 2). While the TMT provides the ability to add custom text fields to diagram elements, there is no way to include these aspects in a clearly structured way in the model. As a result, it can easily happen that assets or assumptions get overlooked in the text fields and are forgotten. Keeping these plain text fields consistent with the model also requires manual effort. In addition, the use of plain text fields to describe countermeasures makes it difficult to adequately analyze whether a countermeasure introduces new threats. For example, a repudiation threat may be mitigated by a logging database, but this database itself may introduce a new threat of logs being manipulated.

OB5: Aggregating risk metrics is a repetitive and tedious task

After determining a threat in our workshop, the next important step is its risk assessment. For this purpose, we play protection poker for determining the two metrics impact and ease of the attack. The risk is then calculated by multiplying these two determined values. This method has the benefit that it can be easily applied by our workshop participants, who are mostly not security experts. For this purpose, we had to add additional fields and calculate the resulting risk value manually, as the TMT does not provide any risk assessment or automatic calculation feature. However, the resulting risk calculation has to be done for each identified threat.

OB6: The participants identify additional system-specific threats

After we modeled the SoI using data flow diagrams, the TMT generates many potential threats. These threats serve as a starting point, as system-specific threats can only be identified by the participants as the stakeholder of the SoI. For this purpose, the TMT provides the important functionality to add custom threats to the list of threats. However, a function to annotate these custom threats to specific DFD elements would be helpful, in order to pinpoint the location of the threat within the SoI. At this point, also often discussion arises to which STRIDE category the threat should be allocated. The fact that one threat can be categorized into multiple categories is the nature of STRIDE and therefore intended, though it is a common point of confusion when the participants are not familiar with STRIDE yet.

OB7: The participants want to integrate threat modeling into their processes

Typically, after the workshop, the participants want to integrate threat modeling as a regular activity into their software development processes. For this purpose, they want to properly document the analyzed threats, their risks, and derived security requirements (i.e., countermeasures). They require the export of the threat modeling results, so that it can be used in their existing processes and tools for requirements management and/or application lifecycle management (RM/ALM). The TMT can only export the threats as CSV or generate an HTML report. However, this is not sufficient, because when changes are made, also the threat model, and as such, the threats are updated. This results in the problem of keeping the threats, which are now spread across different tools, manually up to date. This is also the case for the other way around, i.e., when changes are made to the system, the threat model needs to be updated. There is currently no tool-support by the TMT to propagate detected system design changes made in other tools back to the data flow diagram.

OB8: Safety-relevant elements are important for the risk assessment of threats

In the automotive and automation domain, we observed that in order to assess the risks of the threats, it is important to consider the involved safety aspects of the SoI. However, the TMT does not provide a way for integrating safety concerns into the model. This means, that the participants had to remember and check each time, whether the currently focused threat can also have consequences on the system's safety.

We think, that especially these eight observations show needs for improvement in threat modeling methods, tools, and processes. We also believe that we are not alone with these observations in threat modeling (cf. [14]). We describe in the next section, what we learned from these observations.

4 Lessons Learned

During our threat modeling workshops we observed strengths of the applied methodology as well as shortcomings (see Sect. 3). The observations serve as a neutral basis on which we can identify and formulate underlying problems and derive lessons learned. We list these nine learnings below.

LL1: A threat model is not a security concept
In a security concept the engineers state how they achieve and maintain the desired or required level of security of their system [4]. It therefore includes not only the threats of the system, but also which security objectives and strategies the engineers pursue, which countermeasures were taken, as well as the underlying assumptions on the SoI and its environment. While the TMT helps to identify recurring threats for common components, it currently does not allow to model security objectives, countermeasures, and assumptions explicitly as stand-alone model elements (cf. OB4). Encoding these aspects in the generic text fields provided by the tool complicates their traceability and requires manual maintenance. Furthermore, extensive manual effort is needed to transform the text field contents into a full security concept. In addition, this procedure makes it easy to overlook the assumptions and countermeasures stored in the subtle text fields, so they may not be considered when the development of the system progresses. To address these shortcomings, better tool support is needed that allows to explicitly model and trace above mentioned design aspects thereby reducing the need for manual maintenance.

LL2: The SoI needs to be developed with security goals in mind
STRIDE leads to an extensive list of fine-grained potential threats for concrete DFD elements (OB1). Consequently, it leads to countermeasures that are locally constrained with a focus only on preventing a specific corresponding threat. For example, to prevent manipulation of data on a specific data flow an integrity check is performed. However, to eradicate threats it may be preferable to improve the architecture instead of just fixing threat by threat. For example, the SoI may be redesigned to prevent an attacker from accessing critical parts of the system at all. In this case STRIDE should be performed in combination with a goal-oriented approach. Breaking down broad security goals into smaller sub-goals based on mitigation strategies (e.g., network segmentation) further helps to derive concrete countermeasures more easily.

LL3: Abstracting the analyzed SoI and prioritizing the threat candidates is necessary
Due to the limited modeling space (OB2) and the high number of threat candidates generated for large DFDs (OB1), it is important to keep the DFDs as compact as possible from the beginning. The threat model must remain manageable, otherwise the chance of errors increases and things can easily be overlooked. One way to achieve this is to conceptually divide the system into different isolated parts that can be analyzed more independently. In addition, the user is responsible for finding the right level of abstraction to keep the model manageable. Therefore, to find meaningful threats, the user has to focus on the most important parts of the system, without getting too detailed, but also not staying too general either. Currently, the TMT does not provide any built-in tool support to indicate different levels of abstraction in the same DFD. Therefore, if more detailed descriptions of certain parts of the SoI are needed, one has to create separate sub-diagrams manually.

Nevertheless, even for smaller DFDs, the number of generated threats will remain rather high, including several false positives. To make the participants aware of false positives, we explain how the TMT generates threats (i.e., template-based) and that false positives can therefore occur. Since we aim to explain the basic STRIDE methodology to the participants, the goal of our workshop is not to analyze all threat candidates in detail. Therefore it is also crucial to prioritize the generated threat candidates somehow. To analyze examples of potential threats, we pick out suggestions for threats that we think are potentially relevant to the SoI. Since the participants shall later also be able to find relevant threats from the TMT generated suggestions on their own, the participants are then asked to evaluate these selected threats themselves. However, participants are also free to choose other threats from the list or propose new (user-defined) threats. Letting the participants first select the most relevant threats for each STRIDE category intuitively and then determine their risk using a structured approach (cf. Sect. 2) helped them to get a better feeling for the current level of security of their system.

LL4: Guidance in coming up with countermeasures for the identified threats is needed

When the threats are identified, the next crucial step in making the system more secure is the implementation of suitable countermeasures. However, deciding on the best fitting countermeasures is a daunting task for the participants (OB3) as they are mostly no security experts. As such, it is necessary to guide the participants in choosing fitting countermeasures that on the one hand mitigate sufficiently the identified threats and ensure that their security objective can be accomplished, while on the other hand also managing the other system's requirements. Especially, when architectural changes to the system are necessary, the participants need suggestions on how to proceed best. For this purpose, we first explain different possible countermeasures that might be employed in general. Afterwards, we discuss the participants' requirements and find a suitable solution for their product.

LL5: One goal of the workshop should be to transfer knowledge

Within the limited time of our workshops it is not possible to address all of the many potential threats (OB1). Therefore, one goal of the workshop should be to transfer knowledge to enable the participants to continue working on the threat model on their own. This includes understanding the data flow diagram, the different security objectives, STRIDE, what possible threats might be, as well as the risk assessment.

LL6: Tool-support is needed to exchange threats and security requirements with other tools

In a security by design process where security is considered throughout the whole development lifecycle, results from a threat modeling session (e.g., found threats) need to be utilized in further development steps. Especially if threat modeling is performed in early stages of the development, exchanging threat modeling results with RM/ALM tools is needed to trace security requirements (OB4, OB5). Unfortunately, the TMT does not provide any possibility to export such threat model results other than storing the whole model in an XML-based file format or exporting a list of threats. We found that being able to reuse these artifacts in their existing development toolchain is important to the participants and increases the motivation to perform threat modeling continuously (OB7).

LL7: When the system changes, the threat model has to be updated manually

Keeping the threat model up-to-date is crucial to continuously evaluate the level of security of the system and find new threats when the system is changed. Especially if threat modeling is performed in early stages of the development, the system design is subject to change and changes must be reflected and versioned in the threat model. However, there currently is no tool-support by the TMT to propagate detected system design changes made in other tools back to the data flow diagram (OB7). Therefore, the threat model has to be kept up-to-date manually resulting in a high effort and a high susceptibility to human error. To enable continuous threat modeling, for example in DevOps environments, support for continuous integration is needed.

LL8: Domain-specifics are important for threat modeling

The domains involving cyber-physical systems, e.g., automotive and automation, are typically heavily standardized. This means that specific processes need to be adhered to and it needs to be shown that their process matches to the standard. This is also true for the process of threat modeling, e.g., the IEC 62443 requires that the system is partitioned into zones and conduits. Another requirement is the mandatory inclusion of a specific set of given countermeasures. Some of the domain-specific requirements can be followed using the TMT, as we noticed in OB4, this is still only implicit and there is no support that helps the users in proving compliance. Still, the TMT is a helpful tool in fulfilling the demanded threat analysis as also new threats can be added (OB6). However, it cannot fulfill them completely on its own, but it can be well integrated in a IEC 62443 compliant threat modeling process [2].

LL9: Integrated safety and security analysis would be beneficial

When the risk of threats is assessed, we observed (OB8) that it is important to think about what kind of hazards a threat can lead to, especially in the automotive and automation domain. Typically, in these domains a hazard and risk analysis has to be performed, the results of which can be used for guiding the threats' impact assessment. Whether a threat can lead to hazards concerning the SoI's safety is one of the biggest driving forces behind determining the possible impact of the threat. Therefore, an integrated safety and security analysis would be beneficial to determine the safety & security of the SoI.

5   Threats to Validity

The observations and learnings shared in this report are purely based on the authors' experiences in threat modeling workshops conducted in various organizations and at various times. These workshops were not planned as controlled experiments or case study to test certain hypotheses about threat modeling methodologies, but rather to enable a group of participants to evaluate the security of their software systems. Therefore, our goal is not to claim universal applicability of our findings, but to simply provide our experiences. This section provides an overview of the threats to validity, which may have influenced the results reported in this paper, loosely based on the threats to validity laid out by Wohlin et al. [12].

In terms of conclusion validity, multiple factors could influence the results we have seen within our threat modeling workshops and thus the lessons we heave derived from the observations. First of all, with six conducted workshops, the statistical power is very low and does not allow for statistical analysis, which we therefore cannot provide. In addition, the observations are collected by the moderators of the workshops, i.e., the authors of this paper, who may be biased and could have "fished" for certain results by making false observations. This cannot be fully mitigated, even though through our collaborative process, egregious lies would have been detected. Since no formal measures were taken during the workshops, there are no threats related to their reliability. However, there could be differences in the conduction of the workshops, depending on the specific settings, as well as random circumstances that altered the observations made during the workshops.

The internal validity is concerned with influences that affect the outcome but are not known to the researchers. In our case, we do not know the full history and experiences with threat modeling of the participants, which influences their behavior in the workshops. Additionally, the length of the workshops could lead to tiring of the participants and thus the threat of "maturation", resulting in negative behavior and adverse effects on the results. The selection of the participants also plays a role in the validity of the observations, since volunteering participants are usually more motivated and have a positive attitude towards the workshops.

Construct validity deals with the generalizability of an experiment with regards to a theory. Since we did not design the workshops to test a specific hypotheses, this aspect is not completely relevant. However, the social threat of "experimenter expectancies" could be relevant, since we may have expected certain behaviors from the participants and thus were more likely to observe them.

For the external validity or generalizability to industrial practice, there are a number of threats that are relevant to our experience report. First of all, our subjects were all experienced in the domain of cyber-physical systems, which means they have a certain attitude towards the threats found in the threat model and, e.g., the impact towards safety. This safety-mindedness may not be present in other groups performing threat modeling, which impacts the generalizability of our lessons learned. Additionally, the setting and design of our workshop could have influenced certain observations. For example, we found several inadequacies with the TMT, which could exacerbate a negative attitude, if a participant had experiences with different tools that may lack these shortcomings.

All in all, we acknowledge that the observations and the derived lessons learned are subject to certain threats to their validity. However, they come from real-world threat modeling workshops and may serve as a guide to improve ones approach to threat modeling, the current tooling landscape, and broaden the view of threat modeling practitioners by suggesting the inclusion of safety impact in addition to security aspects.

6 Conclusion

In this paper, we reported our observations and lessons learned from systematic STRIDE-based threat modeling workshops that we conducted with industry partners. We see threat modeling as a must for eliciting security requirements, and our workshops lead to fruitful results. Nevertheless, our observations underpin the conclusion of Yskout et al. that there are many open challenges [14], including a need for better tool-support also claimed by Xiong and Lagerström [13]. In more detail, we see the following challenges that we plan to work on in the future.

Comprehensible Goal-Based Threat Modeling
A major open challenge we see in threat modeling is that the subsequent comprehensibility of decisions (e.g., why a certain countermeasure was chosen) made during threat modeling is limited. These decisions also include the underlying assumptions and pursued goals and strategies on which the decisions can be justified. On the one hand, these limitations can be attributed to the lack of modeling abilities due to limited tool support (LL1). On the other hand, we also recognized methodological improvements, e.g. that a goal-based threat modeling approach can be beneficial (LL2). Engineers should thereby not only be able to model and outline goals and strategies in the data flow diagram, but goals should also actively guide the design process. We are currently focusing on improving the threat modeling tool support to address the shortcomings of the TMT in terms of missing modeling capabilities and to make threat modeling more structured and comprehensible.

Modular and Reusable Threat Models
Because threat modeling is an interactive process in which the exchange between the involved participants is very important, it is usually carried out manually instead of automatically. In order to support this manual work and to reduce the complexity of threat models (LL3), the user should be able to modularize the DFD, identify and isolate recurring structures, and save them for later re-use. To support the reuse and modularization of data flow diagram structures, our objective is to build advanced tool support. This objective integrates well with the proposed goal-based threat modeling approach (see above) using refined goals as a basis for stored recurring diagram structures (i.e., storing diagram patterns according to the security goal they achieve). We thereby want to build up on the approach presented in [8] and put a stronger focus on data flow diagrams.

Suggestions for Countermeasures and Architectural Changes

Choosing good fitting countermeasures is crucial not only for the security but also for the architectural design and thus the future maintainability of the developed software. However, we have learned that the participants, who are typically no security experts but software developers, are overwhelmed in deciding what countermeasure to choose (LL4). Especially, judging the implications on the architecture when choosing a new countermeasure is very difficult for non-security-experts. In order to ease the process in determining the best fitting countermeasures and help the participants to build up necessary knowledge (LL5), we are working on automatically suggesting fitting countermeasures including comprehensible suggestions on necessary architectural changes to the system, which can additionally be automatically applied.

Continuous Threat Modeling

In order for threat modeling to be an effective part of a modern, continuous software development process, its results have to be used in other steps of the process, such as software design, so that countermeasures can be traced and implemented in the software (LL6). While there is a proposal for an open, standardized format for exchanging threat model information called *Open Threat Model (OTM)*[2], released by *IriusRisk* in 2022, no widely agreed upon standard exists for this purpose. Therefore, one challenge is to enable exchangeability of information between threat modeling tools and other parts of the software development tool chain, to allow for the forward integration of threat information into the development process. Conversely, changes in the software, such as realized countermeasures for identified threats, are not automatically represented in the threat model. This causes a lot of manual effort to keep the threat model up to date (LL7). As future work, we aim to develop a mapping between software and threat model, so that it can be kept synchronized with the actual software without the need for manual interference.

Domain Specific Threat Modeling Including Safety

Threat modeling is important for the security of systems, which is more and more recognized by other domains than software engineering as well. This results in standards requiring threat modeling as part of their standard-compliant development process [2]. However, each standard adapts these threat modeling processes with their own slight adaptations and additional concepts, e.g., the zones and conduits in the IEC 62443, which need to be properly handled in these domains (4). A second aspect with the involvement of cyber-physical systems is that safety-relevant components need to be identified for assessing the impact of threats (4). In a recent project with BMW, we developed an integrated safety and security analysis method that supports the analysis of correlations between attacks and hazards on an architectural level [3]. Currently, we simplify the method and develop a tool in order to allow small and medium sized companies an easier integration into their processes. In the future, we plan to adapt this simplified method to specific domains, e.g., the automotive and automation domain.

---

[2] https://github.com/iriusrisk/OpenThreatModel.

Extending Results to Other Domains
Our workshops have been conducted with industry partners working in the domain of cyber-physical systems only. Therefore, our observations are also limited to this domain. A potential topic for future work is to extend the variety of domains and to also conduct these kind of workshops with industry partners from other domains (e.g., web development, mobile applications).

Acknowledgements. This research has been funded by the Federal Ministry of Education and Research (BMBF) under grant 01IS17047 as part of the Software Campus program.

References

1. Fockel, M., Merschjohann, S., Fazal-Baqaie, M.: Threat analysis in practice – systematically deriving security requirements. In: Kuhrmann, M., et al. (eds.) PRO-FES 2018. LNCS, vol. 11271, pp. 355–358. Springer, Cham (2018). https://doi.org/10.1007/978-3-030-03673-7_25
2. Fockel, M., Merschjohann, S., Fazal-Baqaie, M., Förder, T., Hausmann, S., Waldeck, B.: Designing and integrating IEC 62443 compliant threat analysis. In: Walker, A., O'Connor, R.V., Messnarz, R. (eds.) EuroSPI 2019. CCIS, vol. 1060, pp. 57–69. Springer, Cham (2019). https://doi.org/10.1007/978-3-030-28005-5_5
3. Fockel., M., Schubert., D., Trentinaglia., R., Schulz., H., Kirmair., W.: Semi-automatic integrated safety and security analysis for automotive systems. In: Proceedings of the 10th International Conference on Model-Driven Engineering and Software Development - MODELSWARD 2022, pp. 147–154. INSTICC, SciTePress (2022)
4. ISO/SAE: ISO/SAE DIS 21434 Road vehicles - Cybersecurity engineering. Standard 2020. Automotive Security Standard (2020)
5. Scandariato, R., Wuyts, K., Joosen, W.: A descriptive study of microsoft's threat modeling technique. Requir. Eng. **20**(2), 163–180 (2015)
6. Shi, Z., Graffi, K., Starobinski, D., Matyunin, N.: Threat modeling tools: a taxonomy. IEEE Secur. Priv. **20**(4), 29–39 (2022)
7. Shostack, A.: Threat Modeling: Designing for Security. Wiley (2014)
8. Trentinaglia, R.: Deriving model-based safety and security assurance cases from design rationale of countermeasure patterns. In: Proceedings of the 25th International Conference on Model Driven Engineering Languages and Systems: Companion Proceedings, pp. 164–169 (2022)
9. Tuma, K., Scandariato, R.: Two architectural threat analysis techniques compared. In: Cuesta, C.E., Garlan, D., Pérez, J. (eds.) ECSA 2018. LNCS, vol. 11048, pp. 347–363. Springer, Cham (2018). https://doi.org/10.1007/978-3-030-00761-4_23
10. Tuma, K., Scandariato, R., Widman, M., Sandberg, C.: Towards security threats that matter. In: Katsikas, S.K., et al. (eds.) CyberICPS/SECPRE -2017. LNCS, vol. 10683, pp. 47–62. Springer, Cham (2018). https://doi.org/10.1007/978-3-319-72817-9_4
11. Williams, L., Meneely, A., Shipley, G.: Protection poker: the new software security "game". IEEE Secur. Priv. **8**(3), 14–20 (2010)
12. Wohlin, C., Runeson, P., Höst, M., Ohlsson, M.C., Regnell, B., Wesslén, A.: Experimentation in Software Engineering. Springer, Heidelberg (2012). https://doi.org/10.1007/978-3-642-29044-2

13. Xiong, W., Lagerström, R.: Threat modeling - a systematic literature review. Comput. Secur. **84**, 53–69 (2019)
14. Yskout, K., Heyman, T., Van Landuyt, D., Sion, L., Wuyts, K., Joosen, W.: Threat modeling: from infancy to maturity. In: Proceedings of the ACM/IEEE 42nd International Conference on Software Engineering: New Ideas and Emerging Results, ICSE-NIER 2020, pp. 9–12. Association for Computing Machinery, New York, NY, USA (2020)

Author Index

© The Editor(s) (if applicable) and The Author(s), under exclusive license
to Springer Nature Switzerland AG 2023
A. Ferrari and B. Penzenstadler (Eds.): REFSQ 2023, LNCS 13975, pp. 367–368, 2023.
https://doi.org/10.1007/978-3-031-29786-1

Printed in the United States
by Baker & Taylor Publisher Services